FUNDAMENTALS OF NUMERICAL COMPUTING

FUNDAMENTALS OF NUMERICAL COMPUTING

L. F. Shampine

Southern Methodist University

R. C. Allen, Jr.

Sandia National Laboratories

S. Pruess

Colorado School of Mines

JOHN WILEY & SONS, INC.

New York Chichester Brisbane Toronto Singapore

Acquisitions Editor	*Barbara Holland*
Editorial Assistant	*Cindy Rhoads*
Marketing Manager	*Cathy Faduska*
Senior Production Manager	*Lucille Buonocore*
Senior Production Editor	*Nancy Prinz*
Manufacturing Manager	*Mark Cirillo*
Cover Designer	*Steve Jenkins*

This book was set in Times Roman, and was printed and bound by R.R. Donnelley & Sons Company, Crawfordsville. The cover was printed by The Lehigh Press, Inc.

Recognizing the importance of preserving what has been written, it is a policy of John Wiley & Sons, Inc. to have books of enduring value published in the United States printed on acid-free paper, and we exert our best efforts to that end.

The paper in this book was manufactured by a mill whose forest management programs include sustained yield harvesting of its timberlands. Sustained yield harvesting principles ensure that the number of trees cut each year does not exceed the amount of new growth.

Library of Congress Cataloging-in-Publication Data:
Shampine, Lawrence
Fundamentals of numerical computing / Richard Allen, Steve Pruess, Lawrence Shampine.
p. cm.
Includes bibliographical reference and index.
ISBN 0-471-16363-5 (cloth : alk. paper)
1. Numerical analysis—Data processing. I. Pruess, Steven. II. Shampine, Lawrence F. III. Title.
QA297.A52 1997
519.4'0285'51—dc20 96-22074
CIP

Printed in the United States of America

10 9 8 7 6 5 4 3 2 1

PRELIMINARIES

The purpose of this book is to develop the understanding of basic numerical methods and their implementations as software that are necessary for solving fundamental mathematical problems by numerical means. It is designed for the person who wants to do numerical computing. Through the examples and exercises, the reader studies the behavior of solutions of the mathematical problem along with an algorithm for solving the problem. Experience and understanding of the algorithm are gained through hand computation and practice solving problems with a computer implementation. It is essential that the reader understand how the codes provided work, precisely what they do, and what their limitations are. The codes provided are powerful, yet simple enough for pedagogical use. The reader is exposed to the art of numerical computing as well as the science.

The book is intended for a one-semester course, requiring only calculus and a modest acquaintance with FORTRAN, C, C++, or MATLAB. These constraints of background and time have important implications: the book focuses on the problems that are most common in practice and accessible with the background assumed. By concentrating on one effective algorithm for each basic task, it is possible to develop the fundamental theory in a brief, elementary way. There are ample exercises, and codes are provided to reduce the time otherwise required for programming and debugging. The intended audience includes engineers, scientists, and anyone else interested in scientific programming. The level is upper-division undergraduate to beginning graduate and there is adequate material for a one semester to two quarter course.

Numerical analysis blends mathematics, programming, and a considerable amount of art. We provide programs with the book that illustrate this. They are more than mere implementations in a particular language of the algorithms presented, but they are not production-grade software. To appreciate the subject fully, it will be necessary to study the codes provided and gain experience solving problems first with these programs and then with production-grade software.

Many exercises are provided in varying degrees of difficulty. Some are designed to get the reader to think about the text material and to test understanding, while others are purely computational in nature. Problem sets may involve hand calculation, algebraic derivations, straightforward computer solution, or more sophisticated computing exercises.

The algorithms that we study and implement in the book are designed to avoid severe roundoff errors (arising from the finite number of digits available on computers and calculators), estimate truncation errors (arising from mathematical approximations), and give some indication of the sensitivity of the problem to errors in the data. In Chapter 1 we give some basic definitions of errors arising in computations and study roundoff errors through some simple but illuminating computations. Chapter 2 deals with one of the most frequently occurring problems in scientific computation, the solution of linear systems of equations. In Chapter 3 we deal with the problem of interpolation, one of the most fundamental and widely used tools in numerical computation. In Chapter 4 we study methods for finding solutions to nonlinear equations. Numerical integration is taken up in Chapter 5 and the numerical solution of ordinary differential equations is examined in Chapter 6. Each chapter contains a case study that illustrates how to combine analysis with computation for the topic of that chapter.

Before taking up the various mathematical problems and procedures for solving them numerically, we need to discuss briefly programming languages and acquisition of software.

PROGRAMMING LANGUAGES

The FORTRAN language was developed specifically for numerical computation and has evolved continuously to adapt it better to the task. Accordingly, of the widely used programming languages, it is the most natural for the programs of this book. The C language was developed later for rather different purposes, but it can be used for numerical computation.

At present FORTRAN 77 is very widely available and codes conforming to the ANSI standard for the language are highly portable, meaning that they can be moved to another hardware/software configuration with very little change. We have chosen to provide codes in FORTRAN 77 mainly because the newer Fortran 90 is not in wide use at this time. A Fortran 90 compiler will process correctly our FORTRAN 77 programs (with at most trivial changes), but if we were to write the programs so as to exploit fully the new capabilities of the language, a number of the programs would be structured in a fundamentally different way. The situation with C is similar, but in our experience programs written in C have not proven to be nearly as portable as programs written in standard FORTRAN 77. As with FORTRAN, the C language has evolved into C++, and as with Fortran 90 compared to FORTRAN 77, exploiting fully the additional capabilities of C++ (in particular, object oriented programming) would lead to programs that are completely different from those in C. We have opted for a middle ground in our C++ implementations.

In the last decade several computing environments have been developed. Popular ones familiar to us are MATLAB [1] and *Mathematica* [2]. MATLAB is very much in keeping with this book, for it is devoted to the solution of mathematical problems by numerical means. It integrates the formation of a mathematical model, its numerical solution, and graphical display of results into a very convenient package. Many of the tasks we study are implemented as a single command in the MATLAB language. As MATLAB has evolved, it has added symbolic capabilities. *Mathematica* is a similar environment, but it approaches mathematical problems from the other direction. Originally it was primarily devoted to solving mathematical problems by symbolic means, but as it has evolved, it has added significant numerical capabilities. In the book we refer to the numerical methods implemented in these widely used packages, as well as others, but we mention the packages here because they are programming languages in their own right. It is quite possible to implement the algorithms of the text in these languages. Indeed, this is attractive because the environments deal gracefully with a number of issues that are annoying in general computing using languages like FORTRAN or C.

At present we provide programs written in FORTRAN 77, C, C++, and MATLAB that have a high degree of portability. Quite possibly in the future the programs will be made available in other environments (e.g., Fortran 90 or *Mathematica*.)

SOFTWARE

In this section we describe how to obtain the source code for the programs that accompany the book and how to obtain production-grade software. It is assumed that the reader has available a browser for the World Wide Web, although some of the software is available by ftp or gopher.

The programs that accompany this book are currently available by means of anonymous ftp (log in as *anonymous* or as *ftp*) at

ftp.wiley.com

in subdirectories of public/college/math/sapcodes for the various languages discussed in the preceding section.

The best single source of software is the Guide to Available Mathematical Software (GAMS) developed by the National Institute of Standards and Technology (NIST). It is an on-line cross-index of mathematical software and a virtual software repository. Much of the high-quality software is free. For example, GAMS provides a link to netlib, a large collection of public-domain mathematical software. Most of the programs in netlib are written in FORTRAN, although some are in C. A number of the packages found in netlib are state-of-the-art software that are cited in this book. The internet address is

http://gams.nist.gov

for GAMS.

A useful source of microcomputer software and pointers to other sources of software is the Mathematics Archives at

http://archives.math.utk.edu:80/

It is worth remarking that one item listed there is an "Index of resources for numerical computation in C or C++."

There are a number of commercial packages that can be located by means of GAMS. We are experienced with the NAG and IMSL libraries, which are large collections of high-quality mathematical software found in most computing centers. The computing environments MATLAB and *Mathematica* mentioned in the preceding section can also be located through GAMS.

REFERENCES

1. C. Moler, J. Little, S. Bangert, and S. Kleiman, *ProMatlab User's Guide, MathWorks*, Sherborn, Mass., 1987. email: info@mathworks.com

2. S. Wolfram, *Mathematica*, Addison-Wesley, Redwood City, Calif., 1991. email: info@wri.com

ACKNOWLEDGMENTS

The authors are indebted to many individuals who have contributed to the production of this book. Professors Bernard Bialecki and Michael Hosea have been especially sharp-eyed at catching errors in the latest versions. We thank the people at Wiley, Barbara Holland, Cindy Rhoads, and Nancy Prinz, for their contributions. David Richards at the University of Illinois played a critical role in getting the LaTeXmacros functioning for us, and quickly and accurately fixing other LaTeXproblems. We also acknowledge the work of James Otto in checking all solutions and examples, and Hong-sung Jin who generated most of the figures. Last, but certainly not least, we are indeed grateful to the many students, too numerous to mention, who have made valuable suggestions to us over the years.

CONTENTS

CHAPTER 1

ERRORS AND FLOATING POINT ARITHMETIC

Errors in mathematical computation have several sources. One is the modeling that led to the mathematical problem, for example, assuming no wind resistance in studying projectile motion or ignoring finite limits of resources in population and economic growth models. Such errors are not the concern of this book, although it must be kept in mind that the numerical solution of a mathematical problem can be no more meaningful than the underlying model. Another source of error is the measurement of data for the problem. A third source is a kind of mathematical error called discretization or truncation error. It arises from mathematical approximations such as estimating an integral by a sum or a tangent line by a secant line. Still another source of error is the error that arises from the finite number of digits available in the computers and calculators used for the computations. It is called roundoff error. In this book we study the design and implementation of algorithms that aim to avoid severe roundoff errors, estimate truncation errors, and give some indication of the sensitivity of the problem to errors in the data. This chapter is devoted to some fundamental definitions and a study of roundoff by means of simple but illuminating computations.

1.1 BASIC CONCEPTS

How well a quantity is approximated is measured in two ways:

$$\textit{absolute error} = \text{true value} - \text{approximate value}$$

$$\textit{relative error} = \frac{\text{true value} - \text{approximate value}}{\text{true value}}.$$

Relative error is not defined if the true value is zero. In the arithmetic of computers, relative error is the more natural concept, but absolute error may be preferable when studying quantities that are close to zero.

A mathematical problem with input (data) x and output (answer) $y = F(x)$ is said to be *well-conditioned* if "small" changes in x lead to "small" changes in y. If the changes in y are "large," the problem is said to be *ill-conditioned*. Whether a problem is well- or ill-conditioned can depend on how the changes are measured. A concept related to conditioning is stability. It is concerned with the sensitivity of an algorithm for solving a problem with respect to small changes in the data, as opposed to the sensitivity of the problem itself. Roundoff errors are almost inevitable, so the reliability of answers computed by an algorithm depends on whether small roundoff errors might seriously affect the results. An algorithm is *stable* if "small" changes in the input lead to "small" changes in the output. If the changes in the output are "large," the algorithm is *unstable*.

To gain some insight about condition, let us consider a differentiable function $F(x)$ and suppose that its argument, the input, is changed from x to $x+\varepsilon x$. This is a relative change of ε in the input data. According to Theorem 4 of the appendix, the change induces an absolute change in the output value $F(x)$ of

$$F(x) - F(x+\varepsilon x) \approx -\varepsilon x F'(x).$$

The relative change is

$$\frac{F(x) - F(x+\varepsilon x)}{F(x)} \approx -\varepsilon x \frac{F'(x)}{F(x)}.$$

Example 1.1. If, for example, $F(x) = e^x$, the absolute change in the value of the exponential function due to a change εx in its argument x is approximately $-\varepsilon x e^x$, and the relative change is about $-\varepsilon x$. When x is large, the conditioning of the evaluation of this function with respect to a small relative change in the argument depends strongly on whether the change is measured in an absolute or relative sense. ■

Example 1.2. If $F(x) = \cos x$, then near $x = \pi/2$ the absolute error due to perturbing x to $x+\varepsilon x$ is approximately $-\varepsilon x(-\sin x) \approx \pi\varepsilon/2$. The relative error at $x = \pi/2$ is not defined since $\cos(\pi/2) = 0$. However, the accurate values

$$\cos(1.57079) = 0.63267949 \times 10^{-5}$$
$$\cos(1.57078) = 1.63267949 \times 10^{-5}$$

show how a very small change in the argument near $\pi/2$ can lead to a significant (63%) change in the value of the function. In contrast, evaluation of the cosine function is well-conditioned near $x = 0$ (see Exercise 1.4). ■

Example 1.3. A common application of integration by parts in calculus courses is the evaluation of families of integrals by recursion. As an example, consider

$$E_n = \int_0^1 x^n e^{x-1}\, dx \text{ for } n = 1, 2, \ldots.$$

From this definition it is easy to see that

$$E_1 > E_2 > \cdots > E_{n-1} > E_n > \cdots > 0.$$

To obtain a recursion, integrate by parts to get

$$E_n = x^n e^{x-1}\Big|_0^1 - \int_0^1 n x^{n-1} e^{x-1}\,dx$$
$$= 1 - nE_{n-1}.$$

The first member of the family is

$$E_1 = 1 - \int_0^1 e^{x-1}\,dx = e^{-1},$$

and from it we can easily compute any E_n. If this is done in single precision on a PC or workstation (IEEE standard arithmetic), it is found that

$$\begin{array}{lcll}
E_1 & = & 0.367879 & \\
E_2 & = & 0.264241 & \\
 & \vdots & & \\
E_{10} & = & 0.0506744 & \\
E_{11} & = & 0.442581 & \text{(the exact } E_n \text{ decrease!)} \\
E_{12} & = & -4.31097 & \text{(the exact } E_n \text{ are positive!)} \\
 & \vdots & & \\
E_{20} & = & -0.222605 \times 10^{11} & \text{(the exact } E_n \text{ are between 0 and 1!)}
\end{array}$$

This is an example of an unstable algorithm. A little analysis helps us understand what is happening. Suppose we had started with $\widehat{E}_1 = E_1 + \delta$ and made no arithmetic errors when evaluating the recurrence. Then

$$\begin{array}{lcl}
\widehat{E}_2 & = & 1 - 2\widehat{E}_1 = 1 - 2E_1 - 2\delta = E_2 - 2\delta \\
\widehat{E}_3 & = & 1 - 3\widehat{E}_2 = 1 - 3E_2 + 6\delta = E_3 + 3!\delta \\
 & \vdots & \\
\widehat{E}_n & = & E_n \pm n!\delta.
\end{array}$$

A small change in the first value E_1 grows very rapidly in the later E_n. The effect is worse in a relative sense because the desired quantities E_n decrease as n increases.

For this example there is a way to get a stable algorithm. If we could find an approximation $\widehat{E}_N$ to E_N for some N, we could evaluate the recursion in reverse order,

$$E_{n-1} = \frac{1 - E_n}{n}, \qquad n = N, N-1, \ldots, 2,$$

to approximate $E_{N-1}, E_{N-2}, \ldots, E_1$. Studying the stability of this recursion as before, if $\widehat{E}_N = E_N + \varepsilon$, then

$$\widehat{E}_{N-1} = \frac{1 - \widehat{E}_N}{N} = \frac{1 - E_N}{N} - \frac{\varepsilon}{N} = E_{N-1} - \frac{\varepsilon}{N}$$

$$\widehat{E}_{N-2} = E_{N-2} + \frac{\varepsilon}{N(N-1)}$$

$$\vdots$$

$$\widehat{E}_1 = E_1 \pm \frac{\varepsilon}{N!}.$$

The recursion is so cheap and the error damps out so quickly that we can start with a poor approximation $\widehat{E}_N$ for some large N and get accurate answers inexpensively for the E_n that really interest us. Notice that recurring in this direction, the E_n increase, making the relative errors damp out even faster. The inequality

$$0 < E_n < \int_0^1 x^n \, dx = \frac{1}{n+1}$$

shows how to easily get an approximation to E_n with an error that we can bound. For example, if we take $N = 20$, the crude approximation $\widehat{E}_{20} = 0$ has an absolute error less than 1/21 in magnitude. The magnitude of the absolute error in $\widehat{E}_{19}$ is then less than $1/(20 \times 21) = 0.0024, \ldots$, and that in $\widehat{E}_{15}$ is less than 4×10^{-8}. The approximations to $E_{14}, \ldots, E_1$ will be even more accurate.

A stable recurrence like the second algorithm is the standard way to evaluate certain mathematical functions. It can be especially convenient for a series expansion in the functions. For example, evaluation of an expansion in Bessel functions of the first kind,

$$f(x) = \sum_{n=0}^{\infty} a_n J_n(x),$$

requires the evaluation of $J_n(x)$ for many n. Using recurrence on the index n, this is accomplished very inexpensively. ■

Any real number $y \neq 0$ can be written in scientific notation as

$$y = \pm .d_1 d_2 \cdots d_s d_{s+1} \cdots \times 10^e. \tag{1.1}$$

Here there are an infinite number of digits d_i. Each d_i takes on one of the values $0, 1, \ldots, 9$ and we assume the number y is *normalized* so that $d_1 > 0$. The portion $.d_1 d_2 \ldots$ is called the *fraction* or *mantissa* or *significand*; it has the meaning

$$d_1 \times 10^{-1} + d_2 \times 10^{-2} + \cdots + d_s \times 10^{-s} + \cdots.$$

There is an ambiguity in this representation; for example, we must agree that

$$0.24000000 \cdots$$

is the same as

$$0.23999999 \cdots.$$

The quantity e in (1.1) is called the *exponent*; it is a signed integer.

Nearly all numerical computations on a digital computer are done in floating point arithmetic. This is a number system that uses a finite number of digits to approximate the real number system used for exact computation. A system with s digits and base 10 has all of its numbers of the form

$$y = \pm .d_1 d_2 \cdots d_s \times 10^e. \tag{1.2}$$

Again, for nonzero numbers each d_i is one of the digits $0, 1, \ldots, 9$ and $d_1 > 0$ for a normalized number. The exponent e also has only a finite number of digits; we assume the range

$$m \leq e \leq M.$$

The number zero is special; it is written as

$$0.0 \cdots 0 \times 10^m.$$

Example 1.4. If $s = 1$, $m = -1$, and $M = 1$, then the set of floating point numbers is

$$\begin{array}{llll} +0.1 \times 10^{-1}, & +0.2 \times 10^{-1}, & \ldots, & +0.9 \times 10^{-1} \\ +0.1 \times 10^{0}, & +0.2 \times 10^{0}, & \ldots, & +0.9 \times 10^{0} \\ +0.1 \times 10^{1}, & +0.2 \times 10^{1}, & \ldots, & +0.9 \times 10^{1}, \end{array}$$

together with the negative of each of these numbers and 0.0×10^{-1} for zero. There are only 55 numbers in this floating point number system. In floating point arithmetic the numbers are not equally spaced. This is illustrated in Figure 1.1, which is discussed after we consider number bases other than decimal. ■

Because there are only finitely many floating point numbers to represent the real number system, each floating point number must represent many real numbers. When the exponent e in (1.1) is bigger than M, it is not possible to represent y at all. If in the course of some computations a result arises that would need an exponent $e > M$, the computation is said to have *overflowed.* Typical operating systems will terminate the run on overflow. The situation is less clear when $e < m$, because such a y might reasonably be approximated by zero. If such a number arises during a computation, the computation is said to have *underflowed.* In scientific computation it is usually appropriate to set the result to zero and continue. Some operating systems will terminate the run on underflow and others will set the result to zero and continue. Those that continue may report the number of underflows at the end of the run. If the response of the operating system is not to your liking, it is usually possible to change the response by means of a system routine.

Overflows and underflows are not unusual in scientific computation. For example, $\exp(y)$ will overflow for $y > 0$ that are only moderately large, and $\exp(-y)$ will underflow. Our concern should be to prevent going out of range *unnecessarily.*

FORTRAN and C provide for integer arithmetic in addition to floating point arithmetic. Provided that the range of integers allowed is not exceeded, integer arithmetic is exact. It is necessary to beware of overflow because the typical operating system does *not* report an integer overflow; the computation continues with a number that is not related to the correct value in an obvious way.

Both FORTRAN and C provide for two precisions, that is, two arithmetics with different numbers of digits s, called single and double precision. The languages deal with mixing the various modes of arithmetic in a sensible way, but the unwary can get into trouble. This is more likely in FORTRAN than C because by default, constants in C are double precision numbers. In FORTRAN the type of a constant is taken from the way it is written. Thus, an expression like (3/4)*5. in FORTRAN and in C means that the integer 3 is to be divided by the integer 4 and the result converted to a floating point number for multiplication by the floating point number 5. Here the integer division 3/4 results in 0, which might not be what was intended. It is surprising how often users ruin the accuracy of a calculation by providing an inaccurate value for a basic constant like π. Some constants of this kind may be predefined to full accuracy in a compiler or a library, but it should be possible to use intrinsic functions to compute accurately constants like $\pi = \text{acos}(-1.0)$.

Evaluation of an asymptotic expansion for the special function $\text{Ei}(x)$, called the exponential integral, involves computing terms of the form $n!/x^n$. To contrast computations in integer and floating point arithmetic, we computed terms of this form for a range of n and $x = 25$ using both integer and double precision functions for the factorial. Working in C on a PC using IEEE arithmetic, it was found that the results agreed through $n = 7$, but for larger n the results computed with integer arithmetic were useless—the result for $n = 8$ was *negative*! The integer overflows that are responsible for these erroneous results are truly dangerous because there was no indication from the system that the answers might not be reliable.

Example 1.5. In Chapter 4 we study the use of bisection to find a number z such that $f(z) = 0$, that is, we compute a root of $f(x)$. Fundamental to this procedure is the question, Do $f(a)$ and $f(b)$ have opposite signs? If they do, a continuous function $f(x)$ has a root z between a and b. Many books on programming provide illustrative programs that test for $f(a)f(b) < 0$. However, when $f(a)$ and $f(b)$ are sufficiently small, the product underflows and its sign cannot be determined. This is likely to happen because we are interested in a and b that tend to z, causing $f(a)$ and $f(b)$ to tend to zero. It is easy enough to code the test so as to avoid the difficulty; it is just necessary to realize that the floating point number system does not behave quite like the real number system in this test. ■

As we shall see in Chapter 4, finding roots of functions is a context in which underflow is quite common. This is easy to understand because the aim is to find a z that makes $f(z)$ as small as possible.

Example 1.6. Determinants. In Chapter 2 we discuss the solution of a system of linear equations. As a by-product of the algorithm and code presented there, the determinant of a system of n equations can be computed as the product of a set of numbers returned:

$$\text{det} = y_1 y_2 \cdots y_n.$$

Unfortunately, this expression is prone to unnecessary under- and overflows. If, for example, $M = 100$ and $y_1 = 10^{50}$, $y_2 = 10^{60}$, $y_3 = 10^{-30}$, all the numbers are in range and so is the determinant 10^{80}. However, if we form $(y_1 \times y_2) \times y_3$, the partial product $y_1 \times y_2$ overflows. Note that $y_1 \times (y_2 \times y_3)$ *can* be formed. This illustrates the fact that floating point numbers do not always satisfy the associative law of multiplication that is true of real numbers.

The more fundamental issue is that because $\det(cA) = c^n \det(A)$, the determinant is extremely sensitive to the scale of the matrix A when the number of equations n is large. A software remedy used in LINPACK [4] in effect extends the range of exponents available. Another possibility is to use logarithms and exponentials:

$$\ln|\det| = \sum_{i=1}^{n} \ln|y_i|$$

$$|\det| = \exp(\ln|\det|).$$

If this leads to an overflow, it is because the answer cannot be represented in the floating point number system. ■

Example 1.7. Magnitude. When computing the magnitude of a complex number $z = x + iy$,

$$|z| = \sqrt{x^2 + y^2},$$

there is a difficulty when either x or y is large. Suppose that $|x| \geq |y|$. If $|x|$ is sufficiently large, x^2 will overflow and we are not able to compute $|z|$ even when it is a valid floating point number. If the computation is reformulated as

$$|z| = |x|\sqrt{1 + (y/x)^2},$$

the difficulty is avoided. Notice that underflow could occur when $|y| \ll |x|$. This is harmless and setting the ratio y/x to zero results in a computed $|z|$ that has a small relative error.

The evaluation of the Euclidean norm of a vector $\mathbf{v} = (v_1, v_2, \ldots, v_n)$,

$$||v||_2 = \left(\sum_{i=1}^{n} v_i^2 \right)^{0.5},$$

involves exactly the same kind of computations. Some writers of mathematical software have preferred to work with the maximum norm

$$||v||_\infty = \max_{1 \leq i \leq n} |v_i|,$$

because it avoids the unnecessary overflows and underflows that are possible with a straightforward evaluation of the Euclidean norm. ■

If a real number y has an exponent in the allowed range, there are two standard ways to approximate it by a floating point number $fl(y)$. If all digits after the first s in (1.1) are dropped, the result is known as a *chopped* or *truncated* representation. A floating point number that is usually closer to y can be found by adding $5 \times 10^{-(s+1)}$ to the fraction in (1.1) and then chopping. This is called *rounding*.

Example 1.8. If $m = -99$, $M = 99$, $s = 5$, and $\pi = 3.1415926\cdots$, then in chopped arithmetic

$$fl(\pi) = 0.31415 \times 10^1$$

while

$$fl(\pi) = 0.31416 \times 10^1$$

in rounded arithmetic. ■

If the representation (1.1) of y is chopped to s digits, the relative error of $fl(y)$ has magnitude

$$\begin{aligned}\left|\frac{y - fl(y)}{y}\right| &= \frac{0.00\cdots 0d_{s+1}d_{s+2}\cdots \times 10^e}{0.d_1d_2\cdots d_{s+1}d_{s+2}\cdots \times 10^e} \\ &\leq \frac{0.00\cdots 099\cdots}{0.10\cdots 000\cdots} \\ &= \frac{0.00\cdots 100\cdots}{0.10\cdots 000\cdots} = 10^{1-s}.\end{aligned}$$

In decimal arithmetic with s digits the *unit roundoff* u is defined to be 10^{1-s} when chopping is done. In a similar way it is found that

$$\left|\frac{y - fl(y)}{y}\right| \leq \frac{1}{2}10^{1-s}$$

when rounding is done. In this case u is defined to be $\frac{1}{2}10^{1-s}$. In either case, u is a bound on the relative error of representing a nonzero real number as a floating point number.

Because $fl(y)$ is a real number, for theoretical purposes we can work with it like any other real number. In particular, it is often convenient to define a real number δ such that

$$fl(y) = y(1+\delta).$$

In general, all we know about δ is the bound

$$|\delta| \leq u.$$

Example 1.9. Impossible accuracy. Modern codes for the computation of a root of an equation, a definite integral, the solution of a differential equation, and so on, try to obtain a result with an accuracy specified by the user. Clearly it is not possible to compute an answer more accurate than the floating point representation of the true solution. This means that the user cannot be allowed to ask for a relative error smaller than the unit roundoff u. It might seem odd that this would ever happen, but it does. One reason is that the user does not know the value of u and just asks for too much accuracy. A more common reason is that the user specifies an absolute error r. This means that any number y^* will be acceptable as an approximation to y if

$$|y - y^*| \leq r.$$

Such a request corresponds to asking for a relative error of

$$\left| \frac{y - y^*}{y} \right| \leq \frac{r}{|y|}.$$

When $|r/y| < u$, that is, $r < u|y|$, this is an impossible request. If the true solution is unexpectedly large, an absolute error tolerance that seems modest may be impossible in practice. Codes that permit users to specify an absolute error tolerance need to be able to monitor the size of the solution and warn the user when the task posed is impossible. ■

There is a further complication to the floating point number system—most computers do not work with decimal numbers. The common bases are $\beta = 2$, binary arithmetic, and $\beta = 16$, hexadecimal arithmetic, rather than $\beta = 10$, decimal arithmetic. In general, a real number y is written in base β as

$$y = \pm.d_1 d_2 \cdots d_s d_{s+1} \cdots \times \beta^e, \tag{1.3}$$

where each digit is one of $0, 1, \ldots, \beta - 1$ and the number is normalized so that $d_1 > 0$ (as long as $y \neq 0$). This means that

$$\pm(d_1 \times \beta^{-1} + d_2 \times \beta^{-2} + \cdots + d_s \times \beta^{-s} + \cdots) \times \beta^e.$$

All the earlier discussion is easily modified for the other bases. In particular, we have

---

in base β with s digits the *unit roundoff*

$$u = \begin{cases} \beta^{1-s}, & \text{chopped} \\ \frac{1}{2}\beta^{1-s}, & \text{rounded.} \end{cases} \tag{1.4}$$

Likewise,

$$fl(y) = y(1 + \delta), \text{ where } |\delta| \leq u.$$

For most purposes, the fact that computations are not carried out in decimal is inconsequential. It should be kept mind that small rounding errors are made as numbers input are converted from decimal to the base of the machine being used and likewise on output.

Table 1.1 illustrates the variety of machine arithmetics used in the past. Today the IEEE standard [1] described in the last two rows is almost universal. In the table the notation $1.2(-7)$ means 1.2×10^{-7}.

As was noted earlier, both FORTRAN and C specify that there will be two precisions available. The floating point system built into the computer is its single precision arithmetic. Double precision may be provided by either software or hardware. Hardware double precision is not greatly slower than single precision, but software double precision arithmetic is considerably slower.

The IEEE standard uses a normalization different from (1.2). For $y \neq 0$ the leading nonzero digit is immediately to the *left* of the decimal point. Since this digit must be 1, there is no need to store it. The number 0 is distinguished by having its $e = m - 1$.

Table 1.1 Examples of Computer Arithmetics.

machine	β	s	m	M	approximate u
VAX	2	24	-128	127	6.0(−08)
VAX	2	56	-128	127	1.4(−17)
CRAY-1	2	48	-16384	16383	3.6(−15)
IBM 3081	16	6	-64	63	9.5(−07)
IBM 3081	16	14	-64	63	2.2(−16)
IEEE					
Single	2	24	-125	128	6.0(−08)
Double	2	53	-1021	1024	1.1(−16)

It used to be some trouble to find out the unit roundoff, exponent range, and the like, but the situation has improved greatly. In standard C, constants related to floating point arithmetic are available in <float.h>. For example, dbl_epsilon is the unit roundoff in double precision. Similarly, in Fortran 90 the constants are available from intrinsic functions. Because this is not true of FORTRAN 77, several approaches were taken to provide them: some compilers provide the constants as extensions of the language; there are subroutines D1MACH and I1MACH for the machine constants that are widely available because they are public domain. Major libraries like IMSL and NAG include subroutines that are similar to D1MACH and I1MACH.

In Example 1.4 earlier in this section we mentioned that the numbers in the floating point number system were *not* equally spaced. As an illustration, see Figure 1.1 where all 19 floating point numbers are displayed for the system for which $\beta = 4$, $s = 1$, $m = -1$, and $M = 1$.

Arithmetic in the floating point number system is to approximate that in the real number system. We use $\oplus$, $\ominus$, $\otimes$, $\oslash$ to indicate the floating point approximations to the arithmetic operations $+, -, \times, /$. If y and z are floating point numbers of s digits, the product $y \times z$ has $2s$ digits. For example, $0.999 \times 0.999 = 0.998001$. About the best we could hope for is that the arithmetic hardware produce the result $fl(y \times z)$, so that $y \otimes z = (y \times z)(1 + \delta)$ for some real number δ with $|\delta| \leq u$. It is practical to do this for all the basic arithmetic operations. *We assume an idealized arithmetic that for the basic arithmetic operations produces*

$$\begin{aligned} y \oplus z &= fl(y + z) \\ y \ominus z &= fl(y - z) \\ y \otimes z &= fl(y \times z) \\ y \oslash z &= fl(y/z), \end{aligned}$$

provided that the results lie in the range of the floating point system. Hence,

$$y \textcircled{\scriptsize op} z = (y \; op \; z)(1 + \delta)$$

where $op = +, -, \times,$ or $/$ and δ is a real number with $|\delta| \leq u$. This is a reasonable assumption, although hardware considerations may lead to arithmetic for which the bound on δ is a small multiple of u.

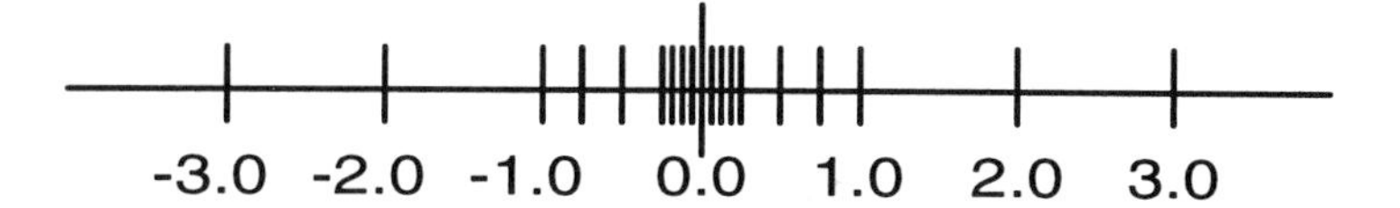

Figure 1.1 Distribution of floating point numbers for $\beta = 4$, $s = 1$, $m = -1$, $M = 1$.

To carry out computations in this model arithmetic by hand, for *each* operation $+, -, \times, /$, perform the operation in exact arithmetic, normalize the result, and round (chop) it to the allotted number of digits. Put differently, for *each* operation, calculate the result and convert it to the machine representation before going on to the next operation.

Because of increasingly sophisticated architectures, the unit roundoff as defined in (1.4) is simplistic. For example, many computers do intermediate computations with more than s digits. They have at least one "guard digit," perhaps several, and as a consequence results can be rather more accurate than expected. (When arithmetic operations are carried out with more than s digits, apparently harmless actions like printing out intermediate results can cause the final result of a computation to change! This happens when the extra digits are shed as numbers are moved from arithmetic units to storage or output devices.) It is interesting to compute $(1+\delta)-1$ for decreasing δ to see how small δ can be made and still get a nonzero result. A number of codes for mathematical computations that are in wide use avoid defining the unit roundoff by coding a test for $u|x| < h$ as

$$\text{if } ((x+h) \neq x) \text{ then } \ldots.$$

On today's computers this is not likely to work properly for two reasons, one being the presence of guard digits just discussed. The other is that modern compilers defeat the test when they "optimize" the coding by converting the test to

$$\text{if } (h \neq 0) \text{ then } \ldots,$$

which is always passed.

EXERCISES

1.1 Solve

$$0.461x_1 + 0.311x_2 = 0.150$$
$$0.209x_1 + 0.141x_2 = 0.068$$

using three-digit chopped decimal arithmetic. The exact answer is $x_1 = 1$, $x_2 = -1$; how does yours compare?

1.2 The following algorithm (due to Cleve Moler) estimates the unit roundoff u by a computable quantity U:

$$A := 4./3.$$
$$B := A - 1.$$
$$C := B + B + B$$
$$U := |C - 1.|$$

(a) What does the above algorithm yield for U in six-digit decimal *rounded* arithmetic?

(b) What does it yield for U in six-digit decimal *chopped* arithmetic?

(c) What are the exact values from (1.4) for u in the arithmetics of (a) and (b)?

(d) Use this algorithm on the machine(s) and calculator(s) you are likely to use. What do you get?

1.3 Consider the following algorithm for generating noise in a quantity x:

$$A := 10^n * x$$
$$B := A + x$$
$$y := B - A$$

(a) Calculate y when $x = 0.123456$ and $n = 3$ using six-digit decimal chopped arithmetic. What is the error $x - y$?

(b) Repeat (a) for $n = 5$.

1.4 Show that the evaluation of $F(x) = \cos x$ is well-conditioned near $x = 0$; that is, for $|x| \leq \delta$ show that the magnitude of the relative error $|[F(x) - F(0)]/F(0)|$ is bounded by a quantity that is not large.

1.5 If $F(x) = (x-1)^2$, what is the exact formula for $[F(x+\varepsilon x) - F(x)]/F(x)$? What does this say about the conditioning of the evaluation of $F(x)$ near $x = 1$?

1.6 Let $S_n := \int_0^\pi (x/\pi)^{2n} \sin x\, dx$ and show that two integrations by parts results in the recursion

$$S_n = 1 - \frac{2n(2n-1)}{\pi^2} S_{n-1}, \quad n = 1, 2, \ldots.$$

Further argue that $S_0 = 2$ and that $S_{n-1} > S_n > 0$ for every n.

(a) Compute S_{15} with this recursion (make sure that you use an accurate value for π).

(b) To analyze what happened in (a), consider the recursion

$$\bar{S}_n = 1 - \frac{2n(2n-1)}{\pi^2} \bar{S}_{n-1}, \quad n = 1, 2, \ldots,$$

with $\bar{S}_0 = 2(1-u)$, that is, the same computation with the starting value perturbed by one digit in the last place. Find a recursion for $S_n - \bar{S}_n$. From this recursion, derive a formula for $S_{15} - \bar{S}_{15}$ in terms of $S_0 - \bar{S}_0$. Use this formula to explain what happened in (a).

(c) Examine the "backwards" recursion

$$\widehat{S}_{n-1} = \frac{(1 - \widehat{S}_n)\pi^2}{2n(2n-1)}$$

starting with $\widehat{S}_{15} = 0$. What is $\widehat{S}_0$? Why?

1.7 For brevity let us write $s = \sin(\theta)$, $c = \cos(\theta)$ for some value of θ. Once c is computed, we can compute s inexpensively from $s = \sqrt{1 - c^2}$. (Either sign of the square root may be needed in general, but let us consider here only the positive root.) Suppose the cosine routine produces $c + \delta c$ instead of c. Ignoring any error made in evaluating the formula for s, show that this absolute error of δc induces an absolute error in s of δs with $\delta s \approx -(c/s)\,\delta c$. For the range $0 \leq \theta \leq \pi/2$, are there θ for which this way of computing $\sin(\theta)$ has an accuracy comparable to the accuracy of $\cos(\theta)$? Are there θ for which it is much less accurate? Repeat for relative errors.

1.2 EXAMPLES OF FLOATING POINT CALCULATIONS

The floating point number system has properties that are similar to those of the real number system, but they are not identical. We have already seen some differences due to the finite range of exponents. It might be thought that because one arithmetic operation can be carried out with small relative error, the same would be true of several operations. Unfortunately this is not true. We shall see that multiplication and division are more satisfactory in this respect than addition and subtraction.

For floating point numbers x, y and z,

$$\begin{aligned} x \otimes y &= xy(1+\delta_1) \\ (x \otimes y) \otimes z &= (xy(1+\delta_1))z(1+\delta_2), \qquad |\delta_2| \leq u \\ &= xyz(1+\delta_1)(1+\delta_2). \end{aligned}$$

The product

$$(1+\delta_1)(1+\delta_2) = 1+\varepsilon,$$

where ε is "small," and can, of course, be explicitly bounded in terms of u. It is more illuminating to note that

$$\begin{aligned}(1+\delta_1)(1+\delta_2) &= 1+\delta_1+\delta_2+\delta_1\delta_2 \\ &\approx 1+\delta_1+\delta_2,\end{aligned}$$

so that

$$\varepsilon \approx \delta_1+\delta_2$$

and an approximate bound for ε is $2u$. Before generalizing this, we observe that it may well be the case that

$$x\otimes(y\otimes z) \neq (x\otimes y)\otimes z,$$

even when the exponent range is not exceeded. However,

$$x\otimes(y\otimes z) = xyz(1+\delta_3)(1+\delta_4),$$

so that

$$\frac{x\otimes(y\otimes z)}{(x\otimes y)\otimes z} = \frac{(1+\delta_3)(1+\delta_4)}{(1+\delta_1)(1+\delta_2)} = 1+\eta,$$

where η is "small." Thus, the associative law for multiplication is *approximately* true.

In general, if we wish to multiply $x_1, x_2, \ldots, x_n$, we might do this by the algorithm

$$\begin{aligned}P_1 &= x_1 \\ P_i &= P_{i-1}\otimes x_i, \qquad i = 2,3,\ldots,n.\end{aligned}$$

Treating these operations in real arithmetic we find that

$$P_i = x_1x_2\cdots x_i(1+\delta_1)(1+\delta_2)\cdots(1+\delta_i),$$

where each $|\delta_i| \leq u$. The relative error of each P_i can be bounded in terms of u without difficulty, but more insight is obtained if we approximate

$$P_i \approx x_1x_2\cdots x_i(1+\delta_1+\delta_2+\cdots+\delta_i),$$

which comes from neglecting products of the δ_i. Then

$$|\delta_1+\delta_2+\cdots+\delta_i| \leq iu.$$

This says that a bound on the approximate relative errors grows additively. Each multiplication could increase the relative error by no more than one unit of roundoff. Division can be analyzed in the same way, and the same conclusion is true concerning the possible growth of the relative error.

Example 1.10. The gamma function, defined as

$$\Gamma(x) = \int_0^\infty t^{x-1}e^{-t}\,dt, \qquad x > 0,$$

generalizes the factorial function for integers to real numbers x (and complex x as well). This follows from the fundamental recursion

$$\Gamma(x) = (x-1)\Gamma(x-1) \tag{1.5}$$

and the fact that $\Gamma(1) = 1$. A standard way of approximating $\Gamma(x)$ for $x \geq 2$ uses the fundamental recursion to reduce the task to approximating $\Gamma(y)$ for $2 \leq y \leq 3$. This is done by letting N be an integer such that $N \leq x < N+1$, letting $y = x - N + 2$, and then noting that repeated applications of (1.5) yield

$$\Gamma(x) = \Gamma(y)(x-N+2)(x-N+3)\cdots(x-2)(x-1).$$

The function $\Gamma(y)$ can be approximated well by the ratio $R(y)$ of two polynomials for $2 \leq y \leq 3$. Hence, we approximate

$$\Gamma(x) \approx R(y)(x-N+2)\cdots(x-1).$$

If x is not too large, little accuracy is lost when these multiplications are performed in floating point arithmetic. However, it is not possible to evaluate $\Gamma(x)$ for large x by this approach because its value grows very quickly as a function of x. This can be seen from the Stirling formula (see Case Study 5)

$$\Gamma(x) \approx \sqrt{2\pi/x}\left(\frac{x}{e}\right)^x.$$

This example makes another point: the virtue of floating point arithmetic is that it automatically deals with numbers of greatly different size. Unfortunately, many of the special functions of mathematical physics grow or decay extremely fast. It is by no means unusual that the exponent range is exceeded. When this happens it is necessary to reformulate the problem to make it better scaled. For example, it is often better to work with the special function $\ln \Gamma(x)$ than with $\Gamma(x)$ because it is better scaled. ■

Addition and subtraction are much less satisfactory in floating point arithmetic than are multiplication and division. It is necessary to be alert for several situations that will be illustrated. When numbers of greatly different magnitudes are added (or subtracted), some information is lost. Suppose, for example, that we want to add $\delta = 0.123456 \times 10^{-4}$ to 0.100000×10^1 in six-digit chopped arithmetic. First, the exponents are adjusted to be the same and then the numbers are added:

$$\begin{array}{rll} & 0.100000 & \times 10^1 \\ + & \underline{0.0000123456} & \times 10^1 \\ & 0.1000123456 & \times 10^1. \end{array}$$

The result is chopped to 0.100012×10^1. Notice that some of the digits did not participate in the addition. Indeed, if $|y| < |x|u$, then $x \oplus y = x$ and the "small" number y plays no role at all. The loss of information does not mean the answer is inaccurate; it is accurate to one unit of roundoff. The problem is that the lost information may be needed for later calculations.

Example 1.11. Difference quotients. Earlier we made use of the fact that for small δ,

$$\frac{F(x+\delta)-F(x)}{\delta} \approx F'(x).$$

In many applications this is used to approximate $F'(x)$. To get an accurate approximation, δ must be "small" compared to x. It had better not be *too* small for the precision, or else we would have $x \oplus \delta = x$ and compute a value of zero for $F'(x)$. If δ is large enough to affect the sum but still "small," some of its digits will not affect the sum in the sense that $x \oplus \delta - x \neq \delta$. In the difference quotient we want to divide by the actual difference of the arguments, not δ itself. A better way to proceed is to define

$$\Delta = (x \oplus \delta) \ominus x$$

and approximate

$$F'(x) \approx \frac{F(x+\Delta)-F(x)}{\Delta}.$$

The two approximations are mathematically equivalent, but computationally different. For example, suppose that $F(x) = x$ and we approximate $F'(x)$ for $x = 1$ using $\delta = 0.123456 \times 10^{-4}$ in six-digit chopped arithmetic. We have just worked out $1 \oplus \delta = 0.100012 \times 10^{1}$; similarly, $\Delta = 0.120000 \times 10^{-4}$ showing the digits of δ that actually affect the sum. The first formula has

$$\frac{(1 \oplus \delta) \ominus 1}{\delta} = \frac{0.120000 \times 10^{-4}}{0.123456 \times 10^{-4}} = 0.972006 \times 10^{0}.$$

The second has

$$\frac{(1 \oplus \Delta) \ominus 1}{\Delta} = \frac{0.120000 \times 10^{-4}}{0.120000 \times 10^{-4}} = 0.100000 \times 10^{1}.$$

Obviously the second form provides a better approximation to $F'(1) = 1$. Quality codes for the numerical approximation of the Jacobian matrices needed for optimization, root solving, and the solution of stiff differential equations make use of this simple device. ■

Example 1.12. Limiting precision. In many of the codes in this book we attempt to recognize when we cannot achieve a desired accuracy with the precision available. The kind of test we make will be illustrated in terms of approximating a definite integral

$$\int_a^b f(x)\,dx.$$

This might be done by splitting the integration interval $[a,b]$ into pieces $[\alpha,\beta]$ and adding up approximations on all the pieces. Suppose that

$$\int_\alpha^\beta f(x)\,dx \approx \frac{\beta-\alpha}{6}\left[f(\alpha)+4f\left(\frac{\alpha+\beta}{2}\right)+f(\beta)\right].$$

The accuracy of this formula improves as the length $\beta - \alpha$ of the piece is reduced, so that, mathematically, any accuracy can be obtained by making this width sufficiently small. However, if $|\beta-\alpha| < 2u|\alpha|$, the floating point numbers α and $\alpha+(\beta-\alpha)/2$ are the same. The details of the test are not important for this chapter; the point is that when the interval is small enough, we cannot ignore the fact that there are only a finite number of digits in floating point arithmetic. If α and β cannot be distinguished in the precision available, the computational results will not behave like mathematical results from the real number system. In this case the user of the software must be warned that the requested accuracy is not feasible. ■

Example 1.13. Summing a divergent series. The sum S of a series

$$\sum_{m=1}^{\infty} a_m$$

is the limit of partial sums

$$S_n = \sum_{m=1}^{n} a_m.$$

There is an obvious algorithm for evaluating S:

$$\begin{aligned} S_1 &= a_1 \\ S_n &= S_{n-1} \oplus a_n, \qquad n = 2,3,\ldots, \end{aligned}$$

continuing until the partial sums stop changing. A classic example of a divergent series is the harmonic series

$$\sum_{m=1}^{\infty} \frac{1}{m}.$$

If the above algorithm is applied to the harmonic series, the computed S_n increase and the a_n decrease until

$$S_n = S_{n-1} \oplus a_n = S_n$$

and the partial sums stop changing. The surprising thing is how small S_n is when this happens—try it and see. In floating point arithmetic this divergent series has a finite sum. The observation that when the terms become small enough, the partial sums stop changing is true of convergent as well as divergent series. Whether the value so obtained is an accurate approximation to S depends on how fast the series converges. It really is necessary to do some mathematical analysis to get reliable results. Later in this chapter we consider how to sum the terms a little more accurately. ■

An acute difficulty with addition and subtraction occurs when some information, lost due to adding numbers of greatly different size, is needed later because of a subtraction. Before going into this, we need to discuss a rather tricky point.

Example 1.14. Cancellation (loss of significance). Subtracting a number y from a number x that agrees with it in one or more leading digits leads to a cancellation of these digits. For example, if $x = 0.123654 \times 10^{-5}$ and $y = 0.123456 \times 10^{-5}$, then

$$\begin{array}{rl} & 0.123654 \quad \times 10^{-5} \\ - & \underline{0.123456} \quad \times 10^{-5} \\ & 0.000198 \quad \times 10^{-5} = 0.198000 \times 10^{-8}. \end{array}$$

The interesting point is that when cancellation takes place, the subtraction is done *exactly*, so that $x \ominus y = x - y$. The difficulty is what is called a *loss of significance*. When cancellation takes place, the result $x - y$ is smaller in size than x and y, so errors *already present* in x and y are *relatively* larger in $x - y$. Suppose that x is an approximation to X and y is an approximation to Y. They might be measured values or the results of some computations. The difference $x - y$ is an approximation to $X - Y$ with the magnitude of its relative error satisfying

$$\left| \frac{(x-y)-(X-Y)}{X-Y} \right| = \left| \frac{(x-X)-(y-Y)}{X-Y} \right| \leq \left| \frac{x-X}{X} \right| \left| \frac{X}{X-Y} \right| + \left| \frac{y-Y}{Y} \right| \left| \frac{Y}{X-Y} \right|.$$

If x is so close to y that there is cancellation, the relative error can be large because the denominator $X - Y$ is small compared to X or Y. For example, if $X = 0.123654700\cdots \times 10^{-5}$, then x agrees with X to a unit roundoff in six-digit arithmetic. With $Y = y$ the value we seek is $X - Y = 0.198700\cdots \times 10^{-8}$. Even though the subtraction $x - y = 0.198000 \times 10^{-8}$ is done exactly, $x - y$ and $X - Y$ differ in the fourth digit. In this example, x and y have at least six significant digits, but their difference has only three significant digits. ■

It is worth remarking that we made use of cancellation in Example 1.11 when we computed

$$\Delta = (x \oplus \delta) \ominus x.$$

Because δ is small compared to x, there is cancellation and $\Delta = (x \oplus \delta) - x$. In this way we obtain in Δ the digits of δ that actually affected the sum.

Example 1.15. Roots of a quadratic. Suppose we wish to compute the roots of

$$x^2 + bx + c = 0.$$

The familiar quadratic formula gives the roots x_1 and x_2 as

$$x_{1,2} = -\frac{b}{2} \pm \sqrt{\left(\frac{b}{2}\right)^2 - c},$$

assuming $b \geq 0$. If c is small compared to b, the square root can be rewritten and approximated using the binomial series to obtain

$$\frac{b}{2}\sqrt{1 - \frac{4c}{b^2}} \approx \frac{b}{2}\left(1 - \frac{2c}{b^2} + \cdots\right).$$

This shows that the true roots

$$\begin{aligned} x_1 &\approx -b \\ x_2 &\approx -c/b. \end{aligned}$$

In finite precision arithmetic some of the digits of c have no effect on the sum $(b/2)^2 - c$. The extreme case is

$$\left(\frac{b}{2}\right)^2 \ominus c = \left(\frac{b}{2}\right)^2.$$

It is important to appreciate that the quantity is computed accurately in a relative sense. However, some information is lost and we shall see that in some circumstances we need it later in the computation. A square root is computed with a small relative error and the same is true of the subtraction that follows. Consequently, the bigger root $x_1 \approx -b$ is computed accurately by the quadratic formula. In the computation of the smaller root, there is cancellation when the square root term is subtracted from $-b/2$. The subtraction itself is done exactly, but the error already present in $(b/2)^2 \ominus c$ becomes important in a relative sense. In the extreme case the formula results in zero as an approximation to x_2.

For this particular task a reformulation of the problem avoids the difficulty. The expression

$$\begin{aligned} (x - x_1)(x - x_2) &= x^2 - (x_1 + x_2) + x_1 x_2 \\ &= x^2 + bx + c \end{aligned}$$

shows that $x_1 x_2 = c$. As we have seen, the bigger root x_1 can be computed accurately using the quadratic formula and then

$$x_2 = c/x_1$$

provides an accurate value for x_2. ■

Example 1.16. Alternating series. As we observed earlier, it is important to know when enough terms have been taken from a series to approximate the limit to a desired accuracy. Alternating series are attractive in this regard. Suppose $a_0 \geq a_1 \geq \cdots \geq a_n \geq a_{n+1} \geq \cdots \geq 0$ and $\lim_{n \to \infty} a_n = 0$. Then the alternating series

$$\sum_{m=0}^{\infty} (-1)^m a_m$$

converges to a limit S and the error of the partial sum

$$S_n = \sum_{m=0}^{n} (-1)^m a_m$$

satisfies

$$|S - S_n| \leq a_{n+1}.$$

To see a specific example, consider the evaluation of $\sin x$ by its Maclaurin series

$$\sin x = x - \frac{x^3}{3!} + \frac{x^5}{5!} - \frac{x^7}{7!} + \cdots.$$

Although this series converges quickly for any given x, there are numerical difficulties when $|x|$ is large. If, say, $x = 10$, the a_m are

$$10, \frac{10^3}{6}, \frac{10^5}{120}, \frac{10^7}{5040}, \ldots.$$

Clearly there are some fairly large terms here that must cancel out to yield a result $\sin 10$ that has magnitude at most 1. The terms a_m are the result of some computation that here can be obtained with small relative error. However, if a_m is large compared to the sum S, a small relative error in a_m will not be small compared to S and S will not be computed accurately.

We programmed the evaluation of this series in a straightforward way, being careful to compute, say,

$$-\frac{x^7}{7!} = -\left(\frac{x^5}{5!}\right)\left(\frac{x}{6}\right)\left(\frac{x}{7}\right),$$

so as to avoid unnecessarily large quantities. Using single precision standard IEEE arithmetic we added terms until the partial sums stopped changing. This produced the value -0.544040 while the exact value should be $\sin x = -0.544021$. Although the series converges quickly for *all* x, some intermediate terms become large when $|x|$ is large. Indeed, we got an overflow due to the small exponent range in IEEE single precision arithmetic when we tried $x = 100$. Clearly floating point arithmetic does not free us from all concerns about scaling.

Series are often used as a way of evaluating functions. If the desired function value is small and if some terms in the series are comparatively large, then there must be cancellation and we must expect that inaccuracies in the computation of the terms will cause the function value to be inaccurate in a relative sense. ■

We have seen examples showing that the sum of several numbers depends on the order in which they are added. Is there a "good" order? We now derive a rule of thumb that can be quite useful. We can form $a_1 + a_2 + \cdots + a_N$ by the algorithm used in Example 1.13. The first computed partial sum is

$$\begin{aligned} fl(S_2) &= a_1 \oplus a_2 = (a_1 + a_2)(1 + \delta_2) \\ &= S_2 + \delta_2 a_1 + \delta_2 a_2, \end{aligned}$$

where $|\delta_2| \leq u$. It is a little special. The general case is represented by the next computed partial sum, which is

$$\begin{aligned} fl(S_3) &= fl(S_2) \oplus a_3 = (fl(S_2) + a_3)(1 + \delta_3) \\ &= S_3 + (\delta_2 + \delta_3)a_1 + (\delta_2 + \delta_3)a_2 + \delta_3 a_3 \\ &\quad + \delta_2\delta_3 a_1 + \delta_2\delta_3 a_2, \end{aligned}$$

where $|\delta_3| \leq u$. To gain insight, we approximate this expression by dropping terms involving the products of small factors so that

$$fl(S_3) \approx S_3 + (\delta_2 + \delta_3)a_1 + (\delta_2 + \delta_3)a_2 + \delta_3 a_3.$$

Continuing in this manner we find that

$$\begin{aligned} fl(a_1+\cdots+a_N) \approx\ & (a_1+\cdots+a_N) \\ & + (\delta_2+\delta_3+\cdots+\delta_N)a_1 \\ & + (\delta_2+\delta_3+\cdots+\delta_N)a_2 \\ & + (\delta_3+\delta_4+\cdots+\delta_N)a_3 \\ & + \cdots + \delta_N a_N. \end{aligned}$$

According to this approximation, the error made when a_k is added to S_k might grow, but its effect in S_N will be no bigger than $(N-k+1)u|a_k|$. This suggests that to reduce the total error, the terms should be added in order of increasing magnitude. A careful bound on the error of repeated summation leads to the same rule of thumb. Adding in order of increasing magnitude is *usually* a good order, but *not necessarily* a good order (because of the complex ways that the individual errors can interact). Much mathematical software makes use of this device to enhance the accuracy of the computations.

The approximate error can be bounded by

$$|fl(S_N) - S_N| \subseteq Nu \sum_{n=0}^{N} |a_n|.$$

Here we use the symbol $\subseteq$ to mean "less than or equal to a quantity that is approximately." (The "less than" is not sharp here.) Further manipulation provides an approximate bound on the magnitude of the sum's relative error

$$\frac{|fl(S_N) - S_N|}{|S_N|} \subseteq Nu \frac{\sum_{n=0}^{N} |a_n|}{\left|\sum_{n=0}^{N} a_n\right|}.$$

The dangerous situation is when $\left|\sum_{n=0}^{N} a_n\right| \ll \sum_{n=0}^{N} |a_n|$, which is when cancellation takes place. An important consequence is that if all the terms have the same sign, the sum will be computed accurately in a relative sense, provided only that the number of terms is not too large for the precision available.

For a convergent series

$$S = \sum_{m=0}^{\infty} a_m,$$

it is necessary that $|a_m| \to 0$ as $m \to \infty$. Rather than sum in the natural order $m = 0, 1, \ldots$, it would often be better to work out mathematically how many terms N are needed to approximate S to the desired accuracy and then calculate S_N in the reverse order $a_N, a_{N-1}, \ldots$.

Example 1.17. There are two ways of interpreting errors that are important in numerical analysis. So far we have been considering a *forward* error analysis. This corresponds to bounding the errors in the answer by bounding at each stage the errors that might arise and their effects. To be specific, recall the expression for the error of summing three numbers:

$$fl(S_3) = S_3 + (\delta_2+\delta_3+\delta_2\delta_3)x_1 + (\delta_2+\delta_3+\delta_2\delta_3)x_2 + \delta_3 x_3.$$

A forward error analysis might bound the absolute error by

$$|fl(S_3) - S_3| \leq \left(2u + u^2\right)(|x_1| + |x_2|) + u|x_3|.$$

(This is a sharp version of the approximate bound given earlier.) A *backward* error analysis views the computed result as the result computed in exact arithmetic of a problem with somewhat different data. Let us reinterpret the expression for $fl(S_3)$ in this light. It is seen that

$$fl(S_3) = y_1 + y_2 + y_3,$$

where

$$\begin{aligned} y_1 &= x_1(1 + \delta_2 + \delta_3 + \delta_2\delta_3) \\ y_2 &= x_2(1 + \delta_2 + \delta_3 + \delta_2\delta_3) \\ y_3 &= x_3(1 + \delta_3). \end{aligned}$$

In the backward error analysis view, the computed sum is the exact sum of terms y_k that are each close in a relative sense to the given data x_k. An algorithm that is stable in the sense of backward error analysis provides the exact solution of a problem with data close to that of the original problem. As to whether the two solutions are close, that is a matter of the conditioning of the problem. A virtue of this way of viewing errors is that it separates the roles of the stability of the algorithm and the condition of the problem. Backward error analysis is particularly attractive when the input data are of limited accuracy, as, for example, when the data are measured or computed. It may well happen that a stable algorithm provides the exact solution to a problem with data that cannot be distinguished from the given data because of their limited accuracy. We really cannot ask more of the numerical scheme in such a situation, but again we must emphasize that how close the solution is to that corresponding to the given data depends on the conditioning of the problem. We shall return to this matter in the next chapter.

A numerical example will help make the point. For $x_1 = 0.12 \times 10^2$, $x_2 = 0.34 \times 10^1$, $x_3 = -0.15 \times 10^2$, the true value of the sum is $S_3 = 0.40 \times 10^0$. When evaluated in two digit decimal chopped arithmetic, $fl(S_3) = 0.00 \times 10^0$, a very inaccurate result. Nevertheless, with $y_1 = 0.116 \times 10^2$, $y_2 = x_2$, and $y_3 = x_3$, we have $fl(S_3) = y_1 + y_2 + y_3$. The computed result is the exact sum of numbers close to the original data. Indeed, two of the numbers are the same as the original data and the remaining one differs by less than a unit of roundoff. ■

For most of the numerical tasks in this book it is not necessary to worry greatly about the effects of finite precision arithmetic. Two exceptions are the subject of the remaining examples. The first is the computation of a root of a continuous function $f(x)$. Naturally we would like to compute a number z for which $f(z)$ is as close to zero as possible in the precision available. Routines for this purpose ask the user to specify a desired accuracy. Even if the user does not request a very accurate root, the routine may "accidentally" produce a number z for which $f(z)$ is very small. Because it is usually not expensive to solve this kind of problem, it is quite reasonable for a user to ask for all the accuracy possible. One way or the other, we must ask what happens when x is very close to a root. An underflow is possible since $f(x) \to 0$ as $x \to z$. If this does not happen, it is usually found that the value of $f(z)$ fluctuates erratically as $x \to z$. Because of this we must devise algorithms that will behave sensibly when the computed value $f(x)$ does not have even the correct sign for x near z. An example will show how the details of evaluation of $f(x)$ are important when x is near a root.

Example 1.18. Let $f(x) = x^2 - 2x + 1$ be evaluated at $x = 1.018$ with three-digit chopped arithmetic and $-100 \le e \le 100$. The exact answer is $f(1.018) = 0.324 \times 10^{-3}$. Because the coefficients of f are small integers, no error arises when they are represented as floating point numbers. However, x is not a floating point number in this arithmetic and there is an error when $\widehat{x} = fl(x) = 0.101 \times 10^1$ is formed. Several algorithms are possible that arise in different ways of writing $f(x)$:

$$\begin{aligned} f(x) &= [(x^2) - (2x)] + 1 \\ &= x(x-2) + 1 \\ &= (x-1)^2. \end{aligned}$$

These forms work out to

$$\begin{aligned} y_1 &= [(\widehat{x} \otimes \widehat{x}) \ominus (2 \otimes \widehat{x})] \oplus 1 \\ &= 0.000 \times 10^{-100} \\ y_2 &= \widehat{x} \otimes (\widehat{x} \ominus 2) \oplus 1 \\ &= 0.100 \times 10^{-2} \\ y_3 &= (\widehat{x} \ominus 1) \otimes (\widehat{x} \ominus 1) \\ &= 0.100 \times 10^{-3}. \end{aligned}$$

All of the results have large relative errors. This should not be too surprising since the problem is poorly conditioned (see Exercise 1.5).

Figure 1.2 is a plot of 281 values of the function $f(x) = (x\exp(x) - 1)^3$ for arguments near $x = 0.567$. Single precision IEEE arithmetic was used for this calculation and the cubed term in the function was expanded out to generate more roundoff. In exact arithmetic $f(x)$ vanishes at only one point α near 0.567, a point that satisfies $\alpha = \exp(-\alpha)$. However, it is clear from the figure that the floating point version is not nearly so well behaved near this α. ■

In Chapter 2 we discuss the numerical solution of a system of linear equations. In contrast to the solution of nonlinear equations, codes based on the method there try to compute an answer as accurately as possible in the precision available. A difficulty with precision arises when we try to assess the accuracy of the result.

Example 1.19. Residual calculation. The simplest system of linear equations is

$$ax = b.$$

The quality of an approximate solution z can be measured by how well it satisfies the equation. The discrepancy is called its residual:

$$r = b - az.$$

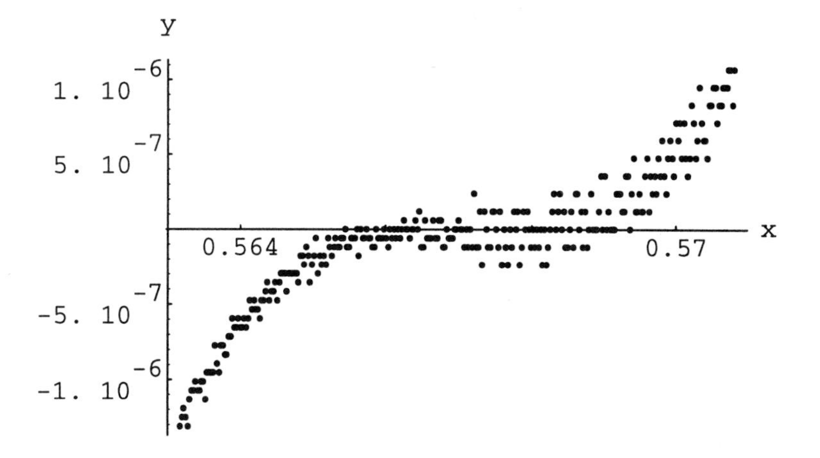

Figure 1.2 Floating point evaluation of $f(x) = x^3 e^{3x} - 3x^2 e^{2x} + 3xe^x - 1$.

If z is a very good solution, its residual r is small and there is cancellation when forming

$$b \ominus (a \otimes z) = b - (a \otimes z).$$

Defining δ by

$$a \otimes z = az(1 + \delta),$$

the computed residual

$$b \ominus (a \otimes z) = b - az - az\delta = r - az\delta.$$

The computed residual differs from the true residual by a quantity that can be as large as $|az|u \approx |b|u$. When r is small because z is a good solution and $|b|$ happens to be large, the computed residual may have few, if any, correct digits (although the relative residual $|r/b|$ is fine). When z is a good solution, it is generally necessary to use double precision to obtain its residual to single precision accuracy. ■

EXERCISES

1.8 Suppose that $z = 0.180 \times 10^2$ is an approximate solution of $ax = b$ for $a = 0.111 \times 10^0$, $b = 0.200 \times 10^1$. Use three-digit decimal chopped arithmetic to compute the residual $r = b - az$. Compute the residual in double precision and in exact arithmetic. Discuss the results.

1.9 For $\alpha = 0.8717$ and $\beta = 0.8719$ calculate the midpoint of the interval $[\alpha, \beta]$ using the formula $(\alpha + \beta)/2$. First use four-digit decimal chopped arithmetic, then four-digit decimal rounded arithmetic. How reasonable are the answers? Find another formula for the midpoint and use four-digit decimal (rounded or chopped) arithmetic to calculate the midpoint of $[0.8717, 0.8719]$. Is your formula better or worse?

1.10 In the model arithmetic, a single operation is carried out with a small relative error. Unfortunately the same is not true of complex arithmetic. To see this, let $z = a + ib$ and $w = c + id$. By definition, $zw = (ac - bd) + i(ad + bc)$. Show how the real part, $ac - bd$, of the product zw might be computed with a large relative error even though all individual calculations are done with a small relative error.

1.11 An approximation S to e^x can be computed by using the Taylor series for the exponential function:

$$\begin{aligned}
&S := 1 \\
&P := 1 \\
&\text{for } k = 1, 2, \ldots \text{ begin} \\
&\qquad P := xP/k \\
&\qquad S := S + P \\
&\text{end } k.
\end{aligned}$$

The loop can be stopped when $S = S + P$ to machine precision.

(a) Try this algorithm with $x = -10$ using single precision arithmetic. What was k when you stopped? What is the relative error in the resulting approximation? Does this appear to be a good way to compute e^{-10} to full machine precision?

(b) Repeat (a) with $x = +10$.

(c) Why are the results so much more reasonable for (b)?

(d) What would be a computationally safe way to compute e^{-10}?

1.12 Many problems in astrodynamics can be approximated by the motion of one body about another under the influence of gravity, for example, the motion of a satellite about the earth. This is a useful approximation because by a combination of analytical and numerical techniques, these two body problems can be solved easily. When a better approximation is desired, for example, we need to account for the effect of the moon or sun on the satellite, it is natural to compute it as a correction to the orbit of the two body problem. This is the basis of Encke's method; for details see Section 9.3 of [2]. A fundamental issue is to calculate accurately the small correction to the orbit. This is reduced to the accurate calculation of a function $f(q)$ for *small* $q \geq 0$. The function is

$$f(q) = \frac{1}{q}[1 - (1 - 2q)^{-3/2}].$$

Explain why $f(q)$ cannot be evaluated accurately in finite precision arithmetic when q is small. In the explanation you should assume that $y^{-3/2}$ can be evaluated with a relative error that is bounded by a small multiple of the unit roundoff. Use the binomial series to show

$$f(q) = -3 - \frac{3 \cdot 5}{2!}q - \frac{3 \cdot 5 \cdot 7}{3!}q^2 - \cdots.$$

Why is this series a better way to evaluate $f(q)$ when q is small?

1.13 Let a regular polygon of N sides be inscribed in a unit circle. If L_N denotes the length of one side, the circumference of the polygon, $N \times L_N$, approximates the circumference of the circle, 2π; hence $\pi \approx NL_N/2$ for large N. Using Pythagoras' theorem it is easy to relate L_{2N} to L_N:

$$L_{2N}^2 = 2\left(1 - \sqrt{1 - L_N^2/4}\right).$$

Starting with $L_4 = \sqrt{2}$ for a square, approximate π by means of this recurrence. Explain why a straightforward implementation of the recurrence in floating point arithmetic does not yield an accurate value for π. (Keep in mind that $L_N \to 0$ as $N \to \infty$.) Show that the recurrence can be rearranged as

$$L_{2N}^2/4 = \frac{L_N^2/4}{2\left(1 + \sqrt{1 - L_N^2/4}\right)}$$

and demonstrate that this form works better.

1.14 A study of the viscous decay of a line vortex leads to an expression for the velocity

$$u_\theta = \frac{\Gamma_0}{2\pi r}\left(1 - \exp\left(-\frac{r^2}{4\nu t}\right)\right)$$

at a distance r from the origin at time $t > 0$. Here Γ_0 is the initial circulation and $\nu > 0$ is the kinematic viscosity. For some purposes the behavior of the velocity at distances $r \ll \sqrt{4\nu t}$ is of particular interest. Why is the form given for the velocity numerically unsatisfactory for such distances? Assuming that you have available a function for the accurate computation of $\sinh(x)$, manipulate the expression into one that can be evaluated in a more accurate way for very small r.

1.3 CASE STUDY 1

Now let us look at a couple of examples that illustrate points made in this chapter. The first considers the evaluation of a special function. The second illustrates the fact that practical computation often requires tools from several chapters of this book. Filon's method for approximating finite Fourier integrals will be developed in Chapter 3 and applied in Chapter 5. An aspect of the method that we take up here is the accurate computation of coefficients for the method.

The representation of the hyperbolic cosine function in terms of exponentials

$$x = \cosh(y) = \frac{\exp(y) + \exp(-y)}{2}$$

makes it easy to verify that for $x > 1$,

$$y = -\ln\left(x - \sqrt{x^2 - 1}\right).$$

Let us consider the evaluation of this expression for $\cosh^{-1}(x)$ in floating point arithmetic when $x \gg 1$. An approximation made earlier in this chapter will help us to understand better what it is that we want to compute. After approximating

$$\sqrt{x^2 - 1} = x\sqrt{1 - \frac{1}{x^2}} \approx x\left(1 - \frac{1}{2x^2} + \cdots\right),$$

we find that

$$y = -\ln\left(x - \sqrt{x^2 - 1}\right) \approx -\ln\left(\frac{1}{2x}\right) = \ln(2x).$$

The first difficulty we encounter in the evaluation is that when x is very large, x^2 overflows. This overflow is unnecessary because the argument we are trying to compute is on scale. If x is large, but not so large that x^2 overflows, the effect of the 1 "falls off the end" in the subtraction, meaning that $fl(x^2 - 1) = fl(x^2)$. This subtraction is carried out with a small relative error, and the same is true of less extreme cases, but there is a loss of information when numbers are of greatly different size. The square root is obtained with a small relative error. The information lost in the subtraction is needed at the next step because there is severe cancellation. Indeed, for large x, we might end up computing $x - x = 0$ as the argument for $\ln(x)$, which would be disastrous.

How might we reformulate the task to avoid the difficulties just noted? A little calculation shows that

$$-\ln\left(x - \sqrt{x^2 - 1}\right) = \ln\left(\frac{1}{x - \sqrt{x^2 - 1}}\right) = \ln\left(x + \sqrt{x^2 - 1}\right),$$

a form that avoids cancellation. The preliminary analysis we did to gain insight suggests a better way of handling the rest of the argument:

$$\sqrt{x^2 - 1} = x\sqrt{1 - \left(\frac{1}{x}\right)^2}.$$

Notice that here we form $(1/x)^2$ instead of $1/x^2$. This rearrangement exchanges a possible overflow when forming x^2 for a harmless underflow, harmless, that is, if the system sets an underflow to zero and continues on. We see now that the expression

$$y = \ln\left(x + x\sqrt{1 - \left(\frac{1}{x}\right)^2}\right)$$

avoids all the difficulties of the original expression for $\cosh^{-1}(x)$. Indeed, it is clear that for large x, evaluation of this expression in floating point arithmetic will lead to an approximation of $\ln(2x)$, as it should.

For our second example we consider Filon's method for approximating finite Fourier integrals, which is developed in Chapter 3:

$$\int_a^b f(x)\cos(\omega x)\,dx \approx h[\alpha(f(b)\sin(\omega b) - f(a)\sin(\omega a)) + \beta C_e + \gamma C_o].$$

Here $\theta = \omega h$ and

$$\alpha = \left(\theta^2 + \theta\sin(\theta)\cos(\theta) - 2\sin^2(\theta)\right)/\theta^3$$
$$\beta = 2\left(\theta\left(1 + \cos^2(\theta)\right) - 2\sin(\theta)\cos(\theta)\right)/\theta^3$$
$$\gamma = 4(\sin(\theta) - \theta\cos(\theta))/\theta^3.$$

The details of the terms C_e and C_o do not concern us here. There is a similar formula for integrals with the sine function in place of the cosine function that involves the same coefficients α, β, γ. It is shown in Case Study 3 that the absolute error of this approximation is bounded by a constant times h^3. To get an accurate integral, it might be necessary to use a small h, meaning that θ is small, but the expressions for the coefficients are unsatisfactory in this case. Each suffers from cancellation in the numerator, and the resulting error is amplified by the division by the small quantity θ^3. To see the cancellation more clearly, let us approximate the sine and cosine terms in, say, α by the leading terms in their Taylor series, $\sin(\theta) \approx \theta$ and $\cos(\theta) \approx 1$, to get

$$\alpha \approx \left(\theta^2 + \theta\cdot\theta\cdot 1 - 2\theta\cdot\theta\right)/\theta^3.$$

Obviously there is perfect cancellation of leading terms in the numerator. This analysis suggests a remedy: for small θ, expand the coefficients in Taylor series and deal with the cancellation and small divisor analytically. The resulting series are

$$\alpha = \frac{2}{45}\theta^3 - \frac{2}{315}\theta^5 + \frac{2}{4725}\theta^7 - \frac{8}{467{,}775}\theta^9 + \frac{4}{8{,}513{,}505}\theta^{11} - \cdots$$

$$\beta = \frac{2}{3} + \frac{2}{15}\theta^2 - \frac{4}{105}\theta^4 + \frac{2}{567}\theta^6 - \frac{4}{22{,}275}\theta^8 + \frac{4}{675{,}675}\theta^{10} - \cdots$$

$$\gamma = \frac{4}{3} - \frac{2}{15}\theta^2 + \frac{1}{210}\theta^4 - \frac{1}{11{,}340}\theta^6 + \frac{1}{997{,}920}\theta^8 - \frac{1}{129{,}729{,}600}\theta^{10} + \cdots.$$

It might be remarked that it was easy to compute these expansions by means of the symbolic capabilities of the Student Edition of MATLAB. In the program used to compute the integral of Case Study 3, these expressions were used for $\theta \leq 0.1$. Because the terms decrease rapidly, nested multiplication is not only an efficient way to evaluate the expressions but is also accurate.

As a numerical illustration of the difficulty we evaluated both forms of α for a range of θ in single precision in FORTRAN. Reference values were computed using the trigonometric form and double precision. This must be done with some care. For instance, if T is a single precision variable and we want a double precision copy DT for computing the reference values, the lines of code

```
T = 0.1E0
DT = 0.1D0
```

are not equivalent to

```
T = 0.1E0
DT = T
```

This is because on a machine with binary or hexadecimal arithmetic, $0.1E0$ agrees with $0.1D0$ only to single precision. For the reference computation we require a double precision version of the actual machine number used in the single precision computations, hence we must use the second code. As we have remarked previously, most computers today perform intermediate computations in higher precision, despite specification of the precision of all quantities. With T, S, and C declared as single precision variables, we found remarkable differences in the result of

```
S = SIN(T)
C = COS(T)
ALPHA = (T**2+T*S*C-2E0*S**2)/T**3
```

and

```
ALPHA = (T**2+T*SIN(T)*COS(T)-2E0*SIN(T)**2)/T**3
```

differences that depended on the machine and compiler used. On a PC with a Pentium chip, the second code gave nearly full single precision accuracy. The first gave the poor results that we expect of computations carried out entirely in single precision.

The coefficient α was computed for a range of θ using the trigonometric definition and single precision arithmetic and its relative error computed using a reference value computed in double precision. Similarly the error of the value computed in single precision from the Taylor series was found. Plotted against θ in Figure 1.3 is the relative error for both methods (on a logarithmic scale). Single precision accuracy corresponds to about seven digits, so the Taylor series approach gives about all the accuracy we could hope for, although for the largest value of θ it appears that another term in the expansion would be needed to get full accuracy. Obviously the trigonometric definition leads to a great loss of accuracy for "small" θ. Indeed, θ is not very small in an absolute sense here; rather, it is small considering its implications for the cost of evaluating the integral when the parameter ω is moderately large.

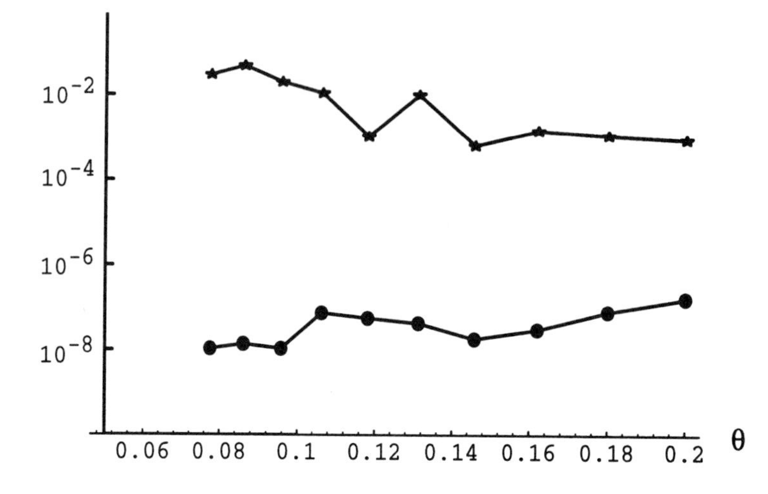

Figure 1.3 Error in series form (•) versus trig form (⋆) for a Filon coefficient.

1.4 FURTHER READING

A very interesting and readable account of the interaction of the floating point number system with the solution of quadratic equations has been given by Forsythe [6]. Henrici [8] gives another elementary treatment of floating point arithmetic that introduces the useful idea of a statistical treatment of errors. To pursue the subject in depth, consult the book *Rounding Errors in Algebraic Processes* by J. H. Wilkinson [10]. Wilkinson's books are unmatched for their blend of theoretical advance, striking examples, practical insight, applications, and readability. For more information on the practical evaluation of special functions, see the books by Cody and Waite [3] or Fike [5]. Other interesting discussions on floating point arithmetic are the books of Goldberg [7] and Higham [9].

REFERENCES

1. ANSI/IEEE, *IEEE Standard for Binary Floating Point Arithmetic*, Std 754-1985, New York, 1985.
2. R. Bate, D. Miller, and J. White, *Fundamentals of Astrophysics*, Dover, New York, 1971.
3. W. Cody and W. Waite, *Software Manual for the Elementary Functions*, Prentice Hall, Englewood Cliffs, N.J., 1980.
4. J. Dongarra, J. Bunch, C. Moler, and G. Stewart, LINPACK *Users' Guide*, SIAM, Philadelphia, 1979.
5. C. Fike, *Computer Evaluation of Mathematical Functions*, Prentice Hall, Englewood Cliffs, N.J., 1968.
6. G. Forsythe, "What is a satisfactory quadratic equation solver?," in *Constructive Aspects of the Fundamental Theorem of Algebra*, B. Dejon and P. Henrici, eds., Wiley, London, 1969.

7. D. Goldberg, "What every computer scientist should know about floating-point arithmetic," *ACM Computing Surveys*, 23 (1991), pp. 5–48.
8. P. Henrici, *Elements of Numerical Analysis*, Wiley, New York, 1964.
9. N. Higham, *Accuracy and Stability of Numerical Algorithms*, SIAM, Philadelphia, 1996.
10. J. Wilkinson, *Rounding Errors in Algebraic Processes*, Dover, Mineola, N.Y., 1994.

MISCELLANEOUS EXERCISES FOR CHAPTER 1

1.15 Use three-digit decimal chopped arithmetic with $m = -100$ and $M = 100$ to construct examples for which

(a) $(x \otimes y) \otimes z \neq x \otimes (y \otimes z)$

(b) $(x \oplus y) \oplus z \neq x \oplus (y \oplus z)$

(c) $x \otimes (y \oplus z) \neq (x \otimes y) \oplus (x \otimes z)$

(d) $(x \oplus y) \oplus z$ has a "large" relative error.

You are allowed to use negative numbers. Examples can be constructed so that either one of the expressions cannot be formed in the arithmetic, or both can be formed but the values are different.

1.16 For a set of measurements $x_1, x_2, \ldots, x_N$, the sample mean $\bar{x}$ is defined to be

$$\bar{x} = \frac{1}{N} \sum_{i=1}^{N} x_i.$$

The sample standard deviation s is defined to be

$$(N-1)s^2 = \sum_{i=1}^{N} (x_i - \bar{x})^2.$$

Another expression,

$$(N-1)s^2 = \sum_{i=1}^{N} x_i^2 - \frac{1}{N} \left(\sum_{i=1}^{N} x_i \right)^2,$$

is often recommended for hand computation of s. Show that these two expressions for s are mathematically equivalent. Explain why one of them may provide better numerical results than the other, and construct an example to illustrate your point.

1.17 Fourier series,

$$a_0 + \sum_{n=1}^{\infty} (a_n \cos nx + \sin nx),$$

are of great practical value. It appears to be necessary to evaluate a large number of sines and cosines if we wish to evaluate such a series, but this can be done cheaply by recursion. For the specific x of interest, for $n = 1, 2, \ldots$ let

$$s_n = \sin nx \text{ and } c_n = \cos nx.$$

Show that for $n = 2, 3, \ldots$

$$s_n = s_1 c_{n-1} + c_1 s_{n-1} \text{ and } c_n = c_1 c_{n-1} - s_1 s_{n-1}.$$

After evaluating $s_1 = \sin x$ and $c_1 = \cos x$ with the intrinsic functions of the programming language, this recursion can be used to evaluate simply and inexpensively all the $\sin nx$ and $\cos nx$ that are needed. To see that the recursion is stable, suppose that for some $m > 1$, s_m and c_m are computed incorrectly as $\widehat{s}_m = s_m + \varepsilon_m$ and $\widehat{c}_m = c_m + \tau_m$. If no further arithmetic errors are made, the errors ε_m and τ_m will propagate in the recurrence so that we compute

$$\widehat{s}_n = s_1 \widehat{c}_{n-1} + c_1 \widehat{s}_{n-1} \text{ and } \widehat{c}_n = c_1 \widehat{c}_{n-1} - s_1 \widehat{s}_{n-1}$$

for $n = m+1, \ldots$. Let ε_n and τ_n be the errors in $\widehat{s}_n$ and $\widehat{c}_n$ so that, by definition,

$$\widehat{s}_n = s_n - \varepsilon_n \text{ and } \widehat{c}_n = c_n - \tau_n.$$

Prove that for all $n \geq m$

$$\varepsilon_n^2 + \tau_n^2 = \varepsilon_m^2 + \tau_m^2,$$

which implies that for all $n \geq m$

$$|\varepsilon_n| \leq \sqrt{\varepsilon_m^2 + \tau_m^2} \text{ and } |\tau_n| \leq \sqrt{\varepsilon_m^2 + \tau_m^2}.$$

In this sense, errors are not amplified and the recurrence is quite stable.

CHAPTER 2

SYSTEMS OF LINEAR EQUATIONS

One of the most frequently encountered problems in scientific computation is that of solving n simultaneous linear equations in n unknowns. If we denote the unknowns by $x_1, x_2, \ldots, x_n$, such a system can be written in the form

$$\begin{array}{ccccccccc} a_{11}x_1 & + & a_{12}x_2 & + & \cdots & + & a_{1n}x_n & = & b_1 \\ a_{21}x_1 & + & a_{22}x_2 & + & \cdots & + & a_{2n}x_n & = & b_2 \\ \vdots & & & & & & \vdots & & \\ a_{n1}x_1 & + & a_{n2}x_2 & + & \cdots & + & a_{nn}x_n & = & b_n. \end{array} \tag{2.1}$$

The given data here are the right-hand sides b_i, $i = 1, 2, \ldots, n$, and the coefficients a_{ij} for $i, j = 1, 2, \ldots, n$. Problems of this nature arise almost everywhere in the applications of mathematics (e.g., the fitting of polynomials and other curves through data and the approximation of differential and integral equations by finite, algebraic systems). Several specific examples are found in the exercises for this chapter (see also [12] or [13]). To talk about (2.1) conveniently, we shall on occasion use some notation from matrix theory. However, we do not presume that the reader has an extensive background in this area. Using matrices, (2.1) can be written compactly as

$$A\mathbf{x} = \mathbf{b}, \tag{2.2}$$

where

$$A = \begin{pmatrix} a_{11} & a_{12} & \cdots & a_{1n} \\ a_{21} & a_{22} & \cdots & a_{2n} \\ \vdots & \vdots & & \vdots \\ a_{n1} & a_{n2} & \cdots & a_{nn} \end{pmatrix}, \quad \mathbf{x} = \begin{pmatrix} x_1 \\ x_2 \\ \vdots \\ x_n \end{pmatrix}, \quad \mathbf{b} = \begin{pmatrix} b_1 \\ b_2 \\ \vdots \\ b_n \end{pmatrix}.$$

Consider for the moment the case $n = 1$ in (2.1),

$$a_{11}x_1 = b_1.$$

If $a_{11} \neq 0$, the equation has a unique solution, namely $x_1 = b_1/a_{11}$. If $a_{11} = 0$, then some problems do not have solutions ($b_1 \neq 0$) while others have many solutions (if $b_1 = 0$, any number x_1 is a solution). The same is true for general n. There are two kinds of matrices, *nonsingular* and *singular*. If the matrix A is nonsingular, there is a unique solution vector $\mathbf{x}$ for any given right-hand side $\mathbf{b}$. If A is singular, there is no solution for some right-hand sides **b** and many solutions for other **b**. In this book we concentrate on systems of linear equations with nonsingular matrices.

Example 2.1. The problem

$$\begin{array}{rcrcr} 2x_1 & + & 3x_2 & = & 8 \\ 5x_1 & + & 4x_2 & = & 13 \end{array} \quad \text{or} \quad \begin{pmatrix} 2 & 3 \\ 5 & 4 \end{pmatrix} \begin{pmatrix} x_1 \\ x_2 \end{pmatrix} = \begin{pmatrix} 8 \\ 13 \end{pmatrix}$$

has a nonsingular coefficient matrix. The linear system has the unique solution

$$x_1 = 1,\ x_2 = 2 \quad \text{or} \quad \mathbf{x} = \begin{pmatrix} 1 \\ 2 \end{pmatrix}.$$

■

Example 2.2. The problem

$$\begin{array}{rcrcr} 2x_1 & + & 3x_2 & = & 4 \\ 4x_1 & + & 6x_2 & = & 7 \end{array} \quad \text{or} \quad \begin{pmatrix} 2 & 3 \\ 4 & 6 \end{pmatrix} \begin{pmatrix} x_1 \\ x_2 \end{pmatrix} = \begin{pmatrix} 4 \\ 7 \end{pmatrix}$$

has a singular coefficient matrix. If

$$\mathbf{b} = \begin{pmatrix} 4 \\ 7 \end{pmatrix},$$

there is no solution, for if x_1 and x_2 were numbers such that $4 = 2x_1 + 3x_2$, then we would have $8 = 2 \times 4 = 2 \times (2x_1 + 3x_2) = 4x_1 + 6x_2$, which is impossible because of the second equation. If

$$\mathbf{b} = \begin{pmatrix} 4 \\ 8 \end{pmatrix},$$

there are many solutions, namely

$$x_1 = \frac{4 - 3c}{2},\ x_2 = c$$

for all real numbers c. ■

In the nonsingular case there exists a matrix called the inverse of A, denoted by A^{-1}, such that the unique solution of (2.2) is given by

$$\mathbf{x} = A^{-1}\mathbf{b}.$$

For $n = 1$, $A^{-1} = (1/a_{11})$. Should we compute A^{-1} and then form the product $A^{-1}\mathbf{b}$ to solve (2.2)? We shall see that the answer is generally no even if we want to solve (2.2) with the same matrix A and many different right-hand sides **b**.

2.1 GAUSSIAN ELIMINATION WITH PARTIAL PIVOTING

The most popular method for solving a nonsingular system of linear equations (2.1) is called Gaussian elimination. It is both simple and effective. In principle it can be used to compute solutions of problems with singular matrices when they have solutions, but there are better ways to do this. The basic idea in elimination is to manipulate the equations of (2.1) so as to obtain an equivalent set of equations that is easy to solve. An equivalent set of equations is one that has the same solutions. There are three basic operations used in elimination: (1) multiplying an equation by a nonzero constant, (2) subtracting a multiple of one equation from another, and (3) interchanging rows. First, if any equation of (2.1) is multiplied by the nonzero constant α, we obtain an equivalent set of equations. To see this, suppose that we multiply the kth equation by α to get

$$\alpha a_{k1}x_1 + \alpha a_{k2}x_2 + \cdots + \alpha a_{kn}x_n = \alpha b_k. \tag{2.3}$$

If $x_1, x_2, \ldots, x_n$ satisfy (2.1), then they obviously satisfy the set of equations that is the same as (2.1) except for the kth equation, which is (2.3). Conversely, because $\alpha \neq 0$, if $x_1, x_2, \ldots, x_n$ satisfy this second set of equations, they obviously satisfy the first. Second, suppose we replace equation i by the result of subtracting the multiple α of equation k from equation i:

$$\begin{aligned} a_{11}x_1 + a_{12}x_2 + \cdots + a_{1n}x_n &= b_1 \\ \vdots \qquad\qquad & \quad \vdots \\ a_{i-1,1}x_1 + a_{i-1,2}x_2 + \cdots + a_{i-1,n}x_n &= b_{i-1} \\ (a_{i1} - \alpha a_{k1})x_1 + (a_{i2} - \alpha a_{k2})x_2 + \cdots + (a_{in} - \alpha a_{kn})x_n &= b_i - \alpha b_k \\ \vdots \qquad\qquad & \quad \vdots \\ a_{n1}x_1 + a_{n2}x_2 + \cdots + a_{nn}x_n &= b_n. \end{aligned} \tag{2.4}$$

If $x_1, x_2, \ldots, x_n$ satisfy (2.1), then by definition

$$a_{i1}x_1 + a_{i2}x_2 + \cdots + a_{in}x_n = b_i$$

and

$$a_{k1}x_1 + a_{k2}x_2 + \cdots + a_{kn}x_n = b_k,$$

so that

$$(a_{i1}x_1 + a_{i2}x_2 + \cdots + a_{in}x_n) - \alpha(a_{k1}x_1 + \cdots + a_{kn}x_n) = b_i - \alpha b_k.$$

Thus $x_1, x_2, \ldots, x_n$ satisfy all the equations of (2.4). To work in reverse, suppose now that $x_1, x_2, \ldots, x_n$ satisfy (2.4). Then in particular they satisfy equations i and k,

$$\begin{aligned} (a_{i1} - \alpha a_{k1})x_1 + \cdots + (a_{in} - \alpha a_{kn})x_n &= b_i - \alpha b_k \\ a_{k1}x_1 + \cdots + a_{kn}x_n &= b_k, \end{aligned}$$

so that

$$\begin{aligned} &[(\alpha a_{i1} - \alpha a_{k1})x_1 + \cdots + (a_{in} - \alpha a_{kn})x_n] \\ &\quad + \alpha[a_{k1}x_1 + \cdots + a_{kn}x_n] = (b_i - \alpha b_k) + \alpha b_k, \end{aligned}$$

which is just

$$a_{i1}x_1 + \cdots + a_{in}x_n = b_i.$$

Thus $x_1, x_2, \ldots, x_n$ also satisfy (2.1). Third, writing the equations in a different order clearly does not affect the solution, so interchanging rows results in an equivalent set of equations.

Example 2.3. Consider the problem

$$\begin{array}{rcrcrcr} 3x_1 & + & 6x_2 & + & 9x_3 & = & 39 \\ 2x_1 & + & 5x_2 & - & 2x_3 & = & 3 \\ x_1 & + & 3x_2 & - & x_3 & = & 2. \end{array} \tag{2.5}$$

If we subtract a multiple α of the first equation from the second, we get

$$(2-3\alpha)x_1 + (5-6\alpha)x_2 + (-2-9\alpha)x_3 = 3-39\alpha.$$

Choosing $\alpha = 2/3$ makes the coefficient of x_1 zero, so that the unknown x_1 no longer appears in this equation:

$$x_2 - 8x_3 = -23.$$

We say that we have "eliminated" the unknown x_1 from the equation. Similarly, we eliminate x_1 from equation (2.5) by subtracting $1/3$ times the first equation from it:

$$x_2 - 4x_3 = -11.$$

The system of equations (2.5)–(2.5) has been reduced to the equivalent system

$$\begin{array}{rcrcrcr} 3x_1 & + & 6x_2 & + & 9x_3 & = & 39 \\ & & x_2 & - & 8x_3 & = & -23 \\ & & x_2 & - & 4x_3 & = & -11. \end{array} \tag{2.6}$$

Now we set aside the first equation and continue the elimination process with the last two equations in the system (2.6) involving only the unknowns x_2 and x_3. Multiply the second equation by 1 and subtract from the third to produce

$$4x_3 = 12, \tag{2.7}$$

a single equation in one unknown. The equations (2.6) have now become the equivalent set of equations

$$\begin{array}{rcrcrcr} 3x_1 & + & 6x_2 & + & 9x_3 & = & 39 \\ & & x_2 & - & 8x_3 & = & -23 \\ & & & & 4x_3 & = & 12. \end{array} \tag{2.8}$$

The system (2.8)–(2.8) is easy to solve. From (2.8) $x_3 = 12/4 = 3$. The known value of x_3 is then used in (2.8) to obtain x_2, that is,

$$x_2 = 8x_3 - 23 = 8 \times 3 - 23 = 1.$$

Finally, the values for x_2 and x_3 are used in (2.8) to obtain x_1,

$$\begin{aligned} x_1 &= (-6x_2 - 9x_3 + 39)/3 \\ &= (-6 \times 1 - 9 \times 3 + 39)/3 = 2. \end{aligned}$$

Because this set of equations is equivalent to the original set of equations, the solution of (2.5)–(2.5) is $x_1 = 2$, $x_2 = 1$, $x_3 = 3$. ■

Let us turn to the general problem (2.1), which we now write with superscripts to help explain what follows:

$$\begin{array}{ccccccccc} a_{11}^{(1)}x_1 & + & a_{12}^{(1)}x_2 & + & \cdots & + & a_{1n}^{(1)}x_n & = & b_1^{(1)} \\ a_{21}^{(1)}x_1 & + & a_{22}^{(1)}x_2 & + & \cdots & + & a_{2n}^{(1)}x_n & = & b_2^{(1)} \\ & & \vdots & & & & \vdots & & \\ a_{n1}^{(1)}x_1 & + & a_{n2}^{(1)}x_2 & + & \cdots & + & a_{nn}^{(1)}x_n & = & b_n^{(1)}. \end{array}$$

If $a_{11}^{(1)} \neq 0$, we can eliminate the unknown x_1 from each of the succeeding equations. A typical step is to subtract from equation i the multiple $a_{i1}^{(1)}/a_{11}^{(1)}$ of the first equation. The results will be denoted with a superscript 2. The step is carried out by first forming

$$m_{i1} = \frac{a_{i1}^{(1)}}{a_{11}^{(1)}},$$

and then forming

$$a_{ij}^{(2)} = a_{ij}^{(1)} - m_{i1}a_{1j}^{(1)}, \; j = 1, 2, \ldots, n$$

and

$$b_i^{(2)} = b_i^{(1)} - m_{i1}b_1^{(1)}.$$

The multiple of the first equation is chosen to make $a_{i1}^{(2)} = 0$, that is, to eliminate the unknown x_1 from equation i. Of course, if $m_{i1} = 0$, the variable x_1 does not appear in equation i, so it does not need to be eliminated. By recognizing this, the arithmetic of elimination can be avoided. Doing this for each $i = 2, \ldots, n$ we arrive at the system

$$\begin{array}{ccccccccc} a_{11}^{(1)}x_1 & + & a_{12}^{(1)}x_2 & + & \cdots & + & a_{1n}^{(1)}x_n & = & b_1^{(1)} \\ & & a_{22}^{(2)}x_2 & + & \cdots & + & a_{2n}^{(2)}x_n & = & b_2^{(2)} \\ & & a_{32}^{(2)}x_2 & + & \cdots & + & a_{3n}^{(2)}x_n & = & b_3^{(2)} \\ & & \vdots & & & & \vdots & & \vdots \\ & & a_{n2}^{(2)}x_2 & + & \cdots & + & a_{nn}^{(2)}x_n & = & b_n^{(2)}. \end{array}$$

Notice that if we start out with A stored in a C or FORTRAN array, we can save a considerable amount of storage by overwriting the $a_{ij}^{(1)}$ with the $a_{ij}^{(2)}$ as they are created. Also, we can save the *multipliers* m_{i1} in the space formerly occupied by the $a_{i1}^{(1)}$ entries of A and just remember that all the elements below the diagonal in the first column are really zero after elimination. Later we shall see why it is useful to save the multipliers. Similarly, the original vector $\mathbf{b}$ can be overwritten with the $b_i^{(k)}$ as they are formed.

Now we set the first equation aside and eliminate x_2 from equations $i = 3, \ldots, n$ in the same way. If $a_{22}^{(2)} \neq 0$, then for $i = 3, 4, \ldots, n$ we first form

$$m_{i2} = \frac{a_{i2}^{(2)}}{a_{22}^{(2)}},$$

and then

$$a_{ij}^{(3)} = a_{ij}^{(2)} - m_{i2} a_{2j}^{(2)}, \; j = 2, 3, \ldots, n$$

and

$$b_i^{(3)} = b_i^{(2)} - m_{i2} b_2^{(2)}.$$

This results in

$$\begin{array}{ccccccccccc}
a_{11}^{(1)} x_1 & + & a_{12}^{(1)} x_2 & + & & & \cdots & + & a_{1n}^{(1)} x_n & = & b_1^{(1)} \\
 & & a_{22}^{(2)} x_2 & + & & & \cdots & + & a_{2n}^{(2)} x_n & = & b_2^{(2)} \\
 & & & & a_{33}^{(3)} x_3 & + & \cdots & + & a_{3n}^{(3)} x_n & = & b_3^{(3)} \\
 & & & & \vdots & & & & \vdots & & \vdots \\
 & & & & a_{n3}^{(3)} x_3 & + & \cdots & + & a_{nn}^{(3)} x_n & = & b_n^{(3)}.
\end{array}$$

As before, we set the first two equations aside and eliminate x_3 from equations $i = 4, \ldots, n$. This can be done as long as $a_{33}^{(3)} \neq 0$. The elements $a_{11}^{(1)}, a_{22}^{(2)}, \ldots$ are called pivot elements or simply *pivots*. Clearly, the process can be continued as long as no pivot vanishes. Assuming this to be the case, we finally arrive at

$$\begin{array}{ccccccccccc}
a_{11}^{(1)} & + & a_{12}^{(1)} x_2 & + & & & \cdots & + & a_{1n}^{(1)} x_n & = & b_1^{(1)} \\
 & & a_{22}^{(2)} x_2 & + & & & \cdots & + & a_{2n}^{(2)} x_n & = & b_2^{(2)} \\
 & & & & a_{33}^{(3)} x_3 & + & \cdots & + & a_{3n}^{(3)} x_n & = & b_3^{(3)} \\
 & & & & & & & & \vdots & & \vdots \\
 & & & & & & & & a_{nn}^{(n)} x_n & = & b_n^{(n)}.
\end{array} \tag{2.9}$$

In the computer implementation, the elements of the original matrix A are successively overwritten by the $a_{ij}^{(k)}$ as they are formed, and the multipliers m_{ij} are saved in the places corresponding to the variables eliminated. The process of reducing the system of equations (2.1) to the form (2.9) is called (forward) *elimination.* The result is a system with a kind of coefficient matrix called *upper triangular.* An upper triangular matrix $U = (u_{ij})$ is one for which

$$u_{ij} = 0 \quad \text{if} \quad i > j.$$

It is easy to solve a system of equations (2.9) with an upper triangular matrix by a process known as *back substitution.* If $a_{nn}^{(n)} \neq 0$, we solve the last equation for x_n,

$$x_n = b_n^{(n)} / a_{nn}^{(n)}.$$

Using the known value of x_n, we then solve equation $n-1$ for x_{n-1}, and so forth. A typical step is to solve equation k,

$$a_{k,k}^{(k)}x_k + a_{k,k+1}^{(k)}x_{k+1} + \cdots + a_{k,n}^{(k)}x_n = b_k^{(k)},$$

for x_k using the previously computed $x_n, x_{n-1}, \ldots, x_{k+1}$:

$$x_k = \left(b_k^{(k)} - \sum_{j=k+1}^{n} a_{kj}^{(k)} x_j \right) / a_{kk}^{(k)}.$$

The only way this process can break down (in principle) is if a pivot element is zero.

Example 2.4. Consider the two examples

$$\begin{aligned} 0 \cdot x_1 + 2x_2 &= 3 \\ 4x_1 + 5x_2 &= 6 \end{aligned}$$

and

$$\begin{aligned} 0 \cdot x_1 + 2x_2 &= 3 \\ 0 \cdot x_1 + 5x_2 &= 6. \end{aligned}$$

The entry $a_{11}^{(1)} = 0$, so it cannot be used as a pivot, but there is a simple remedy for the difficulty. We merely interchange the equations to get the equivalent set

$$\begin{aligned} 4x_1 + 5x_2 &= 6 \\ 0 \cdot x_1 + 2x_2 &= 3. \end{aligned}$$

For this problem the difficulty was easily avoided. This device will not work on the other problem, however, as it is singular. The first equation of this set requires $x_2 = \frac{3}{2}$ and the second requires $x_2 = \frac{6}{5}$, so there is no solution at all. ■

In the general case, suppose we have arrived at

$$\begin{array}{ccccccccccccc} a_{11}^{(1)}x_1 & + & \cdots & + & a_{1k}^{(1)}x_k & + & \cdots & + & a_{1n}^{(1)}x_n & = & b_1^{(1)} \\ & \ddots & & & \vdots & & & & \vdots & & \vdots \\ & & & & a_{kk}^{(k)}x_k & + & \cdots & + & a_{kn}^{(k)}x_n & = & b_k^{(k)} \\ & & & & \vdots & & & & \vdots & & \vdots \\ & & & & a_{nk}^{(k)}x_k & + & \cdots & + & a_{nn}^{(k)}x_n & = & b_n^{(k)} \end{array}$$

and $a_{kk}^{(k)} = 0$. We examine the elements $a_{jk}^{(k)}$ in column k for $j > k$. If for some index l, $a_{lk}^{(k)} \neq 0$, we interchange equations k and l. This does not affect the solution, so we rename the coefficients in the same way as before. The new pivot $a_{kk}^{(k)}$ is the old $a_{lk}^{(k)}$, which was nonzero, so the elimination process can now proceed as usual. If, however, $a_{jk}^{(k)} = 0$ for *all* $j = k, k+1, \ldots, n$, we have a difficulty of another sort: the matrix is singular. We prove this by showing that if a solution exists, it cannot be unique. Assume $\mathbf{x}$ is a solution to the problem. Set $z_{k+1} = x_{k+1}, \ldots, z_n = x_n$ and let z_k be arbitrary. The quantities $z_k, z_{k+1}, \ldots, z_n$ satisfy equations k through n because the unknown x_k does not appear in any of those equations. Now values for $z_1, \ldots, z_{k-1}$ may be determined by back substitution so that equations 1 through $k-1$ are satisfied:

$$a_{11}^{(1)} x_1 + \cdots + a_{1,k-1}^{(1)} x_{k-1} = b_1^{(1)} - \sum_{i=k}^{n} a_{1i}^{(1)} z_i$$

$$\vdots \qquad\qquad \vdots$$

$$a_{k-1,k-1}^{(k-1)} x_{k-1} = b_{k-1}^{(k-1)} - \sum_{i=k}^{n} a_{k-1,i}^{(k-1)} z_i.$$

This can be done because none of these pivot elements vanishes. Since all of the equations are satisfied, we have produced a whole family of solutions, namely z_1, z_2, $\ldots$, z_k, x_{k+1}, $\ldots$, x_n with z_k arbitrary. This shows that the matrix is singular.

Example 2.5. The following problems illustrate how singular systems are revealed during elimination. In the system

$$\begin{array}{rcrcrcl} x_1 & + & 2x_2 & - & x_3 & = & 2 \\ 2x_1 & + & 4x_2 & + & x_3 & = & 7 \\ 3x_1 & + & 6x_2 & - & 2x_3 & = & 7, \end{array}$$

one step of elimination yields

$$\begin{array}{rcrcrcr} x_1 & + & 2x_2 & - & x_3 & = & 27 \\ & & 0x_2 & + & 3x_3 & = & 3 \\ & & 0x_2 & + & x_3 & = & 1. \end{array}$$

Since we cannot find a nonzero pivot for the second elimination step, the system is singular. It is not hard to show that the solutions are

$$\begin{array}{rcl} x_1 & = & 3-2c \\ x_2 & = & c \\ x_3 & = & 1 \end{array}$$

for all real numbers c. The system

$$\begin{array}{rcrcrcr} x_1 & - & x_2 & + & x_3 & = & 0 \\ 2x_1 & + & x_2 & - & x_3 & = & -3 \\ x_1 & + & 2x_2 & - & 2x_2 & = & -2 \end{array}$$

is also singular, since two steps of elimination give

$$\begin{array}{rcrcrcr} x_1 & - & x_2 & + & x_3 & = & 0 \\ & & 3x_2 & - & 3x_3 & = & -3 \\ & & & & 0x_3 & = & 1. \end{array}$$

In this case there is no solution at all. ■

We conclude that by using interchanges, the elimination process has a zero pivot only if the original problem is singular. This statement is fine in theory, but the distinction between singular and nonsingular problems is blurred in practice by roundoff effects. Unless a pivot is exactly zero, interchange of equations is unnecessary in theory. However, it is plausible that working with a pivot that is almost zero will lead to problems of accuracy in finite precision arithmetic, and this turns out to be the case.

Example 2.6. The following example is due to Forsythe and Moler [6]:

$$\begin{aligned} 0.000100x_1 + 1.00x_2 &= 1.00 \\ 1.00x_1 + 1.00x_2 &= 2.00. \end{aligned}$$

Using three-digit decimal rounded floating point arithmetic, one step in the elimination process without interchanging equations yields for the second equation

$$[1.00 - (10{,}000)(1.00)]x_2 = [2.00 - (10{,}000)(1.00)]$$

or

$$-10{,}000x_2 = -10{,}000.$$

Clearly, $x_2 = 1.00$ and, by back substitution, $x_1 = 0.00$. Notice that all information contained in the second equation was lost at this stage. This happened because the small pivot caused a large multiplier and subsequently the subtraction of numbers of very different size. With interchange we have

$$\begin{array}{rcrcl} 1.00x_1 & + & 1.00x_2 & = & 2.00 \\ & & 1.00x_2 & = & 1.00 \end{array}$$

and $x_1 = 1.00$, $x_2 = 1.00$. The true solution is about $x_1 = 1.00010$, $x_2 = 0.99990$. ■

Small pivot elements $a_{kk}^{(k)}$ may lead to inaccurate results. As we saw in the last example, when eliminating the variable x_k in row i, a small pivot element leads to a large multiplier $m_{ik} = a_{ik}^{(k)} / a_{kk}^{(k)}$. When

$$a_{ij}^{(k+1)} = a_{ij}^{(k)} - m_{ik}a_{kj}^{(k)}$$

is formed, there is a loss of information whenever $m_{ik}a_{kj}^{(k)}$ is much larger than $a_{ij}^{(k)}$, information that may be needed later. A large multiplier is also likely to result in a large entry in the upper triangular matrix resulting from elimination. In the solution of the corresponding linear system by back substitution, we compute

$$x_k = \left(b_k^{(k)} - \sum_{j=k+1}^{n} a_{kj}^{(k)} \right) / a_{kk}^{(k)}.$$

If the pivot (the denominator) is small and the true value x_k is of moderate size, it must be the case that the numerator is also small. But if there are entries $a_{kj}^{(k)}$ of the upper triangular matrix that are large, this is possible only if cancellation occurs in the numerator. The large entries might well have been computed with a modest relative error, but because the entries are large this leads to a large absolute error in the numerator after cancellation. The small denominator amplifies this and there is a substantial relative error in x_k.

Partial pivoting is the most popular way of avoiding small pivots and controlling the size of the $a_{ij}^{(k)}$. When we eliminate x_k, we select the *largest* coefficient (in magnitude) of x_k in the last $n-k+1$ equations as the pivot. That is, if $|a_{lk}^{(k)}|$ is the largest of the $|a_{jk}^{(k)}|$ for $j=k,k+1,\ldots,n$, we interchange row k and row l. By renaming the rows we can assume that the pivot $a_{kk}^{(k)}$ has the largest magnitude possible. Partial pivoting avoids small pivots and nicely controls the size of the multipliers

$$|m_{ik}| = \left| \frac{a_{ik}^{(k)}}{a_{kk}^{(k)}} \right| \leq 1.$$

Controlling the size of the multipliers moderates the growth of the entries in the upper triangular matrix resulting from elimination. Let $a = \max_{i,j} |a_{ij}^{(1)}|$. Now

$$|a_{ij}^{(2)}| = |a_{ij}^{(1)} - m_{i1} a_{ij}^{(1)}| \leq 2a$$

and it is easy to go on to show that

$$|a_{ij}^{(k)}| \leq 2^{k-1} a.$$

This implies that

$$\max_{i,j,k} |a_{ij}^{(k)}| \leq 2^{n-1} \max_{i,j} |a_{ij}| \tag{2.10}$$

when partial pivoting is done. The growth that is possible here is very important to bounds on the error of Gaussian elimination. Wilkinson [15] points out that there is equality in this bound for matrices of the form

$$\begin{pmatrix} 1 & 0 & 0 & 0 & 1 \\ -1 & 1 & 0 & 0 & 1 \\ -1 & -1 & 1 & 0 & 1 \\ -1 & -1 & -1 & 1 & 1 \\ -1 & -1 & -1 & -1 & 1 \end{pmatrix}.$$

However, usually the growth is very modest. Research into this matter is surveyed in [11]. There are other ways of selecting pivot elements that lead to better bounds on the error and there are other ways of solving linear systems that have still better bounds. Some details will be mentioned later, but in practice the numerical properties of Gaussian elimination with partial pivoting are so good that it is the method of choice except in special circumstances, and when one speaks of "Gaussian elimination" it is assumed that partial pivoting is done unless something to the contrary is said. Gaussian elimination with partial pivoting is the basic method used in the popular computing environments MATLAB, *Mathematica*, and MATHCAD.

Example 2.7. Using exact arithmetic and elimination with partial pivoting, solve the following system:

$$\begin{pmatrix} 1 & 2 & 1 \\ 2 & 2 & 3 \\ -1 & -3 & 0 \end{pmatrix} \begin{pmatrix} x_1 \\ x_2 \\ x_3 \end{pmatrix} = \begin{pmatrix} 0 \\ 3 \\ 2 \end{pmatrix}.$$

Since $|2| > |1|$, we interchange the first and second equations to get

$$\begin{pmatrix} 2 & 2 & 3 \\ 1 & 2 & 1 \\ -1 & -3 & 0 \end{pmatrix} \begin{pmatrix} x_1 \\ x_2 \\ x_3 \end{pmatrix} = \begin{pmatrix} 3 \\ 0 \\ 2 \end{pmatrix}.$$

Using 2 as a pivot, we eliminate the coefficients of x_1 in equations two and three to get

$$\begin{pmatrix} 2 & 2 & 3 \\ 0 & 1 & -1/2 \\ 0 & -2 & 3/2 \end{pmatrix} \begin{pmatrix} x_1 \\ x_2 \\ x_3 \end{pmatrix} = \begin{pmatrix} 3 \\ -3/2 \\ 7/2 \end{pmatrix}.$$

Since $|-2| > |1|$, we interchange equations two and three,

$$\begin{pmatrix} 2 & 2 & 3 \\ 0 & -2 & 3/2 \\ 0 & 1 & -1/2 \end{pmatrix} \begin{pmatrix} x_1 \\ x_2 \\ x_3 \end{pmatrix} = \begin{pmatrix} 3 \\ 7/2 \\ -3/2 \end{pmatrix},$$

and using -2 as pivot obtain

$$\begin{pmatrix} 2 & 2 & 3 \\ 0 & -2 & 3/2 \\ 0 & 0 & 1/4 \end{pmatrix} \begin{pmatrix} x_1 \\ x_2 \\ x_3 \end{pmatrix} = \begin{pmatrix} 3 \\ 7/2 \\ 1/4 \end{pmatrix}.$$

Back substitution then gives

$$\begin{aligned} x_3 &= (1/4)/(1/4) = 1 \\ x_2 &= [(7/2) - (3/2)(1)]/(-2) = -1 \\ x_1 &= [(3) - (3)(1) - (2)(-1)]/(2) = 1 \end{aligned}$$

or

$$\mathbf{x} = \begin{pmatrix} 1 \\ -1 \\ 1 \end{pmatrix}.$$

■

The algorithm for elimination is quite compact:
Elimination, modification of **b**.

for $k = 1, \ldots, n-1$ begin

interchange rows so that $|a_{kk}| = \max_{k \le i \le n} |a_{ik}|$
if $|a_{kk}| = 0$, set singularity indicator, return
for $i = k+1, \ldots, n$ begin

$t := a_{ik}/a_{kk}$
for $j = k+1, \ldots, n$ begin
$a_{ij} := a_{ij} - t * a_{kj}$
end j
$b_i := b_i - t * b_k$
end i
end k
if $|a_{nn}| = 0$, set singularity indicator.

Back substitution

for $i = n, \ldots, 1$ begin
$x_i := b_i$
for $j = i+1, \ldots, n$ begin
$x_i := x_i - a_{ij} * x_j$
end j
$x_i := x_i / a_{ii}$
end i.

Sometimes we are interested in solving problems involving one matrix A and several right-hand sides $\mathbf{b}$. Examples are given in the exercises of problems with the right-hand sides corresponding to different data sets. Also, if we should want to compute the inverse of an $n \times n$ matrix A, this can be done a column at a time. It is left as an exercise to show that column i of A^{-1} is the result of solving the system of equations with column i of the identity matrix as $\mathbf{b}$. If we know all the right-hand sides in advance, it is clear that we can process them simultaneously. It is not always the case that they are all known in advance. The residual correction process we take up later is an example. For such problems it is important to observe that if we save the multipliers and record how the rows are interchanged when processing A, we can process $\mathbf{b}$ separately. To understand how this can be valuable, we first need to look at the costs of the various portions of this algorithm.

As a measure of work we count arithmetic operations. Since the number of additions and subtractions equals the number of multiplications, only the latter (as well as divisions) are counted. It is easy enough to see that elimination requires

$$n(n-1)(2n-1)/6 \text{ multiplications and } n(n-1)/2 \text{ divisions.}$$

Modification of $\mathbf{b}$ requires

$$n(n-1)/2 \text{ multiplications.}$$

Back substitution requires

$$n(n-1)/2 \text{ multiplications and } n \text{ divisions.}$$

For large n the multiplications dominate the cost, both because there are more of them and because they are relatively expensive. The most important point is that processing the matrix A is the bulk of the cost of solving a system of linear equations of even moderate size.

Several designs are seen in popular codes. The most straightforward is to input A and $\mathbf{b}$ and have the code compute the solution $\mathbf{x}$ and return it. It is quite easy to modify such a code to accept input of m right-hand sides, process all the right-hand sides along with A, and return all the solutions in an array. This is considerably cheaper than solving the problems one after another because A is processed only once and this is the most expensive part of the computation. In detail, solving m systems with the same A simultaneously requires

$$\frac{n(n-1)(2n-1)}{6} + m\left[\frac{n(n-1)}{2} + \frac{n(n-1)}{2}\right] = \frac{n^3}{3} - \frac{n^2}{2} + \frac{n}{6} + m(n^2 - n)$$

multiplications. Solving them independently requires

$$m\left[\frac{n(n-1)(2n-1)}{6} + \frac{n(n-1)}{2} + \frac{n(n-1)}{2}\right] = m\frac{n^3}{3} + \frac{mn^2}{2} - \frac{5mn}{6}$$

multiplications. If, for example, we wished to invert A, we would have $m = n$ and the cost would be

$$\frac{n^4}{3} + \frac{n^3}{2} - \frac{5n^2}{6}$$

multiplications. For large n, there is a considerable difference. The most flexible design separates the two phases of the computation. By saving the information necessary for processing the right-hand side, systems involving the same matrix A can be solved independently and just as inexpensively as if all the right-hand sides were available to begin with. This is the design found in production-grade software and in the programs of this chapter. Because it is a little more trouble to use than the simplest design, it is not unusual for libraries to have both. The computing environment MATLAB is an example of this.

EXERCISES

2.1 Using elimination with partial pivoting, determine which of the following systems are singular and which are nonsingular. For the nonsingular problems, find solutions. Use exact arithmetic.

(a)

$$\begin{aligned} -x_1 + 2x_2 + x_3 &= 5 \\ x_1 + 4x_2 - 3x_3 &= -8 \\ -2x_1 \qquad\quad + x_3 &= 5 \end{aligned}$$

(b)

$$\begin{aligned} x_1 - x_2 - 2x_3 &= -1 \\ -2x_1 - 2x_2 + 4x_3 &= 4 \\ 3x_1 + 3x_2 + x_3 &= 1 \end{aligned}$$

(c)

$$\begin{aligned} x_1 + 2x_2 - x_3 &= 2 \\ 2x_1 + 4x_2 + x_3 &= 7 \\ 3x_1 + 6x_2 - 2x_3 &= 7 \end{aligned}$$

(d)

$$\begin{aligned} x_1 - x_2 + x_3 &= 0 \\ 2x_1 + x_2 - x_3 &= -3 \\ x_1 + 2x_2 - 2x_3 &= -2 \end{aligned}$$

(e)

$$\begin{aligned} x_1 + x_2 + x_3 &= 0 \\ 2x_1 + x_2 - x_3 &= -3 \\ 2x_1 - 4x_3 &= -6 \end{aligned}$$

(f)

$$\begin{aligned} 2x_1 - 3x_2 + 2x_3 + 5x_4 &= 3 \\ x_1 - x_2 + x_3 + 2x_4 &= 1 \\ 3x_1 + 2x_2 + 2x_3 + x_4 &= 0 \\ x_1 + x_2 - 3x_3 - x_4 &= 0 \end{aligned}$$

2.2 Four loads applied to a three-legged table yield the following system for the reactions on the legs:

$$\begin{aligned} R_1 + R_2 + R_3 &= 110.00 \\ R_1 + R_2 &= 78.33 \\ R_2 + R_3 &= 58.33. \end{aligned}$$

Solve for R_1, R_2, and R_3 by hand.

2.3 The following set of equations arises in analyzing loads on an A-frame:

$$\begin{aligned} 8.00R_E - 1784.00 &= 0.00 \\ -8.00R_A + 1416.00 &= 0.00 \\ C_h + D_h &= 0.00 \\ C_v + D_v + 223.00 &= 0.00 \\ -5.18C_v - 5.18C_h + 446.00 &= 0.00 \\ -5.77D_v - 1456.00 &= 0.00 \\ -5.77B_v - 852.00 &= 0.00 \\ B_h + D_h &= 0.00. \end{aligned}$$

Solve the equations by hand.

2.4 Consider the linear system

$$\begin{aligned} x_1 + \tfrac{1}{2}x_2 + \tfrac{1}{3}x_3 &= 1 \\ \tfrac{1}{2}x_1 + \tfrac{1}{3}x_2 + \tfrac{1}{4}x_3 &= 0 \\ \tfrac{1}{3}x_1 + \tfrac{1}{4}x_2 + \tfrac{1}{5}x_3 &= 0. \end{aligned}$$

(a) Solve the system using *exact* arithmetic (any method).

(b) Put the system in matrix form using a two-digit decimal chopped representation.

(c) Solve the system in (b) *without* partial pivoting [same arithmetic as (b)].

(d) Solve the system in (b) *with* partial pivoting [same arithmetic as (b)].

(e) Solve the system in (b) using exact arithmetic.

2.2 MATRIX FACTORIZATION

Because it is illuminating, more advanced books invariably study elimination by viewing it as a matrix factorization. In this section we shall see that if no pivoting is done, the elimination algorithm factors, or decomposes, the matrix A into the product LU of a lower triangular matrix $L = (\ell_{ij})$, where

$$\ell_{ij} = 0 \quad \text{if} \quad i < j,$$

and an upper triangular matrix $U = (u_{ij})$, where

$$u_{ij} = 0 \quad \text{if} \quad i > j.$$

When partial pivoting is done, it is a version of A with its rows interchanged that is decomposed. Rows can be interchanged in A by multiplication with a matrix P called a permutation matrix. It is easy to construct P. If PA is to be the result of interchanging some rows of A, all we need do is take P to be the result of interchanging these rows in the identity matrix I. For example, to interchange rows 1 and 3 of the 3×3 matrix A in PA, we use

$$P = \begin{pmatrix} 0 & 0 & 1 \\ 0 & 1 & 0 \\ 1 & 0 & 0 \end{pmatrix}.$$

The entire elimination process with partial pivoting can be written as the

---

LU factorization

$$PA = LU.$$

Rather than sort out the permutations, we concentrate here on the factorization without pivoting to show how the "LU decomposition" arises. The remainder of this section provides the details of this factorization and it may be skipped by the reader unfamiliar with linear algebra. Looking back at the elimination described in the preceding section, we see that if $a_{11}^{(1)} \neq 0$, we could multiply row 1 by $m_{i1} = a_{i1}^{(1)} / a_{11}^{(1)}$ and subtract it from row i to eliminate the first unknown from row i. This is done for rows

$i=2,3,\ldots,n$. It is easily verified that when the matrix

$$M_1 = \begin{pmatrix} 1 & & & & \\ -m_{21} & 1 & & & \\ -m_{31} & 0 & 1 & & \\ \vdots & \vdots & & \ddots & \\ -m_{n1} & 0 & & 0 & 1 \end{pmatrix}$$

multiplies any matrix A on the left, it has the effect of multiplying row 1 of A by m_{i1} and subtracting the result from row i of A for $i=2,\ldots,n$. As with permutation matrices, this kind of matrix is found by performing the operations on the identity matrix. With the multipliers m_{i1} chosen as specified, the product M_1A has the form

$$M_1A = \begin{pmatrix} a_{11}^{(1)} & a_{12}^{(1)} & a_{13}^{(1)} & \cdots & a_{1n}^{(1)} \\ 0 & a_{22}^{(2)} & a_{23}^{(2)} & \cdots & a_{2n}^{(2)} \\ \vdots & \vdots & & & \\ 0 & a_{n2}^{(2)} & a_{n3}^{(2)} & \cdots & a_{nn}^{(2)} \end{pmatrix}.$$

For later use we note that multiplication by the inverse of a matrix "undoes" a multiplication by the matrix. To "undo" multiplication by M_1, we need to multiply row 1 by m_{i1} and *add* the result to row i for $i=2,\ldots,n$. In this way, we see that

$$M_1^{-1} = \begin{pmatrix} 1 & & & & \\ m_{21} & 1 & & & \\ m_{31} & 0 & 1 & & \\ \vdots & \vdots & & \ddots & \\ m_{n1} & 0 & & 0 & 1 \end{pmatrix}.$$

It is also easy to verify directly that $M_1^{-1}M_1 = I$. Suppose we have formed

$$M_{k-1}M_{k-2}\cdots M_2M_1A = \begin{pmatrix} a_{11}^{(1)} & a_{12}^{(1)} & \cdots & a_{1k}^{(1)} & \cdots & a_{1n}^{(1)} \\ 0 & a_{22}^{(2)} & \cdots & a_{2k}^{(2)} & \cdots & a_{2n}^{(2)} \\ & & \ddots & & & \\ & & & a_{kk}^{(k)} & \cdots & a_{kn}^{(k)} \\ & & & & \vdots & \\ & & & a_{nk}^{(k)} & \cdots & a_{nn}^{(k)} \end{pmatrix}.$$

If $a_{kk}^{(k)} \neq 0$, we want to multiply row k of this matrix by $m_{ik} = a_{ik}^{(k)}/a_{kk}^{(k)}$ and subtract it from row i of the matrix for $i=k+1,\ldots,n$. This is done by multiplying the matrix by

$$M_k = \begin{pmatrix} 1 & & & & \\ & \ddots & & & \\ & & 1 & & \\ & & -m_{k+1,k} & & \\ & & \vdots & \ddots & \\ & & -m_{n,k} & & 1 \end{pmatrix}.$$

Then

$$M_k M_{k-1} \cdots M_2 M_1 A = \begin{pmatrix} a_{11}^{(1)} & & & \cdots & & a_{1n}^{(1)} \\ & \ddots & & & & \\ & & a_{kk}^{(k)} & \cdots & & a_{kn}^{(k)} \\ & & 0 & a_{k+1,k+1}^{(k+1)} & & a_{k+1,n}^{(k+1)} \\ & & \vdots & \vdots & & \\ & & 0 & a_{n,k+1}^{(k+1)} & \cdots & a_{n,n}^{(k+1)} \end{pmatrix}$$

and

$$M_k^{-1} = \begin{pmatrix} 1 & & & & \\ & \ddots & & & \\ & & 1 & & \\ & & m_{k+1,k} & & \\ & & \vdots & \ddots & \\ & & m_{n,k} & & 1 \end{pmatrix}.$$

Elimination without pivoting results after $n-1$ steps in

$$M_{n-1} M_{n-2} \cdots M_1 A = \begin{pmatrix} a_{11}^{(1)} & \cdots & & a_{1n}^{(1)} \\ & a_{22}^{(2)} & \cdots & a_{2n}^{(2)} \\ & & \ddots & \\ & & & a_{n,n}^{(n)} \end{pmatrix},$$

which is an upper triangular matrix that we shall call U. Multiplication of this equation on the left by M_{n-1}^{-1}, then $M_{n-2}^{-1}, \ldots$, results in

$$A = M_1^{-1} \cdots M_{n-2}^{-1} M_{n-1}^{-1} U.$$

Earlier we saw the simple form of these inverse matrices. It is a delightful fact that their product is also extremely simple. Now

$$M_{n-1}^{-1} = \begin{pmatrix} 1 & & & & \\ & \ddots & & & \\ & & 1 & & \\ & & 0 & 1 & \\ & & 0 & m_{n,n-1} & 1 \end{pmatrix}.$$

Multiplication by M_{n-2}^{-1} means to multiply row $n-2$ by $m_{i,n-2}$ and add it to row i for $i = n-1, n$. In the special case of M_{n-2}^{-1} times M_{n-1}^{-1} this clearly results in

$$M_{n-2}^{-1} M_{n-1}^{-1} = \begin{pmatrix} 1 & & & & \\ & \ddots & & & \\ & & 1 & & \\ & & m_{n-1,n-2} & 1 & \\ & & m_{n,n-2} & m_{n,n-1} & 1 \end{pmatrix}.$$

Repetition of this argument shows that

$$M_1^{-1} \cdots M_{n-2}^{-1} M_{n-1}^{-1} = \begin{pmatrix} 1 & & & & & \\ m_{2,1} & 1 & & & & \\ m_{3,1} & m_{3,2} & 1 & & & \\ \vdots & \vdots & \vdots & & 1 & \\ m_{n,1} & m_{n,2} & m_{n,3} & \cdots & m_{n,n-1} & 1 \end{pmatrix},$$

which is a lower triangular matrix that we shall call L. Finally, then, we see that $A = LU$, where the L and U arise in an simple way during elimination. Because the diagonal elements of L are all ones, we do not need to store them. The matrix L is formed a column at a time and the elements can be written in the space occupied by elements of A that are set to zero. As the scheme was described in the preceding section, the elements of U are written over the elements of A as they are formed. One of the virtues of describing elimination in terms of a matrix factorization is that it is clear how to handle more than one vector $\mathbf{b}$ in solving $A\mathbf{x} = LU\mathbf{x} = \mathbf{b}$. For any given $\mathbf{b}$ we first solve

$$L\mathbf{y} = \mathbf{b}$$

and then

$$U\mathbf{x} = \mathbf{y}.$$

This yields the desired $\mathbf{x}$, for substitution shows that

$$L\mathbf{y} = L(U\mathbf{x}) = (LU)\mathbf{x} = A\mathbf{x} = \mathbf{b}.$$

Forward substitution to solve the lower triangular system $L\mathbf{y} = \mathbf{b}$, or

$$\begin{aligned} y_1 &= b_1 \\ m_{2,1}y_1 + y_2 &= b_2 \\ &\vdots \\ m_{n,1}y_1 + m_{n,2}y_2 + \cdots + y_n &= b_n, \end{aligned}$$

is just

$$\begin{aligned} y_1 &= b_1 \\ y_2 &= b_2 - m_{2,1}y_1 \\ &\vdots \\ y_n &= b_n - m_{n,1}y_1 - m_{n,2}y_2 - \cdots - m_{n,n-1}y_{n-1}. \end{aligned}$$

Back substitution is used to solve $U\mathbf{x} = \mathbf{y}$, or

$$\begin{array}{ccccccccc} u_{11} & + & u_{12}x_2 & + & \cdots & + & u_{1n}x_n & = & y_1 \\ & & u_{22}x_2 & + & \cdots & + & u_{2n}x_n & = & y_2 \\ & & & \ddots & & & \vdots & & \\ & & & & & & u_{nn}x_n & = & y_n, \end{array}$$

but now the order is $x_n, x_{n-1}, \ldots, x_1$:

$$\begin{aligned} x_n &= y_n / u_{n,n} \\ x_{n-1} &= (y_{n-1} - u_{n-1,n} x_n) / u_{n-1,n-1} \\ &\vdots \\ x_1 &= (y_1 - u_{12} x_2 - u_{13} x_3 - \cdots - u_{1,n} x_n) / u_{1,1}. \end{aligned}$$

There is another important decomposition of A that arises naturally in a discussion of least squares fitting of data. The reader should turn to the advanced texts cited for details, but a little perspective is useful. If the Gram–Schmidt process is used to form a set of orthonormal vectors from the columns of A, a decomposition $A = QR$ is obtained, where Q is an orthogonal matrix and R is an upper triangular matrix. An orthogonal matrix Q is one for which $Q^{-1} = Q^T$, so to solve $A\mathbf{x} = QR\mathbf{x} = \mathbf{b}$, all we have to do is form $R\mathbf{x} = Q^T\mathbf{b}$ and solve it by backward substitution. The Gram–Schmidt process in its classic form is not numerically stable, but there is a modification that is. A more popular way to obtain a QR decomposition stably is by means of Householder transformations. Error *bounds* for this way of solving systems of linear equations are much better than those for Gaussian elimination because there is no growth factor. However, the method is about twice as expensive. In another section dealing with matrices with special structure we take up important circumstances that favor Gaussian elimination over QR decomposition. Because the accuracy of Gaussian elimination is almost always satisfactory in practice, it is preferred except in special circumstances. One exception is in the solution of least squares problems where the QR decomposition is especially convenient and problems are often very ill-conditioned.

EXERCISES

2.5 Find the L and U in the LU decomposition (no pivoting) for the coefficient matrices in

(a) Exercise 2.1a;

(b) Exercise 2.1b;

(c) Exercise 2.1f.

2.6 Find an LU decomposition for the singular coefficient matrix in Exercise 2.1d. Is the decomposition unique?

2.3 ACCURACY

There are two main sources of error in the computed solution $\mathbf{z}$ of the linear system $A\mathbf{x} = \mathbf{b}$. The data A and $\mathbf{b}$ may not be measured exactly, and even if they are, errors are generally made in representing them as floating point numbers. Roundoff errors occur in the elimination and forward/backward substitution algorithms. It seems obvious that we should study the error

$$\mathbf{e} = \mathbf{x} - \mathbf{z},$$

but it turns out that a different way of approaching the issue of accuracy is illuminating. A backward error analysis views $\mathbf{z}$ as the exact solution of a perturbed problem

$$(A + \Delta A)\mathbf{z} = \mathbf{b} + \Delta\mathbf{b}.$$

If the perturbations ΔA and $\Delta \mathbf{b}$ are comparable to the measurement errors or the roundoff in the entries of A or $\mathbf{b}$, then it is reasonable to say that $\mathbf{z}$ is about as good a solution as we might hope to get.

A BACKWARD ERROR ANALYSIS

A floating point error analysis of a simple system of linear equations will be illuminating. Suppose that the system

$$\begin{aligned} u_{11}x_1 + u_{12}x_2 &= b_1 \\ u_{22}x_2 &= b_2 \end{aligned}$$

has arisen directly or as the result of applying Gaussian elimination to a more general system. In our development of elimination we discussed how a small pivot, here u_{11} and u_{22}, could be dangerous both for its direct effects and because it might lead to large elements in the upper triangular matrix, here u_{12}. Analyzing the error in this simple case will help us to understand this. Backward substitution in exact arithmetic produces the true solution as

$$\begin{aligned} x_2 &= \frac{b_2}{u_{22}} \\ x_1 &= \frac{b_1 - x_2 u_{12}}{u_{11}}. \end{aligned}$$

In floating point arithmetic,

$$x_2^* = b_2 \oslash u_{22} = \frac{b_2}{u_{22}}(1+\delta_1) = x_2(1+\delta_1).$$

Computation of the other component involves several steps. First we compute

$$x_2^* \otimes u_{12} = x_2^* u_{12}(1+\delta_2) = x_2 u_{12}(1+\delta_1)(1+\delta_2),$$

then

$$b_1 \ominus (x_2^* \otimes u_{12}) = (b_1 - (x_2^* \otimes u_{12}))(1+\delta_3),$$

and finally

$$\begin{aligned} x_1^* &= (b_1 \ominus (x_2^* \otimes u_{12})) \oslash u_{11} \\ &= \frac{(b_1 \ominus (x_2^* \otimes u_{12}))}{u_{11}}(1+\delta_4) \\ &= \frac{(b_1 - x_2^* u_{12}(1+\delta_2))}{u_{11}}(1+\delta_3)(1+\delta_4). \end{aligned}$$

In a backward error analysis, we express the solution x_1^*, x_2^* computed in floating point arithmetic as the solution in exact arithmetic of a perturbed problem:

$$\begin{aligned} u_{11}^* x_1^* + u_{12}^* x_2^* &= b_1 \\ u_{22}^* x_2^* &= b_2. \end{aligned}$$

As the notation suggests, this can be done without perturbing the right-hand side of the equation. The equation

$$\begin{aligned} x_2^* &= \frac{b_2}{u_{22}^*} \\ &= \frac{b_2}{u_{22}}(1+\delta_1) \end{aligned}$$

will be valid if we define

$$u_{22}^* = \frac{u_{22}}{(1+\delta_1)} \approx u_{22}(1-\delta_1).$$

Similarly, the equation

$$\begin{aligned} x_1^* &= \frac{b_1 - x_2^* u_{12}^*}{u_{11}^*} \\ &= \frac{(b_1 - x_2^* u_{12}(1+\delta_2))}{u_{11}}(1+\delta_3)(1+\delta_4) \end{aligned}$$

will be valid if we define

$$\begin{aligned} u_{12}^* &= u_{12}(1+\delta_2) \\ u_{11}^* &= \frac{u_{11}}{(1+\delta_3)(1+\delta_4)} \approx u_{11}(1-\delta_3-\delta_4). \end{aligned}$$

With these definitions we have expressed the computed solution of the given problem as the exact solution of a problem with perturbed matrix. It is seen that none of the coefficients of the matrix is perturbed by more than about two units of roundoff.

This analysis tells us that the backward substitution algorithm is sure to produce a good result in the sense that the computed solution is the exact solution of a problem close to the one given. However, that is not the same as saying that the computed solution is close to the true solution. A forward error analysis bounds directly the difference between the computed and true solutions.

Our basic assumption about floating point arithmetic is that a single operation is carried out with a relative error bounded by the unit of roundoff u, so we have

$$\left|\frac{x_2^* - x_2}{x_2}\right| = |\delta_1| \leq u.$$

Substitution of the expressions developed earlier and a little manipulation shows that

$$\frac{x_1^* - x_1}{x_1} = \sigma_2 - \frac{x_2 u_{12}}{x_1 u_{11}}\sigma_1(1+\sigma_2),$$

where

$$\begin{aligned} \sigma_1 &= \delta_1 + \delta_2 + \delta_1\delta_2 \\ \sigma_2 &= \delta_3 + \delta_4 + \delta_3\delta_4. \end{aligned}$$

This implies that

$$\left|\frac{x_1^* - x_1}{x_1}\right| \leq \left(2u+u^2\right)\left[1+\left|\frac{x_2 u_{12}}{x_1 u_{11}}\right|\left(1+2u+u^2\right)\right].$$

According to this bound, the relative error is generally small. A large relative error is possible only when $|x_2 u_{12}| \gg |x_1 u_{11}|$. If the solution is such that both components are of comparable size, a large relative error is possible only when the pivot u_{11} is small and/or the entry u_{12} in the upper triangular matrix is large. Large relative errors are more likely when $|x_2| \gg |x_1|$. The denominator can be written in the form

$$x_1 u_{11} = b_1 - x_2 u_{12},$$

showing that the relative error can be large when the numerator is large and the denominator is small because of cancellation.

ROUNDOFF ANALYSIS

A natural way to measure the quality of an approximate solution $\mathbf{z}$ is by how well it satisfies the equation. A virtue of this is that it is easy to compute the residual

$$\mathbf{r} = \mathbf{b} - A\mathbf{z}.$$

In this measure, a good solution $\mathbf{z}$ has a small residual. Because of cancellation (see Example 1.10), if we should want an accurate residual for a good solution, it will be necessary to compute it in higher precision arithmetic, and this may not be available. The residual provides a $\Delta\mathbf{b}$ for the backward error analysis, namely,

$$\Delta\mathbf{b} = -\mathbf{r}.$$

The residual $\mathbf{r}$ is connected to the error $\mathbf{e}$ by

$$\mathbf{r} = \mathbf{b} - A\mathbf{z} = A\mathbf{x} - A\mathbf{z} = A(\mathbf{x} - \mathbf{z}) = A\mathbf{e}$$

or $\mathbf{e} = A^{-1}\mathbf{r}$. A small residual $\mathbf{r}$, hence a small $\Delta\mathbf{b}$, may be perfectly satisfactory from the point of view of backward error analysis even when the corresponding error $\mathbf{e}$ is not small.

Example 2.8. To illustrate the distinction between the two points of view, consider the system

$$\begin{pmatrix} 0.747 & 0.547 \\ 0.623 & 0.457 \end{pmatrix} \begin{pmatrix} x_1 \\ x_2 \end{pmatrix} = \begin{pmatrix} 0.200 \\ 0.166 \end{pmatrix}. \tag{2.11}$$

We carry out the elimination process using three-digit chopped decimal arithmetic. After the first step we have

$$\begin{pmatrix} 0.747 & 0.547 \\ 0 & 0.001 \end{pmatrix} \begin{pmatrix} x_1 \\ x_2 \end{pmatrix} = \begin{pmatrix} 0.200 \\ 0.000 \end{pmatrix}.$$

It then follows that

$$z_2 = \frac{0.000}{0.001} = 0.000,$$

$$z_1 = (0.200 - 0.547 z_2)/0.747 = 0.267,$$

so the computed solution is

$$\mathbf{z} = \begin{pmatrix} 0.267 \\ 0.000 \end{pmatrix}.$$

The exact solution to (2.11) is easily found to be $x_1 = 1$ and $x_2 = -1$. Therefore the error (in exact arithmetic) is

$$\mathbf{e} = \mathbf{x} - \mathbf{z} = \begin{pmatrix} 1 - 0.267 \\ -1 - 0.000 \end{pmatrix} = \begin{pmatrix} 0.733 \\ -1 \end{pmatrix}.$$

In contrast, the residual (in exact arithmetic) is

$$\begin{aligned} \mathbf{r} &= \mathbf{b} - A\mathbf{z} \\ &= \begin{pmatrix} 0.200 - [(0.747 \times 0.267) + (0.547 \times 0.000)] \\ 0.166 - [(0.623 \times 0.267) + (0.457 \times 0.000)] \end{pmatrix} \\ &= \begin{pmatrix} 0.000551 \\ -0.000341 \end{pmatrix}. \end{aligned}$$

This says that $\mathbf{z}$ is the exact solution of $A\mathbf{z} = \mathbf{b} + \Delta\mathbf{b}$, where $b_1 = 0.200$ is perturbed to 0.199449 and b_2 is perturbed to 0.166341. Thus, $\mathbf{z}$ is the solution of a problem very close to the one posed, even though it differs considerably from the solution $\mathbf{x}$ of the original problem. ■

The fundamental difficulty in Example 2.8 is that the matrix in the system (2.11) is nearly singular. In fact, the first equation is, to within roundoff error, 1.2 times the second. If we examine the elimination process we see that z_2 was computed from two quantities that were themselves on the order of roundoff error. Carrying more digits in our arithmetic would have produced a totally different z_2. The error in z_2 propagates to an error in z_1. This accounts for the computed solution being in error. Why then are the residuals small? Regardless of z_2, the number z_1 was computed to make the residual for the first equation as nearly zero as possible in the arithmetic being used. The residual for the second equation should also be small because the system is close to singular: the first equation is approximately a multiple of the second. In Section 2.2 we observed that any matrix A could have its rows interchanged to obtain a matrix PA, which can be decomposed as the product of a lower triangular matrix L and an upper triangular matrix U. For simplicity we ignore the permutation matrix P in what follows. An error analysis of elimination using floating point arithmetic shows that L and U are computed with errors ΔL and ΔU, respectively. Then A is not exactly equal to the product $(L+\Delta L)(U+\Delta U)$. Let ΔA be defined so that

$$\begin{aligned} A + \Delta A &= (L+\Delta L)(U+\Delta U) \\ &= LU + (\Delta L)U + L(\Delta U) + (\Delta L)(\Delta U), \end{aligned}$$

that is,

$$\Delta A = (\Delta L)U + L(\Delta U) + (\Delta L)(\Delta U).$$

We might reasonably hope to compute L with errors ΔL that are small relative to L, and the same for U. However, the expression for ΔA shows that the sizes of L and U play important roles in how well A is represented by the computed factors. Partial pivoting keeps the elements of L less than or equal to 1 in magnitude. We also saw in (2.10) that the size of elements of U, the $a_{ij}^{(k)}$, was moderated with partial pivoting. In particular, they cannot exceed $2^{n-1} \max_{ij} |a_{ij}|$ for an $n \times n$ matrix. It can be shown rigorously, on taking into account the errors of decomposition and of forward/backward substitution, that the computed solution $\mathbf{z}$ of $A\mathbf{x} = \mathbf{b}$ satisfies

$$(A + \Delta A)\mathbf{z} = \mathbf{b}, \tag{2.12}$$

where the entries of ΔA are usually small. To make precise how small these entries are, we need a way of measuring the sizes of vectors and matrices. One way to measure the size of a vector $\mathbf{x}$ of n components is by its norm, which is denoted by $||\mathbf{x}||$. Several definitions of norm are common in numerical analysis. One that is likely to be familiar is the Euclidean length of $\mathbf{x}$, $\left(\sum_{i=1}^{n} x_i^2\right)^{1/2}$. All vector norms possess many of the properties of length. The norm used in this chapter is the maximum norm

$$||\mathbf{x}|| = \max_{1 \le i \le n} |x_i|. \tag{2.13}$$

If A is an $n \times n$ matrix and $\mathbf{x}$ is an n-vector, then $A\mathbf{x}$ is also an n-vector. A matrix norm can be defined in terms of a vector norm by

$$||A|| = \max_{\mathbf{x} \neq \mathbf{0}} \frac{||A\mathbf{x}||}{||\mathbf{x}||}. \tag{2.14}$$

Geometrically, this says that $||A||$ is the maximum relative distortion that the matrix A creates when it multiplies a vector $\mathbf{x} \neq \mathbf{0}$. It is not easy to evaluate $||A||$ directly from (2.14), but it can be shown that for the maximum norm (2.13)

$$||A|| = \max_i \sum_j |a_{ij}|, \tag{2.15}$$

which is easy enough to evaluate. An important inequality connects norms of vectors and matrices:

$$||A\mathbf{x}|| \le ||A|| \, ||\mathbf{x}||. \tag{2.16}$$

For $\mathbf{x} \neq \mathbf{0}$ this follows immediately from the definition (2.14). For $\mathbf{x} = \mathbf{0}$ we note that $A\mathbf{x} = \mathbf{0}$ and that $||\mathbf{x}|| = 0$, from which the inequality is seen to hold.

Example 2.9. Let $\mathbf{x} = \begin{pmatrix} -1 \\ 2 \\ 3 \end{pmatrix}$. Then

$$||\mathbf{x}|| = \max[|-1|, |2|, |3|] = 3.$$

Let $A = \begin{pmatrix} 1 & -1 & 0 \\ 2 & -2 & 3 \\ -4 & 1 & -1 \end{pmatrix}$. Then

$$\begin{aligned} \|A\| &= \max[(|1|+|-1|+|0|),(|2|+|-2|+|3|),(|-4|+|1|+|-1|)] \\ &= \max[(2),(7),(6)] = 7. \end{aligned}$$

■

Returning to the roundoff analysis for Gaussian elimination, it can be shown rigorously [11] that the computed solution z satisfies the perturbed equation (2.12) where

$$\|\Delta A\| \leq \gamma_n u \|A\|. \tag{2.17}$$

As usual, u is the unit roundoff. The factor γ_n depends on n and can grow as fast as 2^{n-1}. To put this in perspective, suppose that ΔA arises from rounding A to form machine numbers. Then $|\Delta a_{ij}|$ could be as large as $u|a_{ij}|$ and $\|\Delta A\|$ could be as large as

$$\max_i u \sum_{j=1}^{n} |a_{ij}| = u\|A\|.$$

According to the *bounds*, the perturbations due to the decomposition and forward/backward substitution process are at worst a factor of γ_n times the error made in the initial rounding of the entries of A. If the rigorous bound 2^{n-1} on γ_n truly reflected practice, we would have to resort to another algorithm for large n. Fortunately, for most problems γ_n is more like 10, independent of the size of n.

From this it can be concluded that Gaussian elimination practically always produces a solution $\mathbf{z}$ that is the exact solution of a problem close to the one posed. Since $A\mathbf{z} - \mathbf{b} = -\Delta A\mathbf{z}$, the residual $\mathbf{r}$ satisfies

$$\|\mathbf{r}\| = \|A\mathbf{z} - \mathbf{b}\| \leq \|\Delta A\|\,\|\mathbf{z}\| \leq \gamma_n u \|A\|\|\mathbf{z}\|.$$

This says that the size of the residual is nearly always small relative to the sizes of A and $\mathbf{z}$. However, recall that this does not imply that the actual error $\mathbf{e}$ is small.

For additional insight as to why Gaussian elimination tends to produce solutions with small residuals, think of the LU factorization of A discussed in Section 2.2. The forward substitution process used to solve the lower triangular system $L\mathbf{y} = \mathbf{b}$ successively computes $y_1, y_2, \ldots, y_n$ so as to make the residual zero. For example, regardless of the errors in y_1 and $m_{2,1}$ the value of y_2 is computed so that

$$m_{2,1}y_1 + y_2 = b_2,$$

that is, the residual of this equation is zero (in exact arithmetic) with this value of y_2. The same thing happens in the back substitution process to compute $x_n, x_{n-1}, \ldots, x_1$ that satisfy $U\mathbf{x} = \mathbf{y}$. Thus, the very nature of the process responds to errors in the data in such a way as to yield a small residual. This is not at all true when $\mathbf{x}$ is computed by first calculating the inverse A^{-1} and then forming $A^{-1}\mathbf{b}$. With a little extra work it is possible to make Gaussian elimination stable in a very strong sense.

Suppose that we have solved $A\mathbf{x} = \mathbf{b}$ to obtain an approximate solution $\mathbf{z}$. We can expect it to have some accuracy, although perhaps not all the accuracy possible in the precision used. A little manipulation shows that the error $\mathbf{e} = \mathbf{x} - \mathbf{z}$ satisfies $A\mathbf{e} = \mathbf{r}$, where $\mathbf{r}$ is the residual of the approximate solution $\mathbf{z}$. We have seen that if Gaussian elimination is organized properly, it is inexpensive to solve this additional system of equations. Of course, we do not expect to solve it exactly either, but we do expect that the computed approximation $\mathbf{d}$ to the error in $\mathbf{z}$ will have some accuracy. If it does, $\mathbf{w} = \mathbf{z} + \mathbf{d}$ will approximate $\mathbf{x}$ better than $\mathbf{z}$ does. In principle this process, called *iterative refinement*, can be repeated to obtain an approximation to $\mathbf{x}$ correct in all its digits. The trouble in practice is that for the process to work as described, we have to have an accurate residual, and the better the approximate solution, the more difficult this is to obtain. Skeel [14] has shown that just one step of iterative refinement with the residual computed in the working precision will provide a computed solution that is very satisfactory. This solution will have a small residual and will satisfy exactly a system of equations with each coefficient differing slightly from that of the given system. This is much better than the result for $\mathbf{z}$ that states that the perturbation in a coefficient is small compared to the norm of the whole matrix, not that it is small compared to the coefficient itself. So, if we are concerned about the reliability of Gaussian elimination with partial pivoting, we could save copies of the matrix and right-hand side and perform one step of iterative refinement in the working precision to correct the result as necessary.

NORM BOUNDS FOR THE ERROR

In the preceding subsection, we found that roundoff errors in the algorithm could be considered equivalent to errors in the data A. We now study the effect of such perturbations, as well as errors in the given data, on the error $\mathbf{e}$. For simplicity, let us first consider the case where only the data $\mathbf{b}$ is in error. Let $\mathbf{x} + \Delta\mathbf{x}$ be the solution of

$$A(\mathbf{x} + \Delta\mathbf{x}) = \mathbf{b} + \Delta\mathbf{b}.$$

Multiply this by A^{-1} and use the fact that $\mathbf{x} = A^{-1}\mathbf{b}$ to get

$$\Delta\mathbf{x} = A^{-1}\Delta\mathbf{b}. \tag{2.18}$$

Norm inequalities say that

$$||\Delta\mathbf{x}|| \leq ||A^{-1}||\,||\Delta\mathbf{b}||. \tag{2.19}$$

But $\mathbf{b} = A\mathbf{x}$ similarly implies $||\mathbf{b}|| \leq ||A||\,||\mathbf{x}||$, hence

$$\frac{||\Delta\mathbf{x}||}{||\mathbf{x}||} \leq ||A||\,||A^{-1}||\frac{||\Delta\mathbf{b}||}{||\mathbf{b}||}. \tag{2.20}$$

Inequality (2.19) says that in an absolute sense, input errors $||\Delta\mathbf{b}||$ can be amplified by as much as $||A^{-1}||$ in the solution. In contrast, (2.20) says that in a relative sense, input errors can be magnified by as much as $||A||\,||A^{-1}||$. The important quantity $||A||\,||A^{-1}||$ denoted by $\text{cond}(A)$ is called the

---

condition number of A

$$\text{cond}(A) = ||A||\,||A^{-1}||.$$

A theorem that helps us understand the condition number is

$$\min_{\det(S)=0} \frac{||S-A||}{||A||} = \frac{1}{\kappa(A)}.$$

In words this says that there is a singular matrix S that differs from A in a relative sense by the reciprocal of the condition number of A. Put differently, if A has a "large" condition number, it is "close" to a singular matrix.

The condition number clearly depends on the norm, but in this book we consider only the maximum norm (2.15).

Example 2.10. For $A = \begin{pmatrix} 1 & 2 \\ 3 & 4 \end{pmatrix}$, find $||A||, ||A^{-1}||, \text{cond}(A)$. First,

$$||A|| = \max[(|1|+|2|), (|3|+|4|)] = \max[(3), (7)] = 7.$$

For 2×2 matrices, the inverse matrix is easy to work out. If $A = \begin{pmatrix} a_{11} & a_{12} \\ a_{21} & a_{22} \end{pmatrix}$, then

$$A^{-1} = \begin{pmatrix} a_{22} & -a_{12} \\ -a_{21} & a_{11} \end{pmatrix} \Big/ (a_{11}a_{22} - a_{12}a_{21}). \tag{2.21}$$

So, in our case

$$A^{-1} = \begin{pmatrix} -2 & 1 \\ \frac{3}{2} & -\frac{1}{2} \end{pmatrix}$$

and

$$\begin{aligned} ||A^{-1}|| &= \max[(|-2|+|1|), (|3/2|+|-1/2|)] \\ &= \max[(3), (2)] = 3. \end{aligned}$$

Then

$$\text{cond}(A) = ||A|| \cdot ||A^{-1}|| = 7 \times 3 = 21.$$

■

Example 2.11. The matrix

$$A = \begin{pmatrix} 1 & -1 \\ 1 & -1+10^{-5} \end{pmatrix}$$

is much closer to being singular than the matrix in Example 2.10 since

$$A^{-1} = \begin{pmatrix} 1-10^5 & 10^5 \\ -10^5 & 10^5 \end{pmatrix}$$

and

$$||A|| = 2,\ ||A^{-1}|| = 2 \times 10^5,\ \mathrm{cond}(A) = 4 \times 10^5.$$

The theorem about the condition number says that there is a singular matrix that is within $1/\mathrm{cond}(A) = 2.5 \times 10^{-6}$ of A. Although not quite this close, the simple matrix

$$S = \begin{pmatrix} 1 & -1 \\ 1 & -1 \end{pmatrix}$$

is obviously singular and

$$\frac{||S - A||}{||A||} = 5 \times 10^{-6}.$$

■

The effect of perturbing A is more complicated because it is possible that $A + \Delta A$ is singular. However, if the perturbation is sufficiently small, say $||A^{-1}||||\Delta A|| < 1$, then it can be shown [9] that $A + \Delta A$ is nonsingular and further that if $(A + \Delta A)(\mathbf{x} + \Delta \mathbf{x}) = \mathbf{b} + \Delta \mathbf{b}$, then we have the so-called

---

condition number inequality

$$\frac{||\Delta \mathbf{x}||}{||\mathbf{x}||} \le \frac{\mathrm{cond}(A)}{1 - \mathrm{cond}(A)\frac{||\Delta A||}{||A||}} \left(\frac{||\Delta A||}{||A||} + \frac{||\Delta \mathbf{b}||}{||\mathbf{b}||} \right), \tag{2.22}$$

valid for $||A^{-1}||\,||\Delta A|| < 1$.

Inequality (2.17) and the related discussion say that rounding errors in the course of Gaussian elimination are equivalent to solving a perturbed system for which we usually have

$$\frac{||\Delta A||}{||A||} \approx 10u, \tag{2.23}$$

where u is the unit roundoff. In some applications data errors may be much larger than this, and they must be used in (2.22).

Example 2.12. Suppose we solve $A\mathbf{x} = \mathbf{b}$ on a machine with $u = 5 \times 10^{-11}$ and obtain

$$\mathbf{z} = \begin{pmatrix} 6.23415 \\ 18.6243 \end{pmatrix},\ \mathrm{cond}(A) = 1.0 \times 10^4.$$

Then from (2.22), assuming exact data so that $||\Delta A||/||A|| \approx 5 \times 10^{-10}$, the bound on the relative error is

$$\frac{||\Delta \mathbf{x}||}{||\mathbf{x}||} \le \frac{10^4}{1 - 10^4(5 \times 10^{-10})}(5 \times 10^{-10}) \approx 5 \times 10^{-6}.$$

On the other hand, if the data are known to be in error, say $||\Delta A||/||A|| \approx 10^{-6}$ and $||\Delta \mathbf{b}||/||\mathbf{b}|| \approx 10^{-6}$, then

$$\frac{||\Delta \mathbf{x}||}{||\mathbf{x}||} \leq \frac{10^4}{1 - 10^4(10^{-6})}(2 \times 10^{-6}) \approx 0.02\,.$$

With $||\mathbf{x}|| \approx ||\mathbf{z}|| \approx 18.6$ the absolute error bound is 0.37, so this analysis says that

$$x_1 = 6.23 \pm 0.37$$
$$x_2 = 18.62 \pm 0.37.$$

■

One criticism of the analysis based on norms is that it is a worst-case analysis. The condition number does not depend on **b**, so the inequality (2.22) must allow for the worst choice of **b** and $\Delta \mathbf{b}$. A large condition number is cause for concern, but it does not mean a given problem will be solved inaccurately. Furthermore, the fact that $||\Delta \mathbf{b}||/||\mathbf{b}||$ is small does *not* mean that for each i, $|\Delta b_i / b_i|$ is small.

OTHER ERROR EXPRESSIONS AND APPROXIMATIONS

A better understanding of the size of the error can be obtained with a more careful analysis. Again, it is simpler to consider first only changes in **b**. If the entries of A^{-1} are denoted by α_{ij}, then $\Delta \mathbf{x} = A^{-1} \Delta \mathbf{b}$ in component form is

$$\Delta x_i = \sum_{p=1}^{n} \alpha_{ip} \Delta b_p, \qquad i = 1, \ldots, n. \tag{2.24}$$

Hence, for $x_i \neq 0$

$$\frac{\Delta x_i}{x_i} = \sum_{p=1}^{n} \alpha_{ip} \frac{b_p}{x_i} \frac{\Delta b_p}{b_p}$$

is the exact formula for the relative change in x_i as a function of the relative changes in the b_p. The special case of a change in only one entry of **b** is

$$\Delta x_i = \alpha_{ip} \Delta b_p \tag{2.25}$$

and

$$\frac{\Delta x_i}{x_i} = \alpha_{ip} \frac{b_p}{x_i} \frac{\Delta b_p}{b_p}. \tag{2.26}$$

This says that the relative error of a solution component x_i will be sensitive to the relative error in a data component b_p whenever the factor $\alpha_{ip} b_p / x_i$ is large. The results using norms told us that "large" components in the inverse matrix indicate possible sensitivity to errors in the data. This result goes much further. It brings to our attention that "small" components in the solution vector indicate possible sensitivity to errors in the data. More to the point, it shows which components are sensitive and how much.

What if A is perturbed? For simplicity we study the effect on a solution component x_i when only a single entry a_{pq} of A is altered. In component form $A\mathbf{x} = \mathbf{b}$ is

$$b_i = \sum_{j=1}^{n} a_{ij} x_j, \qquad i = 1, \ldots, n.$$

Taking the partial derivative with respect to a_{pq} leads to

$$0 = \sum_{j=1}^{n} a_{ij} \frac{\partial x_j}{\partial a_{pq}} \quad \text{for} \quad i \neq p$$

and

$$0 = x_q + \sum_{j=1}^{n} a_{pj} \frac{\partial x_j}{\partial a_{pq}} \quad \text{for} \quad i = q.$$

In terms of the vectors

$$\mathbf{v} = \left(\frac{\partial x_i}{\partial a_{pq}} \right)$$

$$\mathbf{w} = (w_i), \quad \text{where} \quad w_i = 0 \quad \text{for} \quad i \neq p,\ w_p = -x_q,$$

this is a system of equations $A\mathbf{v} = \mathbf{w}$. Then $\mathbf{v} = A^{-1}\mathbf{w}$, or in component form,

$$\frac{\partial x_i}{\partial a_{pq}} = v_i = \sum_{j=1}^{n} \alpha_{ij} w_j = -\alpha_{ip} x_q \quad \text{for} \quad i = 1, \ldots, n.$$

We conclude that for "small" perturbations Δa_{pq} of a_{pq}, the solution component x_i is perturbed to $x_i + \Delta x_i$, where

$$\Delta x_i \approx \frac{\partial x_i}{\partial a_{pq}} \Delta a_{pq} = -\alpha_{ip} x_q \Delta a_{pq}.$$

In terms of relative changes, this is

$$\frac{\Delta x_i}{x_i} \approx -\alpha_{ip} a_{pq} \frac{x_q}{x_i} \left(\frac{\Delta a_{pq}}{a_{pq}} \right). \tag{2.27}$$

This *approximate* result shows much more detail than the *bound* (2.22). In particular, it is clear that if there is a solution component x_q that is large compared to a component x_i, then x_i can be sensitive to perturbations in column q of A.

ESTIMATING CONDITION

Although there is far more information in the equality (2.26) and the approximation (2.27) than in the condition number inequality (2.22), they require knowledge of A^{-1}, which is relatively expensive to compute. An efficient algorithm for the calculation of A^{-1} requires roughly three times the work needed for elimination. To compute cond(A) exactly requires A^{-1}, but for our purposes an estimate for $\|A^{-1}\|$

would suffice. An adequate estimate can be obtained with little more work than a forward/backward substitution. The calculation of $||A||$ using the maximum norm is easy. To estimate $||A^{-1}||$ the basic observation is that if $A\mathbf{y} = \mathbf{d}$, then from (2.16)

$$||\mathbf{y}|| = ||A^{-1}\mathbf{d}|| \leq ||A^{-1}|| \, ||\mathbf{d}||.$$

For any vector $\mathbf{y}$ we can form $\mathbf{d}$ and then obtain a lower bound for $||A^{-1}||$ from

$$||A^{-1}|| \geq ||\mathbf{y}|| / ||\mathbf{d}||.$$

Using a factorization of A, Cline, Moler, Stewart, and Wilkinson [3] construct $\mathbf{y}$ and $\mathbf{d}$ that result in a "large" ratio $||\mathbf{y}||/||\mathbf{d}||$. A lower bound for $||A^{-1}||$ is obtained in any case, and often the algorithm results in a value comparable to $||A^{-1}||$. The code Factor discussed below uses this approach to estimating cond(A). LAPACK [1] refines this estimate by an iterative procedure that involves repeated solutions of linear systems involving A.

The idea behind the condition estimate explains a common way that ill-conditioning is revealed in a computation. Let $\mathbf{y}$ be the computed solution of $A\mathbf{x} = \mathbf{d}$. In the inequality

$$||A|| \, ||A^{-1}|| \geq \frac{||A|| \, ||\mathbf{x}||}{||\mathbf{d}||}$$

let us approximate $||\mathbf{x}||$ by $||\mathbf{y}||$. Suppose we find that the right-hand side of the inequality is then "large." If the computed solution is comparable to the size of the true solution, this says that the problem is ill-conditioned. If the computed solution is not even comparable in size to the true solution, it is very inaccurate. Either way, a large value of this quantity is a warning. Often a problem is scaled naturally so that $||A||$ and $||\mathbf{d}||$ are about the same size, and in this common situation a large value of the quantity corresponds to a large computed solution. With this in mind, if you should compute a solution that seems "large," you should question the conditioning of the problem and/or the accuracy of the solution.

Often a problem is scaled so that $||A||$ and $||\mathbf{d}||$ are about 1. If this is the case and $\mathbf{y}$ should turn out to be large, the inequality shows that the matrix must be ill-conditioned.

Example 2.13. An example of Wilkinson [15] illustrates this and makes the point that a matrix can be very ill-conditioned without any small pivots arising in the elimination process. Because the matrix of the set of 100 equations

$$\begin{pmatrix} 0.501 & -1 & & & \\ & 0.502 & -1 & & \\ & & \ddots & \ddots & \\ & & & 0.599 & -1 \\ & & & & 0.600 \end{pmatrix} x = \begin{pmatrix} 0 \\ 0 \\ \vdots \\ 0 \\ 1 \end{pmatrix}$$

is upper triangular, the pivot elements are on the diagonal, and obviously none is "small." Back substitution shows that

$$x_1 = 1/(0.600 \times 0.599 \times \cdots \times 0.502 \times 0.501) > (0.6)^{-100} > 10^{22},$$

hence this matrix is very ill-conditioned. ■

The purpose of the condition number estimator in Factor is not to get an accurate value for the condition number, rather to recognize when it is "large." The inequality (2.22) is a worst-case bound. If the condition estimate of Factor indicates that the problem is ill-conditioned, you might well feel justified in computing A^{-1} so that you can use the more informative (2.26) and (2.27) to assess the effects of perturbations to the data.

EXERCISES

2.7 For the linear system in Exercise 1.1, let

$$\mathbf{y} = \begin{pmatrix} 0.999 \\ -1.001 \end{pmatrix}, \quad \mathbf{z} = \begin{pmatrix} 0.463 \\ -0.204 \end{pmatrix}.$$

In exact arithmetic, calculate the residuals $\mathbf{r} = \mathbf{b} - A\mathbf{y}$ and $\mathbf{s} = \mathbf{b} - A\mathbf{z}$. Does the better approximation have the smaller residual?

2.8 Consider the system

$$\begin{aligned} x_1 + x_2 &= 2 \\ 10x_1 + 10^{18}x_2 &= 10 + 10^{18}. \end{aligned}$$

Do not use more than 15-digit decimal arithmetic in the computations of parts (a) and (b). This will, for example, result in $10 + 10^{-18}$ becoming just 10.

(a) Solve using Gaussian elimination with partial pivoting.

(b) Divide each row by its largest $|a_{ij}|$ and then use Gaussian elimination with partial pivoting.

(c) Solve by hand using any method and exact arithmetic.

(d) Use exact arithmetic to calculate residuals for each solution. Which method seems better [compare with part (c)]? Do the residuals indicate this?

(e) Using the formula (2.21), compute A^{-1} for this system, and cond(A).

2.9 Assume that the computed solution to a nonsingular linear system is

$$(-10.4631,\ 0.00318429,\ 3.79144,\ -0.000422790)$$

and the condition number is 1200.

(a) What is the uncertainty ($\pm$?) in each component of the computed solution? Assume exact data and a unit roundoff of 10^{-6}.

(b) Repeat (a) but for the case when $||\Delta A||/||A||$ and $||\Delta\mathbf{b}||/||\mathbf{b}||$ are each 10^{-5}.

2.10 On a machine with unit roundoff 10^{-17} with $\mathbf{b}$ exact but $||\Delta A||/||A|| = 10^{-10}$, how large a condition number can we tolerate if we want $||\Delta\mathbf{x}||/||\mathbf{x}|| \le 10^{-5}$?

2.4 ROUTINES FACTOR AND SOLVE

In this section we describe routines Factor and Solve that can be used to solve the system $A\mathbf{x} = \mathbf{b}$. Routine Factor performs Gaussian elimination with partial pivoting on the matrix A and estimates its condition number. It saves the multipliers and the pivot information to be used by Solve to obtain the solution $\mathbf{x}$ for any right-hand side $\mathbf{b}$. A typical call to Factor in FORTRAN is

```
CALL FACTOR(A,MAXROW,NEQ,COND,PVTIDX,FLAG,TEMP)
```

while a typical function evaluation of Factor is

```
flag = Factor(a, neq, cond, pivot_index);
```

in the C++ version. The parameter cond must be passed by reference since its value is set by the function Factor; in C the address of cond must be explicitly passed so that its call looks like

flag = Factor(a, neq, &cond, pivot_index);

Input variables in FORTRAN are A, the array containing the coefficient matrix A; MAXROW, the declared row dimension of A in the program that calls Factor; and NEQ, the number of equations to be solved. Output variables are A, containing the upper triangular matrix U in positions $a_{ij}, i \leq j$, and the lower triangular matrix L in positions $a_{ij}, i > j$ (as long as FLAG is zero); FLAG, an integer variable that indicates whether or not zero pivots were encountered; COND, an estimate of the condition number of A in the maximum norm; PVTIDX, an array that records the row interchanges; and in FORTRAN 77 we need TEMP, an array used for temporary storage.

In the C and C++ versions corresponding variables are a for A, neq for NEQ, cond for COND, and pivot_index for PVTIDX. Note that in the C and C++ versions, (1) instead of declaring the matrix A to have two indices, we use a pointer to a vector consisting of the rows of A; (2) there is no need to reserve space for the temporary index TEMP because this allocation can be made dynamically when needed (as can be done for a and for pivot_index); (3) the output flag is the return variable for the function Factor. Because arrays are typically indexed starting with zero in C and C++, we have modified the algorithm accordingly. Some further comments are in order about the variables for Factor

MAXROW and NEQ (or neq). If, for example, the array A is declared as A(10,10) in the calling program and we are solving three equations in three unknowns, then MAXROW would have the value 10 in the FORTRAN version and NEQ (or neq in the C version) the value 3.

COND. COND is a lower bound for cond(A) in the maximum norm and is often a good approximation to cond(A). If $fl(\text{COND} + 1) = \text{COND}$, the matrix A is singular to working precision. Because we are working with arithmetic of limited precision, exact singularity is difficult to detect. In particular, the occurrence of a zero pivot does not necessarily mean that the matrix is singular nor does a singular matrix necessarily produce a zero pivot (see Exercise 2.16). When FLAG is nonzero, the output COND is meaningless.

PVTIDX (or pivot_index). When the elimination process has been completed, the kth component of PVTIDX (pivot_index) is the index of the kth pivot row and the nth component is set to $(-1)^m$ where m is the number of interchanges. Computation of the determinant of A requires PVTIDX(NEQ) in FORTRAN or pivot_index[neq-1] in C and C++ (see Exercise 2.23).

The argument list for Solve is

SOLVE(A,MAXROW,NEQ,PVTIDX,B)

in FORTRAN; in C and C++ it is

```
Solve(a,neq,pivot_index,b);
```

The variables A, MAXROW, NEQ, PVTIDX are as specified in the Factor list. Routine Solve uses the arrays A and PVTIDX as output from Factor and the right-hand side contained in the vector B to solve $A\mathbf{x} = \mathbf{b}$. The solution $\mathbf{x}$ is returned in B.

Example 2.14. To illustrate the codes, we solve the problem

$$\begin{pmatrix} 3 & 6 & 9 \\ 2 & 5 & -2 \\ 1 & 3 & -1 \end{pmatrix} \begin{pmatrix} x_1 \\ x_2 \\ x_3 \end{pmatrix} = \mathbf{b},$$

for two right sides $\mathbf{b} = (39,3,2)^T$ and $\mathbf{b} = (6,7,-12)^T$. The main program sets up the problem and calls the routines Factor/Solve to obtain a solution. Note that after the call to Factor, the variable FLAG is checked to see if any pivots are zero. If FLAG $>$ 0 and Solve were used, a zero divide error would result. When FLAG $= 0$ the routine Solve is used once for each right-hand side to obtain the solutions. Note that Factor is not (and should not be) used twice since it does not act on $\mathbf{b}$. The output is as follows (the floating point values may vary slightly from one computer to another).

```
Condition number =  106.642857142857100
Solution of the first system
   2.000000000000000     1.000000000000000    3.000000000000000
Solution of the second system
   76.75000000000000    -31.00000000000000   -4.250000000000000
```

■

EXERCISES

2.11 Solve the system in Exercise 2.2 using Factor/Solve. Compare to the true solution.

2.12 Solve the system in Exercise 2.3 using Factor/Solve. Compute the residual.

2.13 The codes Factor/Solve can be used to find the elements of A^{-1}. The inverse is used to estimate certain statistical parameters. It is also used to study the sensitivity of the solution to errors in the data. Let $\mathbf{x}^i$ denote the solution of

$$A\mathbf{x} = \mathbf{b}^i,\ i = 1,2,\ldots,n,$$

where the ith right-hand side $\mathbf{b}^i$ is

$$b^i_j = \begin{cases} 0 & \text{if } i \neq j \\ 1 & \text{if } i = j. \end{cases}$$

If we form the matrix

$$X = \begin{bmatrix} x^1_1 & x^2_1 & \cdots & x^n_1 \\ x^1_2 & x^2_2 & \cdots & x^n_2 \\ \vdots & \vdots & & \vdots \\ x^1_n & x^2_n & \cdots & x^n_n \end{bmatrix},$$

it is easy to show that $X = A^{-1}$. Do so. Use Factor/Solve to find the inverse of the matrix

$$A = \begin{pmatrix} 1 & 2 & 3 \\ 4 & 5 & 6 \\ 7 & 8 & 9.01 \end{pmatrix}.$$

2.14 Consider the linear system of Exercise 2.4.

(a) Using the method of Exercise 2.13, find the inverse of the original linear system. Calculate the exact condition number.

(b) Use Factor/Solve on the system in Exercise 2.4b. What is COND? Is the condition number inequality (2.22) valid for this problem? (Use $||\Delta A|| = 0.003333$.)

2.15 Suppose we are given the electrical network shown in Figure 2.1, and we desire to find the potentials at junctions (1) through (6). The potential applied between A and B is V volts. Denoting the potentials by $v_1, v_2, \ldots, v_6$, application of Ohm's law and Kirchhoff's current law yield the following set of linear equations for the v_i:

$$11v_1 - 5v_2 - v_6 = 5V$$
$$-20v_1 + 41v_2 - 15v_3 - 6v_5 = 0$$
$$-3v_2 + 7v_3 - 4v_4 = 0$$
$$-v_3 + 2v_4 - v_5 = 0$$
$$-3v_2 - 10v_4 + 28v_5 - 15v_6 = 0$$
$$-2v_1 - 15v_5 + 47v_6 = 0.$$

Solve when $V = 50$.

2.16 The system

$$\begin{aligned} 0.473x_1 - 0.115x_2 & & &= b_1 \\ 0.731x_1 - 0.391x_2 &+ 0.267x_3 &&= b_2 \\ -0.782x_2 &+ 0.979x_3 &&= b_3 \end{aligned}$$

is singular.

(a) Apply Factor to the coefficient matrix. What is the smallest pivot? Is it near the unit roundoff u? Is it near underflow? Is COND large? The results you obtain will depend on the hardware and software that you use. If Factor turns up a pivot that is exactly zero, perturb the coefficient 0.979 by a very small amount so as to get a system that is computationally nonsingular for the rest of this problem. Adding 10^{-14} to the coefficient will suffice for a number of configurations.

(b) Use Factor/Solve to compute $\mathbf{x}$ for $\mathbf{b} = (0.084, 0.357, 0.833)$. Is there any indication of singularity in the answer?

(c) Use Factor/Solve to compute $\mathbf{x}$ for $\mathbf{b} = (0.566, 0.404, 0.178)$. Is there any indication of singularity in the answer?

(d) Compute the residuals for (b) and for (c). Do they give any indication of singularity?

2.17 In analyzing environmental samples taken from the atmosphere, a simple model with m samples and n sources and chemicals produces $AX = B$, where a_{ik} is the average concentration of element i from source k, x_{kj} is the mass of particles from source k contributing to sample j, b_{ij} is the concentration of element i in sample j, and $1 \le i \le n$, $1 \le k \le n$, $1 \le j \le m$. If $m = 4$, $n = 3$, then

$$A = \begin{bmatrix} 0.172 & 0.013 & 0.144 \\ 0.368 & 0.681 & 0.271 \\ 0.099 & 0.510 & 0.329 \end{bmatrix},$$

$$B = \begin{bmatrix} 1.44 & 4.35 & 1.32 & 3.95 \\ 2.84 & 9.30 & 2.90 & 8.29 \\ 2.36 & 3.45 & 3.25 & 7.35 \end{bmatrix}.$$

(a) What is X? What does COND tell you about the reliability of this result? First assume exact data, then that the entries of A and B are rounded to the displayed values.

(b) Use the method of Exercise 2.13 to compute A^{-1}. What is the exact cond(A)?

(c) What does (2.24) tell you about the sensitivity of x_{21} to changes in b_{11}? Replace b_{11} by 1.43 and recalculate x_{21}. Do the numerical answers confirm the theory? Here you are to consider relative changes to the data and the solution.

2.18 Consider the linear system

$$\begin{bmatrix} 0.217 & 0.732 & 0.414 \\ 0.508 & 0.809 & 0.376 \\ 0.795 & 0.886 & 0.338 \end{bmatrix} \mathbf{x} = \begin{bmatrix} 0.741 \\ 0.613 \\ 0.485 \end{bmatrix}.$$

(a) Solve for $\mathbf{x}$ using Factor/Solve.

(b) If each entry in A and $\mathbf{b}$ might have an error of ± 0.0005, how reliable is $\mathbf{x}$?

(c) Make arbitrary changes of ± 0.0005 in the elements of A to get $A + \Delta A$ and in the elements of $\mathbf{b}$ to get $\mathbf{b} + \Delta\mathbf{b}$. Solve $(A + \Delta A)(\mathbf{x} + \Delta\mathbf{x}) = \mathbf{b} + \Delta\mathbf{b}$ to get $\mathbf{x} + \Delta\mathbf{x}$. Calculate $||\Delta\mathbf{x}||/||\mathbf{x}||$. Is this consistent with (b)? What is the relative change in each x_i?

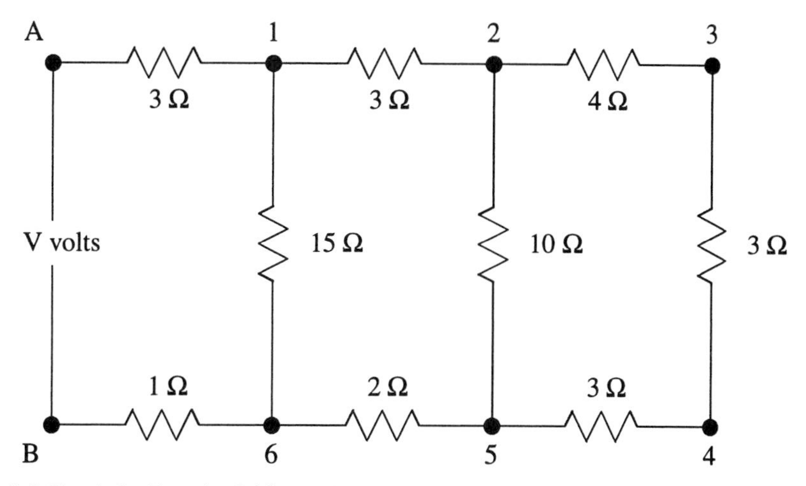

Figure 2.1 Circuit for Exercise 2.15.

2.5 MATRICES WITH SPECIAL STRUCTURE

Most general-purpose linear system solvers are based on Gaussian elimination with partial pivoting. When the matrix A has special properties, it is possible to reduce the storage and the cost of solving linear systems very substantially. This section takes up important examples.

When it is possible to factor the matrix without pivoting, this is both faster and reduces the storage required. There are two kinds of matrices that are common for which it is not necessary to pivot. An $n \times n$ matrix A is said to be *diagonally dominant* (by columns) if for each column

$$|A_{ii}| \geq \sum_{i \neq j} |A_{ij}| .$$

This says that the entry on the diagonal is the biggest one in the column, and by some margin. An induction argument can be used to show that Gaussian elimination applied to a nonsingular, diagonally dominant matrix will always select the entry on the diagonal; hence do no row interchanges. The matrix A is said to be symmetric if $A^T = A$. A symmetric matrix is *positive definite* if for any vector $\mathbf{v} \neq \mathbf{0}$, the quantity $\mathbf{v}^T A \mathbf{v} > 0$. It is not only possible to dispense with pivoting for positive definite matrices, but even to exploit symmetry by working with a variant of Gaussian elimination.

BAND MATRICES

Recall that in the basic elimination algorithm, the innermost loop can be omitted when the multiplier $t = 0$. This reflects the fact that the variable is not present in this equation and so does not need to be eliminated. When the matrix A is already "close" to an upper triangular matrix, testing for a zero multiplier can save quite a bit of work. A kind of matrix that is extremely important in several areas of computation is a *banded* matrix. A matrix $A = (a_{ij})$ is said to be banded when all the nonzero elements lie in a band about the diagonal. Specifically, when $a_{ij} = 0$ if $i - j > m_\ell$ and $j - i > m_u$, the matrix is said to have the lower band width m_ℓ, upper band width m_u, and band width $m = m_\ell + m_u + 1$. An example of a matrix with $m_\ell = 2$ and $m_u = 1$ is

$$\begin{bmatrix} x & x & 0 & 0 & 0 \\ x & x & x & 0 & 0 \\ x & x & x & x & 0 \\ 0 & x & x & x & x \\ 0 & 0 & x & x & x \end{bmatrix}.$$

Here x indicates an entry that might not be zero. When elimination is performed on such a matrix, at most m_ℓ elements have to be eliminated at each stage. Examination of the elements above the diagonal shows that many zero elements remain zero. Indeed, partial pivoting will leave zeros in $a_{ij}^{(k)}$ for $j - i > m_u + m_\ell$. As with zero multipliers, we can speed up the computation by recognizing elements that are zero and stay zero. Another important observation is that by using a special storage scheme, there is no need to provide storage for $a_{ij}^{(k)}$ with $i - j > m_\ell$ or $j - i > m_u + m_\ell$. Codes implementing a special version of Gaussian elimination for banded matrices can be found in either LINPACK [5] or LAPACK [1]. Although it is a little more trouble to set up A in the special storage format used, the advantages can be great. The numerical results are identical, but the storage in the banded case is roughly $n(2m_\ell + m_u)$ instead of n^2. The operation count for the decomposition is comparable to $nm_\ell(m_\ell + m_u)$ instead of $n^3/3$ and there is a similar advantage for the forward and back substitution. Complete details are found in [5], [1], or [9]. The main point is that when n is large and m_ℓ and m_u are small, tremendous savings are possible. This is what makes solution of systems with $n = 10^3$, and, say, $m_\ell = m_u = 5$, a routine matter when solving differential equations.

It will be convenient now to derive an alternative form of the Gaussian elimination algorithm. Assume the decomposition $A = LU$ exists with L lower triangular and U upper triangular. First note that

$$a_{11} = \sum_{m=1}^{n} \ell_{1m} u_{m1} = \ell_{11} u_{11}$$

because the matrices are triangular. Choose $\ell_{11} \neq 0$ and then

$$u_{11} = a_{11}/\ell_{11}.$$

For $i > 1$

$$a_{i1} = \sum_{m=1}^{n} \ell_{im} u_{m1} = \ell_{i1} u_{11},$$

so

$$\ell_{i1} = a_{i1}/u_{11} \quad \text{for} \quad i = 2, \ldots, n.$$

Also, for $j > 1$,

$$a_{1j} = \sum_{m=1}^{n} \ell_{1m} u_{mj} = \ell_{11} u_{1j},$$

so

$$u_{ij} = a_{1j}/\ell_{11} \quad \text{for} \quad j = 2,\ldots,n.$$

In general we form a column of L and a row of U at a time. Suppose we have computed columns $1,\ldots,k-1$ of L and rows $1,\ldots,k-1$ of U. Then

$$a_{kk} = \sum_{m=1}^{n} \ell_{km}u_{mk} = \ell_{kk}u_{kk} + \sum_{m=1}^{k-1} \ell_{km}u_{mk}.$$

The terms in the sum on the right are known. Choose $\ell_{kk} \neq 0$, and then

$$u_{kk} = \left(a_{kk} - \sum_{m=1}^{k-1} \ell_{km}u_{mk} \right) \bigg/ \ell_{kk}.$$

Now for $i > k$,

$$a_{ik} = \sum_{m=1}^{n} \ell_{im}u_{mk} = \ell_{ik}u_{kk} + \sum_{m=1}^{k-1} \ell_{im}u_{mk}.$$

The terms in the sum on the right are known, so

$$\ell_{ik} = \left(a_{ik} - \sum_{m=1}^{k-1} \ell_{im}u_{mk} \right) \bigg/ u_{kk} \quad \text{for} \quad i = k+1,\ldots,n.$$

Similarly, for $j > k$,

$$a_{kj} = \sum_{m=1}^{n} \ell_{km}u_{mj} = \ell_{kk}u_{kj} + \sum_{m=1}^{k-1} \ell_{km}u_{mj}$$

and

$$u_{kj} = \left(a_{kj} - \sum_{m=1}^{k-1} \ell_{km}u_{mj} \right) \bigg/ \ell_{kk} \quad \text{for} \quad j = k+1,\ldots,n.$$

If all the diagonal elements of L are taken to be 1, this algorithm is Gaussian elimination without pivoting. Later it will prove useful to choose other values for these elements. In our discussion of elimination applied to a band matrix A, we observed that quite a lot of storage and work could be saved. The situation is even better when no pivoting is done. If A is a band matrix with lower band width m_ℓ and upper band width m_u, then L is also a band matrix with lower band width m_ℓ and U is a band matrix with upper band width m_u. If we choose the diagonal elements of L to be 1, it is not necessary to store them and as is the case for full matrices, the factors L and U can be written over A as they are computed. These statements about the form of L and U follow by examination of the recipe for their computation when the form of A is taken into account. Of course, a special storage scheme is needed to exploit the fact that only elements in a band about the diagonal can be nonzero. This approach to solving banded linear systems is of great practical importance. In Chapter 3 matrices with $m_\ell = 1 = m_u$ arise in the fitting of splines to data. They are examples of nonsingular diagonally dominant matrices for which we can be sure that pivoting is not needed for stability of the algorithm. Fortunately, many banded systems arising in practice are diagonally dominant.

TRIDIAGONAL LINEAR SYSTEMS

Some special cases of m_ℓ and m_u occur sufficiently often that special codes are written for them. For example, when $m_\ell = m_u = 1$ the coefficient matrix is called *tridiagonal*. For the numerical solution of partial differential equations by a number of important methods, it is necessary to solve a great many tridiagonal systems involving a great many unknowns, perhaps thousands of each. This would be impractical with Factor/Solve, but with a special-purpose algorithm it is not that difficult. Let us assume that the tridiagonal system is written as

$$\begin{pmatrix} a_1 & c_1 & & & & \\ b_2 & a_2 & c_2 & & & 0 \\ & \ddots & \ddots & \ddots & & \\ 0 & & & b_{n-1} & a_{n-1} & c_{n-1} \\ & & & & b_n & a_n \end{pmatrix} \begin{pmatrix} x_1 \\ x_2 \\ \vdots \\ x_{n-1} \\ x_n \end{pmatrix} = \begin{pmatrix} d_1 \\ d_2 \\ \vdots \\ d_{n-1} \\ d_n \end{pmatrix}.$$

When *no* pivoting is done, elimination zeros out the b_i to lead to a structure like

$$\begin{pmatrix} f_1 & c_1 & & & \\ & f_2 & c_2 & & 0 \\ & \ddots & \ddots & \ddots & \\ 0 & & & f_{n-1} & c_{n-1} \\ & & & & f_n \end{pmatrix} \begin{pmatrix} x_1 \\ x_2 \\ \vdots \\ x_{n-1} \\ x_n \end{pmatrix} = \begin{pmatrix} e_1 \\ e_2 \\ \vdots \\ e_{n-1} \\ e_n \end{pmatrix}.$$

As we shall see, the c_i are unchanged from the original system. To show this and to derive formulas for f_i and e_i, first observe that $f_1 = a_1$ and $e_1 = d_1$ since Gaussian elimination without pivoting does not change the first equation. To eliminate b_2 the multiplier is $m_2 = b_2/f_1$. Hence,

$$\begin{aligned} f_2 &= a_2 - m_2 c_1 \\ c_2 &= c_2 - m_2 \cdot 0 = c_2 \qquad \text{(as we stated)} \\ e_2 &= d_2 - m_2 d_1. \end{aligned}$$

Notice that x_1 does not appear in any other equation, so this completes the first stage of the elimination. To complete the derivation we use induction. Assume that we have derived the f_i and e_i through row k. Then we have the pattern

$$\left(\begin{array}{cccc|c} 0 & f_k & c_k & 0 & e_k \\ 0 & b_{k+1} & a_{k+1} & c_{k+1} & d_{k+1} \end{array}\right)$$

in rows k and $k+1$. Clearly the multiplier is $m_{k+1} = b_{k+1}/f_k$; then eliminate on row $k+1$ to get

$$\begin{aligned} f_{k+1} &= a_{k+1} - m_{k+1} c_k \\ c_{k+1} &= c_{k+1} - m_{k+1} \cdot 0 = c_{k+1} \\ e_{k+1} &= d_{k+1} - m_{k+1} d_k, \end{aligned}$$

which finishes the elimination of variable x_k. In algorithmic form this is

$f_1 = a_1$
for $k = 1, \ldots, n-1$ begin

$$m_{k+1} = b_{k+1}/f_k$$
$$f_{k+1} = a_{k+1} - m_{k+1}c_k$$

end k.

For the solution of many systems with the same coefficient matrix, we save the multipliers m_k as well as the f_k, c_k and solve separately for the data $d_1, \dots, d_n$. Forward elimination is

$e_1 = d_1$
for $k = 1, \dots, n-1$ begin

$$e_{k+1} = d_{k+1} - m_{k+1}e_k$$

end k.

Back substitution is simply

$x_n = e_n/f_n$
for $k = n-1, n-2, \dots, 1$ begin

$$x_k = (e_k - c_k x_{k+1})/f_k$$

end k.

Storage can be managed extremely efficiently. A general matrix requires storage for n^2 entries, but a tridiagonal matrix requires storage for only $3n-2$ nonzero entries. A natural scheme is to store the three diagonal bands with a_k, b_k, and c_k as three vectors of length n. We can write m_k over b_k and f_k over a_k as they are computed; during the forward and backward substitution stages e_k and x_k can overwrite d_k so that only one additional n vector is needed. We leave as an exercise the task of counting arithmetic operations to see that they are dramatically less than for a general matrix (see Exercise 2.15). The above algorithm assumed *no* pivoting, but as was seen earlier in this chapter, it is not always possible to solve linear systems without pivoting, and even when it is possible, the numerical results may be poor. For tridiagonal systems there is a simple condition that guarantees that all goes well, a condition often satisfied by systems that arise in practice. First note that if any c_k or b_k vanishes, the system can be broken into smaller systems that are also tridiagonal. Hence, we can assume that $c_k \neq 0$ and $b_k \neq 0$ for all k. The key assumption is that

$$\begin{aligned}
|a_1| &> |b_2| \\
|a_k| &\geq |b_{k+1}| + |c_{k-1}|, \qquad k = 2, \dots, n-1. \\
|a_n| &> |c_{n-1}|.
\end{aligned}$$

This condition is a little stronger than being diagonally dominant, enough that we can prove that the matrix is nonsingular. The argument is by induction. By assumption $|m_2| = |b_2/a_1| < 1$. Supposing now that $|m_j| < 1$ for $j = 2, \dots, k$,

$$|m_{k+1}| = \left| \frac{b_{k+1}}{f_k} \right|,$$

but

$$\begin{aligned}
|f_k| &= |a_k - m_k c_{k-1}| \geq |a_k| - |m_k||c_{k-1}| \\
&> |a_k| - |c_{k-1}| \geq |b_{k+1}| > 0.
\end{aligned}$$

This implies that $|m_{k+1}| < 1$ as desired. From the above inequality we have $|f_k| > |b_{k+1}| > 0$, so

$$\begin{aligned} |f_k| &= |a_k - m_k c_{k-1}| \leq |a_k| + |m_k||c_{k-1}| \\ &< |a_k| + |c_{k-1}|. \end{aligned}$$

Thus, under these assumptions all the quantities computed in the elimination are well defined and they are nicely bounded in terms of the data. In particular, the matrix must be nonsingular.

SYMMETRIC MATRICES

So far we have not considered how we might take advantage of a symmetric matrix A. It might seem possible to decompose A into the product of an upper triangular matrix U and its transpose U^T, which is a lower triangular matrix. However, it is *not* always possible to decompose a symmetric matrix in this way. This follows from the fact that such a decomposition implies that the matrix must be positive definite and not all symmetric matrices are positive definite. To see this, if $A = U^T U$ for a nonsingular matrix U, then

$$\mathbf{v}^T A \mathbf{v} = \mathbf{v}^T U^T U \mathbf{v} = \mathbf{y}^T \mathbf{y},$$

where $\mathbf{y} = U\mathbf{v}$. Because U is nonsingular, if $\mathbf{v} \neq 0$, then $\mathbf{y} \neq 0$ and

$$\mathbf{y}^T \mathbf{y} = \sum_{i=1}^{n} y_i^2 > 0,$$

showing that A is positive definite. Although we shall not prove it, any positive definite, symmetric A *can* be decomposed as $A = U^T U$ for a nonsingular upper triangular matrix U. Symmetric, positive definite matrices are very important. In applications such as the least squares fitting of data and the variational formulation of the finite element solution of partial differential equations, the quantity $\mathbf{v}^T A \mathbf{v}$ is a kind of "energy" and is naturally positive. We shall discuss a very effective way of solving problems with A that are symmetric, positive definite. There are more complicated ways of dealing with symmetric matrices that are not positive definite that approach the efficiency of the definite case for large systems, but they do not cope with the storage nearly so well. Codes can be found in LINPACK [5] or LAPACK [1] for both cases. Supposing that A is symmetric, positive definite and using the fact stated that it can be factored as $A = U^T U$, we can obtain U by specializing the recipe given earlier. Now we are to have $L^T = U$. Thus

$$a_{11} = \ell_{11} u_{11} = u_{11}^2,$$

so

$$u_{11} = \sqrt{a_{11}},$$

and as before

$$u_{1j} = a_{1j} / u_{11} \quad \text{for} \quad j = 2, \ldots, n.$$

Now

$$a_{kk} = \sum_{m=1}^{n} \ell_{km} u_{mk} = \sum_{m=1}^{k} u_{mk}^2, \tag{2.28}$$

from which we find

$$u_{kk} = \left(a_{kk} - \sum_{m=1}^{k-1} u_{mk}^2 \right)^{1/2}.$$

Then, as before,

$$\begin{aligned} u_{kj} &= \left(a_{kj} - \sum_{m=1}^{k-1} \ell_{km} u_{mj} \right) \Big/ \ell_{kk} \\ &= \left(a_{kj} - \sum_{m=1}^{k-1} u_{mk} u_{mj} \right) \Big/ u_{kk}, \quad j = k+1, \ldots, n. \end{aligned}$$

This decomposition has excellent numerical properties. From (2.28) we see that

$$a_{kk} \geq u_{mk}^2;$$

hence

$$|u_{mk}| \leq \sqrt{a_{kk}} \text{ all } m \geq k, \text{ all } k,$$

which says the multipliers in Gaussian elimination cannot get large relative to A. This decomposition is called the Cholesky or square root decomposition. The square roots in this algorithm can be avoided if we use a factorization of the form LDL^T, where D is diagonal and L is lower triangular with ones on the diagonal. As in the case of the LU decomposition, when A is a band matrix, so is U. The Cholesky decomposition preserves more than the band structure of a matrix. By examination of its recipe it is seen that as one goes down a column of U, the first (possibly) nonzero element occurs in the same place as the first nonzero element of A. This says that the "profile" or "envelope" or "skyline" of A is preserved. Obviously it is more trouble to work with a data structure that takes advantage of this fact than with one suitable for a band, but it is not much more trouble and the storage can be reduced quite a lot. Renumbering the unknowns alters the band width and the envelope of a matrix. There are algorithms that attempt to find the best numbering in the sense of minimizing the storage and cost of computing the Cholesky factorization. Many techniques have been developed for the solution of large, symmetric, positive definite systems when most of the components of A are zero. The monograph [8] explains the methods and presents codes; see [9] also. It is possible to solve efficiently systems arising in many areas of scientific computation that involve thousands of unknowns.

EXERCISES

2.19 Count the arithmetic operations required by the algorithm in Section 2.5.2 for a linear system of n equations with a tridiagonal coefficient matrix and m right-hand sides. Compare this with what is required for a general system.

2.6 CASE STUDY 2

All the standard methods of approximating the solution of elliptic partial differential equations, PDEs, involve the solution of a system of linear equations. The numerical solution of PDEs is a large subject and there are many books devoted to the solution of particular classes of problems and to the use of particular kinds of methods. The Galerkin method is a very important one that we illustrate here with a simple example from [4]. After some preparation, the velocity $w(x,y)$ of the steady flow of a viscous fluid in a duct with square cross section can be found as the solution of

$$1+\frac{\partial^2 w}{\partial x^2}+\frac{\partial^2 w}{\partial y^2}=1+\nabla^2 w=0$$

subject to no-slip boundary conditions

$$w(x,y)=0 \text{ on } |x|=1 \text{ and } |y|=1.$$

The same mathematical problem arises in the torsion of a bar of square cross section. Galerkin methods approximate $w(x,y)$ by an expansion

$$w_N(x,y)=\sum_{j=1}^{N} a_j\phi_j(x,y).$$

The choice of the basis functions ϕ_j is of fundamental importance. For the example we shall use

$$\phi_j(x,y)=\left[(1-x^2)(1-y^2)\right]^j.$$

Notice that each ϕ_j satisfies the boundary conditions. Also, the problem is such that $w(x,y)$ has certain symmetry properties, properties shared by these basis functions. Because the ϕ_j reflect well the qualitative behavior of the solution, we can hope to get a reasonable approximation with just a few terms in the expansion. This is a global Galerkin method because each ϕ_j approximates the solution over the entire domain. We shall see in Chapter 3 that approximating functions by piecing together polynomial approximations over subdomains can be very effective. In the present context the subdomains are called elements, and a Galerkin method based on piecewise polynomial basis functions is called a finite element method. In any case, when w_N is substituted into the differential equation, there is a residual

$$R(x,y)=1+\nabla^2 w_N.$$

The idea is to find coefficients a_j that make this residual small in some sense. Generally there is a residual arising from boundary conditions, too, but for simplicity we discuss only the case of basis functions that satisfy the boundary conditions exactly. To quantify the notion of being small, we make use of an inner product of two functions $f(x,y)$ and $g(x,y)$:

$$(f,g) = \int_{-1}^{1}\int_{-1}^{1} f(x,y)g(x,y)\,dxdy.$$

The Galerkin method requires that the residual of the approximation be small in the sense that it is orthogonal to all the basis functions, that is, $(\phi_i, R) = 0$ for $i = 1,\dots,N$. If we substitute the expansion into the definition of the residual, this becomes

$$0 = (\phi_i, R) = \left(\phi_i, 1 + \sum_{j=1}^{N} a_j \nabla^2 \phi_j\right) = (\phi_i, 1) + \sum_{j=1}^{N} a_j \left(\phi_i, \nabla^2 \phi_j\right).$$

This is a system of linear equations $\mathbf{Ca} = \mathbf{b}$ for the coefficients a_i, where $C_{ij} = \left(\phi_i, \nabla^2 \phi_j\right)$ and $b_i = -(\phi_i, 1)$. When there is more than one independent variable, it is not so easy to piece together polynomial approximations over elements so that they connect smoothly, and it is usual to reduce the order of differentiation in the inner product of C_{ij} from two to one by means of integration by parts (Green's theorem). With the global polynomial approximation of this example, there is no difficulty forming directly the Laplacian of the basis functions.

In the classic Galerkin method the integrals of the inner products are computed exactly. This is possible for our simple example, but in practice integrals are approximated numerically. A common procedure for general functions is based on Gaussian quadrature. As we shall see in Chapter 5, a Gaussian quadrature formula of M points consists of a set of nodes η_i and weights A_i such that

$$\int_{-1}^{1} u(x)\,dx \approx \sum_{i=1}^{M} A_i u(\eta_i)$$

provides the best possible approximation in a certain sense. In particular, the formula is exact for any polynomial $u(x)$ of degree no more than $2M - 1$. In Section 5.6 we discuss how such a formula can be used to obtain an approximation to the integral of a function of two variables:

$$\int_{-1}^{1}\int_{-1}^{1} f(x,y)\,dxdy \approx \sum_{i=1}^{M}\sum_{j=1}^{M} A_i A_j f(\eta_i, \eta_j).$$

Generally M is quite small, but for our example we take it to be 5. The basis functions we use are polynomials and inspection of the integrands for the cases $N = 1$, 2, 3 that we solve numerically shows that with $M = 5$, the integrals we need are computed exactly.

Often when solving PDEs, quantities computed from the solution are at least as interesting as the solution itself. Such a quantity for this example is the nondimensional flow rate

$$\dot{q} = \int_{-1}^{1}\int_{-1}^{1} w(x,y)\,dxdy.$$

The Galerkin approximations computed in the manner described gave

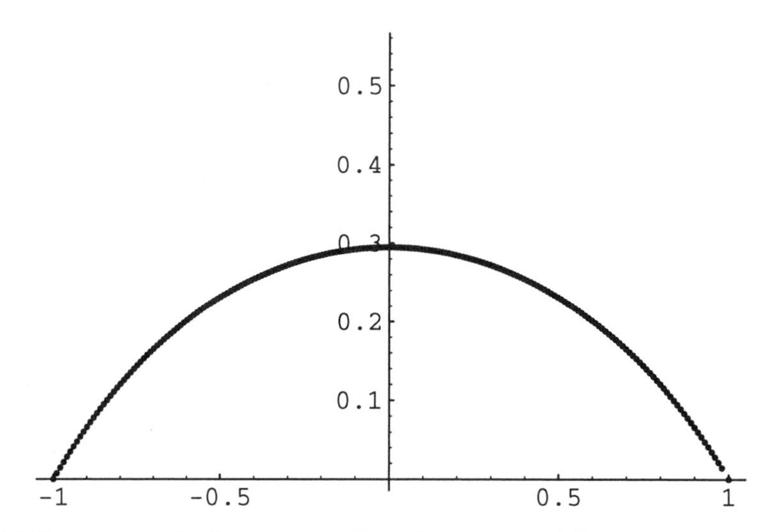

Figure 2.2 The flow rate $w(x,0)$ approximated by Galerkin's method (3 terms).

	$N=1$	$N=2$	$N=3$	Exact
$w(0,0)$	0.3125	0.2927	0.2947	0.2947
$\dot{q}$	0.5556	0.5607	0.5608	0.5623

The "exact" values here were obtained from a series solution for $w(x,y)$. Figure 2.2 shows $w_3(x,0)$, which exhibits the kind of velocity distribution across the duct that we expect on physical grounds. These are exceptionally good approximations because we have incorporated so much information about the solution into the basis functions. It should be appreciated that if the solution did not have the symmetry properties of the basis functions, we could not expect such a good answer. Moreover, consistent results found on adding terms to the expansion might just be consistently poor results because it is not possible to represent well the solution with the basis functions chosen. Because finite element methods approximate the solution locally, they are better suited for a general problem.

Finite differences represent another approach to solving PDEs that leads to large systems of linear equations. In this approach $w(x,y)$ is approximated only on a mesh. For example, if we choose an integer N and define a mesh spacing $h = 1/N$, we might approximate $w(ih, jh)$ by w_{ij} for $i, j = -N, -N+1, \ldots, N-1, N$. To satisfy the boundary conditions we take $w_{ij} = 0$ when $i = \pm N$ and $j = \pm N$. Taylor series expansion of a smooth solution $w(x,y)$ shows that

$$\frac{w((i+1)h, jh) - 2w(ih, jh) + w((i-1)h, jh)}{h^2} = \frac{\partial^2 w}{\partial x^2}(ih, jh) + \tau.$$

Further, there is a constant γ such that $|\tau| \leq \gamma h^2$ for all i, j and all sufficiently small h. We say that this is a difference approximation to the second derivative that is of order two. All we wish to do here is make plausible that the expression on the left imitates the partial derivative for small h. This expression and the differential equation suggest that we define the approximation w_{ij} at mesh points (ih, jh) interior to the square by the equation

$$\frac{w_{i+1,j} - 2w_{i,j} + w_{i-1,j}}{h^2} + \frac{w_{i,j+1} - 2w_{i,j} + w_{i,j-1}}{h^2} + 1 = 0.$$

The set of $(2N-1)^2$ equations can be rewritten as

$$w_{i,j-1} + w_{i+1,j} - 4w_{i,j} + w_{i-1,j} + w_{i,j+1} = -h^2.$$

For this discrete approximation to the PDE to be a reasonable one, the mesh spacing h must be small, but then we must have a large set of linear equations. It is of the utmost practical importance that the equations are sparse. Indeed, there are only five nonzero coefficients in each row of the matrix. Historically such equations were solved by iterative methods. The methods used are very simple and self-correcting, so that a mistake does not prevent computing the correct answer. Both characteristics are very important to hand computation. Also important is that the methods require storage only for the nonzero coefficients, the solution itself, and perhaps another vector of the same size as the solution. For a problem as simple as this example, it is not even necessary to store the coefficients of the matrix. Iterative methods take advantage of the fact that we do not require an accurate solution. After all, $w_{i,j}$ is only an approximation to $w(ih, jh)$ of modest accuracy, so there is little point to computing it very accurately.

The classic iterative procedures for solving equations like those arising in this example are most naturally described in terms of the equation itself. First let us rewrite the equation as

$$w_{i,j} = \left(h^2 + w_{i,j-1} + w_{i+1,j} + w_{i-1,j} + w_{i,j+1}\right)/4.$$

The Jacobi iteration improves an approximation $\mathbf{w}^{(k)}$ by computing

$$w_{i,j}^{(k+1)} = \left(h^2 + w_{i,j-1}^{(k)} + w_{i+1,j}^{(k)} + w_{i-1,j}^{(k)} + w_{i,j+1}^{(k)}\right)/4$$

for all i, j. This amounts to saying that we define $w_{i,j}^{(k+1)}$ so as to make the residual in equation j equal to zero. This is a very simple and inexpensive iteration that requires only storage for the current and next approximations. A refinement suggests itself: Would it not be better to use $w_{i,j}^{(k+1)}$ instead of $w_{i,j}^{(k)}$ for the rest of the computation? Besides, if we do this, we would halve the storage required. This is called the Gauss–Seidel iteration. There is a complication with this method. So far we have not written out the matrix explicitly, so we have not specified the order of the components $w_{i,j}^{(k)}$ in the vector $\mathbf{w}^{(k)}$. If we are to start using $w_{i,j}^{(k+1)}$ as soon as it is computed, the order in which components are improved matters. A question that arises in the use of any iterative method is when to quit. A natural way is to measure the residual of the current solution and quit when it is less than a tolerance.

For the example we used the Gauss–Seidel iteration and improved the unknowns from left to right, bottom to top, that is, for each $i = -N+1, \ldots, N-1$, we improved $w_{i,j}$ for $j = -N+1, \ldots, N-1$. We took $N = 5$, which led to a system of 81 equations for unknowns at points interior to the region. It was convenient to start with $w_{i,j}^{(0)} = 0$ for all i, j. With a tolerance of 5×10^{-4}, convergence was achieved in 89 iterations.

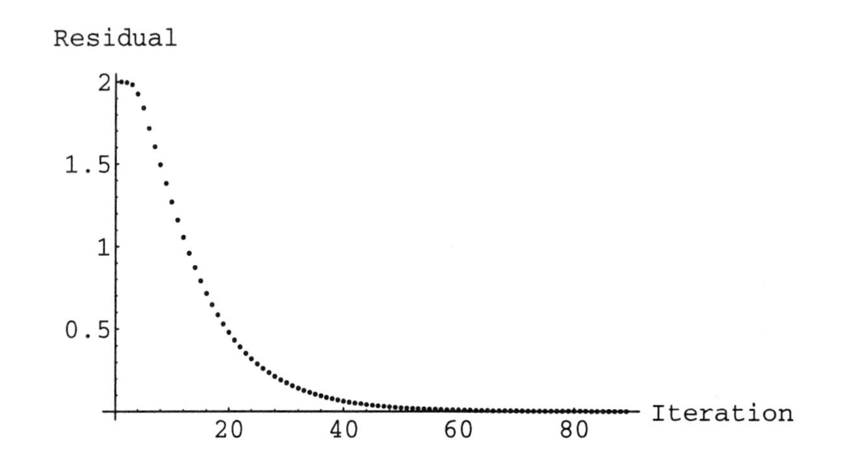

Figure 2.3 Residual versus iteration number for Gauss–Seidel iteration.

The behavior of the residuals is displayed in Figure 2.3. Except for the first few iterates, the residual was reduced by a factor close to 0.9 at each iteration. It might be remarked that the residual measured was that of the finite difference equation in the form with the 1 present because it provides a natural measure of scale. The approximation to $w(0,0)$ found in this way was 0.2923. A plot of the approximations to $w(ih,0)$ was in good agreement with that of Figure 2.2. We have had to solve a very much larger system of equations to get an approximation comparable to that of the global Galerkin method. On the other hand, the finite difference approximation does not depend on the special form of the solution. Another distinction appears when we ask about the flow rate, for we now have to compute the integral of a function defined only on a mesh. How this might be done is discussed in Chapter 5, but for the sake of simplicity, we do not go into the matter here.

Let us now consider the Jacobi and Gauss–Seidel iterations for a general problem $A\mathbf{x} = \mathbf{b}$ and prove a simple result about convergence. One approach to iterative methods supposes that it is relatively easy to solve systems of the form $M\mathbf{y} = \mathbf{c}$ for a matrix M that is "close" to A. The idea is to rewrite the given problem as

$$M\mathbf{x} = \mathbf{b} + (M - A)\mathbf{x}$$

and calculate a sequence of approximate solutions $\mathbf{x}^{(k)}$ by

$$M\mathbf{x}^{(k+1)} = \mathbf{b} + (M - A)\mathbf{x}^{(k)}.$$

The Jacobi iteration arises in this form when we take M to be the diagonal of A. Similarly, the Gauss-Seidel iteration arises when we take M to be the diagonal of A and all the elements below the diagonal. Clearly it is very easy to solve linear systems involving M in either form. We have also seen another example in this chapter. Generally the factors L and U resulting from Gaussian elimination applied to A are very accurate in the sense that $M = LU$ is very nearly A. By virtue of having a factorization of this M, it is easy to compute the iterates by forward and back substitution in the usual fashion. With this choice of M, the iteration is the iterative refinement discussed in Section 2.3.2.

Passing to the limit on both sides of the equation defining the iteration, we see that if the approximations converge at all, they must converge to $\mathbf{x}$. A little manipulation of the equations for $\mathbf{x}$ and $\mathbf{x}^{(k+1)}$ shows that error $\mathbf{e}^{(k)} = \mathbf{x} - \mathbf{x}^{(k)}$ satisfies

$$\mathbf{e}^{(k+1)} = M^{-1}(M-A)\mathbf{e}^{(k)}.$$

This then implies that

$$||\mathbf{e}^{(k+1)}|| \leq ||M^{-1}(M-A)||\,||\mathbf{e}^{(k)}||.$$

If the number $\rho = \left\|M^{-1}(M-A)\right\| < 1$, this inequality implies that the process converges for any starting guess $\mathbf{x}^{(0)}$. The error decreases by a factor of ρ at each iteration, so if M is close to A in the sense that ρ is small, the process converges quickly. Notice that the crucial quantity ρ is a kind of relative error of the approximation of A by M. Sharp convergence results are known, but they are beyond the scope of this book; see, for example, [2] or [10].

The situation with iterative refinement is special because we want a very accurate value for $\mathbf{x}$. For reasons taken up in Chapter 1, it would be better then to compute the iterate by means of the difference of successive iterates, $\delta^{(k+1)} = \mathbf{x}^{(k+1)} - \mathbf{x}^{(k)}$. This leads to the form suggested in Section 2.3.2,

$$\begin{aligned} M\delta^{(k+1)} &= \mathbf{b} - A\mathbf{x}^{(k)} = \mathbf{r}^{(k)} \\ \mathbf{x}^{(k+1)} &= \mathbf{x}^{(k)} + \delta^{(k+1)}. \end{aligned}$$

Although we have not provided all the details, it is not surprising that even when A is extraordinarily ill-conditioned, M is sufficiently close to A that the result just established guarantees convergence.

The Jacobi iteration is more easily understood by looking at components. Equation j is

$$a_{j1}x_1 + \cdots + a_{jj}x_j + \cdots + a_{jn}x_n = b_j.$$

A little manipulation results in

$$x_j^{(k+1)} = \left(b_j - a_{j1}x_1^{(k)} - \cdots - a_{j,j-1}x_{j-1}^{(k)} - a_{j,j+1}x_{j+1}^{(k)} - \cdots - a_{jn}x_n^{(k)}\right)/a_{jj}.$$

We cannot expect this iteration to work unless the entries off the diagonal of A are small compared to those on the diagonal. To obtain a sufficient condition for convergence, let us suppose that A is strictly diagonally dominant by rows, meaning that there is a number $\rho < 1$ such that for all j,

$$\sum_{\substack{i=1 \\ i \neq j}}^{n} |a_{ji}| \leq \rho\,|a_{jj}|.$$

A little calculation shows that the error $e_j^{(k+1)} = x_j - x_j^{(k+1)}$ satisfies

$$\left| e_j^{(k+1)} \right| \le \left| \sum_{\substack{i=1 \\ i \ne j}}^{n} -a_{ji} e_j^{(k)} \right| \Big/ \left| a_{jj} \right| \le \|\mathbf{e}^{(k)}\| \sum_{\substack{i=1 \\ i \ne j}}^{n} \left| a_{ji} \right| / \left| a_{jj} \right| \le \rho \|\mathbf{e}^{(k)}\|.$$

This holds for each j, so

$$\|\mathbf{e}^{(k+1)}\| \le \rho \|\mathbf{e}^{(k)}\|,$$

telling us that the worst error is reduced by a factor of ρ at each iteration. A (possibly) large system of linear equations arises in Chapter 3 when fitting data by smooth cubic splines. Equation j has the form

$$h_{j-1} x_{j-1} + 2(h_{j-1} + h_j) x_j + h_j x_{j+1} = b_j,$$

and the first and last equations have the form

$$2h_1 x_1 + h_1 = b_1$$
$$h_{n-1} x_{n-1} + 2h_n x_n = b_n.$$

The quantities h_j appearing in these equations are all positive. Evidently this system of equations is strictly diagonally dominant by rows with $\rho = 1/2$. The Jacobi iteration would be a reasonable way to solve these systems, but A is a symmetric tridiagonal matrix and the special version of elimination from Section 2.5.2 is so effective that there is little point to iterative methods. In components the Gauss–Seidel iteration is

$$x_j^{(k+1)} = \left(b_j - a_{j1} x_1^{(k+1)} - \cdots - a_{j,j-1} x_{j-1}^{(k+1)} - a_{j,j+1} x_{j+1}^{(k)} - \cdots - a_{jn} x_n^{(k)} \right) / a_{jj}.$$

A modification of the proof given for the Jacobi iteration shows that this iteration also converges whenever A is strictly diagonally dominant by rows. The finite difference equations of the example do not satisfy the convergence criterion just developed, but they do have a kind of diagonal dominance for which it is possible to prove convergence of both the Jacobi and Gauss–Seidel methods.

The Jacobi iteration and the Gauss–Seidel iteration are too slow to be practical for most problems arising in practice. However, they are still important as preconditioners for more elaborate iterative procedures. For more information about iterative methods, see [2] or [10]. The latter provides some substantial applications to the numerical solution of PDEs.

REFERENCES

1. E. Anderson, et al., LAPACK *User's Guide*, SIAM, Philadelphia, 1992.
2. O. Axelsson, *Iterative Solution Methods*, Cambridge University Press, New York, 1994.
3. A. Cline, C. Moler, G. Stewart, and J. Wilkinson, "An estimate for the condition number of a matrix," *SIAM J. Numer. Anal.*, 16 (1979), pp. 368–375.
4. C. A. J. Fletcher, *Computational Galerkin Methods*, Springer-Verlag, New York, 1984.
5. J. Dongarra, J. Bunch, C. Moler, and G. Stewart, LINPACK *User's Guide*, SIAM, Philadelphia, 1979.

6. G. Forsythe and C. Moler, *Computer Solutions of Linear Algebraic Systems*, Prentice Hall, Englewood Cliffs, N.J., 1969.
7. K. Gallivan and R. Plemmons, "Parallel algorithms for dense linear algebra computations," *SIAM Review*, 32 (1990), pp. 54–135.
8. A. George and J. Liu, *Computer Solution of Large Sparse Positive Definite Systems*, Prentice Hall, Englewood Cliffs, N.J., 1981.
9. G. Golub and C. Van Loan, *Matrix Computations*, 2nd ed., The Johns Hopkins University Press, Baltimore, M.D., 1989.
10. L. A. Hageman and D. M. Young, *Applied Iterative Methods*, Academic Press, Orlando, Fla., 1981.
11. N. Higham, *Accuracy and Stability of Numerical Algorithms*, SIAM, Philadelphia, 1996.
12. B. Noble, *Applications of Undergraduate Mathematics in Engineering*, Macmillan, New York, 1967.
13. B. Noble and J. Daniel, *Applied Linear Algebra*, 3rd ed., Prentice Hall, Englewood Cliffs, N.J., 1988.
14. R. Skeel, "Iterative refinement implies numerical stability for Gaussian elimination," *Math. Comp.*, 35 (1980), pp. 817–832.
15. J. Wilkinson, *The Algebraic Eigenvalue Problem*, Oxford University Press, Oxford, England, 1988.

MISCELLANEOUS EXERCISES FOR CHAPTER 2

2.20 A finite element analysis of a certain load bearing frame yields the stiffness equations

$$\begin{bmatrix} \alpha & 0 & 0 & 0 & \beta & -\beta \\ 0 & \alpha & 0 & -\beta & 0 & -\beta \\ 0 & 0 & \alpha & \beta & \beta & 0 \\ 0 & -\beta & \beta & \gamma & 0 & 0 \\ \beta & 0 & \beta & 0 & \gamma & 0 \\ -\beta & -\beta & 0 & 0 & 0 & \gamma \end{bmatrix} \mathbf{x} = \begin{bmatrix} 15 \\ 0 \\ -15 \\ 0 \\ 25 \\ 0 \end{bmatrix},$$

where $\alpha = 482,317.$, $\beta = 2,196.05$, and $\gamma = 6,708.43$. Here x_1, x_2, x_3 are the lateral and x_4, x_5, x_6 the rotational (three-dimensional) displacements corresponding to the applied force (the right-hand side).

(a) Solve for $\mathbf{x}$.

(b) How reliable is the computation? First assume exact data, then $||\Delta A||/||A|| = 5 \times 10^{-7}$.

2.21 Wang (*Matrix Methods of Structural Analysis*, International Textbook Company, Scranton, Pa., 1966) considers a statically indeterminate pin-jointed truss. With this problem is associated a statics matrix A that defines the configuration of the framework, a member stiffness matrix S that relates the elastic properties of the constituent members, and an external force vector $\mathbf{p}$ that describes the applied forces at the joints. A displacement vector bfx that accounts for the displacement at each degree of freedom and an internal force vector **f** acting on each member satisfies

$$K\mathbf{x} = \mathbf{p},\ K = ASA^T,\ \mathbf{f} = (SA^T)\mathbf{x}.$$

For one example

$$A = \begin{bmatrix} 0.6 & -1.0 & 0.0 & 0.0 & 0.0 & 0.0 & 0.0 & -0.6 & 0.0 & 0.0 \\ 0.8 & 0.0 & 0.0 & 0.0 & 0.0 & 0.0 & 1.0 & 0.8 & 0.0 & 0.0 \\ 0.0 & 1.0 & -0.6 & 0.0 & 0.0 & 0.0 & 0.0 & 0.0 & 0.6 & 0.0 \\ 0.0 & 0.0 & 0.8 & 0.0 & 0.0 & 0.0 & 0.0 & 0.0 & 0.8 & 1.0 \\ 0.0 & 0.0 & 0.0 & 1.0 & -1.0 & 0.0 & 0.0 & 0.0 & -0.6 & 0.0 \\ 0.0 & 0.0 & 0.0 & 0.0 & 0.0 & 0.0 & -1.0 & 0.0 & -0.8 & 0.0 \\ 0.0 & 0.0 & 0.0 & 0.0 & 1.0 & -1.0 & 0.0 & 0.6 & 0.0 & 0.0 \\ 0.0 & 0.0 & 0.6 & 0.0 & 0.0 & 1.0 & 0.0 & 0.0 & 0.0 & 0.0 \end{bmatrix},$$

The matrix S has all zero entries except along the diagonal where the entries are

$$\{4800, 10000, 4800, 10000, 10000, 10000, 3000, 4800, 4800, 3000\}.$$

Write a program to form matrix products and determine the elements of K. Solve for **x** using the three **p** vectors

$$\mathbf{p} = \begin{bmatrix} 0.0 \\ -1.0 \\ 0.0 \\ 0.0 \\ 0.0 \\ 0.0 \\ 0.0 \\ 0.0 \end{bmatrix}, \begin{bmatrix} 0.0 \\ 0.0 \\ 0.0 \\ -1.0 \\ 0.0 \\ 0.0 \\ 0.0 \\ 0.0 \end{bmatrix}, \begin{bmatrix} 0.0 \\ 0.0 \\ 0.0 \\ 0.0 \\ 0.0 \\ -1.0 \\ 0.0 \\ 0.0 \end{bmatrix}.$$

Find the corresponding vectors **f**.

2.22 This exercise assumes a familiarity with matrix multiplication. An appropriate organization of Gaussian elimination as in Factor/Solve makes it efficient to solve systems of equations with different right-hand sides but the same coefficient matrix A. It is more complicated to deal with changes in A, but a formula called the Sherman–Morrison formula makes it possible to deal with certain modifications efficiently. Assume that we have already factored A into LU and we want to solve $(A + \mathbf{u}\mathbf{v}^T)\mathbf{x} = \mathbf{b}$ for given column vectors **u**, **v**, and **b**. Show that this can be done by first solving $A\mathbf{z} = \mathbf{u}$ and $A\mathbf{y} = \mathbf{b}$, then forming

$$\mathbf{x} = \mathbf{y} - \frac{\mathbf{v}^T\mathbf{y}}{1 + \mathbf{v}^T\mathbf{z}}\mathbf{z}.$$

A proper choice of **u** and **v** handles the change of one row or one column of A. For example, if row i of A is to be changed by adding to it a given row vector $\mathbf{v}^T$, just take the column vector **u** to be zero in all entries except the ith, which is 1.

(a) How do you change column j in A so as to add a given column vector **u**?

(b) How do you choose **u** and **v** in order to change a_{ij} into $a_{ij} + \delta$?

(c) A change in A may make it singular. How would this be revealed when using the Sherman–Morrison formula? Note that this approach may not be an accurate way to solve for $\mathbf{x}$ when A is poorly conditioned because the elimination is done on A. Still for *small* changes to A, the accuracy should be acceptable and this is an inexpensive way to study the effect of changes to the data of A.

2.23 Occasionally it is desirable to compute the determinant of a matrix A with n rows and columns. Using the factorization $PA = LU$ discussed in Section 2.2, it can be shown that

$$\det A = (-1)^{\text{number of row interchanges}} \times \text{ product of pivots.}$$

In terms of the output from Factor (FORTRAN version) this is

$$\det A = \text{PVTIDX}(n) * A(1,1) * \cdots * A(n,n).$$

In the C or C++ versions this would be

$$\det A = pivot\_index[n-1] * a[0][0] * \cdots * a[n-1][n-1].$$

Use this formula to compute the determinant of

(a) A in Exercise 2.13 and

(b) the matrix in Exercise 2.15.

CHAPTER 3

INTERPOLATION

Often we need to approximate a function $f(x)$ by another "more convenient" function $F(x)$. This arises in physical processes where $f(x)$ is known only through its values at certain sample points x and $F(x)$ is needed to approximate maximum or minimum values, to estimate integrals or derivatives, or merely to generate values for $f(x)$ at points where experimental data are not available. This also arises when $f(x)$ is known but is difficult to evaluate, integrate, or differentiate. A familiar example is the function $f(x) = \sin x$ that has to be approximated in a way that can be evaluated by calculators and computers. This is a fundamental principle of numerical analysis: if we cannot carry out a basic computation with the function of interest, we approximate it by a function for which we can do the computation.

In this chapter a function $f(x)$ is approximated by an interpolant $F(x)$, a function that agrees with $f(x)$ at certain points. A formal definition of the verb interpolate is as follows.

Definition. interpolation: A function $F(x)$ is said to *interpolate* $f(x)$ at the points $\{x_1, \ldots, x_N\}$ if

$$F(x_j) = f(x_j), \qquad j = 1, 2, \ldots, N.$$

The process of constructing such a function $F(x)$ is called *interpolation.*

There are many types of approximating functions $F(x)$ and which one to use depends to a large extent on the nature of the data and the intended use of the approximation. Perhaps the simplest approximating functions are polynomials. It can be shown that any continuous function can be approximated arbitrarily well over a finite interval by a polynomial. More to the point, polynomials and their ratios (called rational functions) are the only functions that can be evaluated directly on a computer. For this reason polynomials are used not only for interpolation but also as a foundation for most of the methods in the remaining chapters of the book. Polynomial splines, that is, piecewise polynomial functions, are a very powerful tool for approximating functions and are the main object of study in this chapter. In many applications the appearance of the graph of $F(x)$ is of great importance. For this reason it is very helpful to have a graphing package for visualization of the approximating functions derived in this chapter.

For a more thorough treatment of the theory of polynomial interpolation see [15, Chapters 5 and 6] and for more about approximation theory see [4]. The book [3] is an excellent introduction to polynomial splines and contains many FORTRAN codes.

3.1 POLYNOMIAL INTERPOLATION

In this section the approximation $F(x)$ is a polynomial and it is traditional to use the notation P_N instead of F. The interpolation problem, formally stated, is as follows. Given the ordered pairs (x_j, f_j) for $j = 1, 2, \ldots, N$, where each $f_j = f(x_j)$ for some probably unknown function $f(x)$,

$$\text{find a polynomial } P_N(x) \text{ such that } P_N(x_j) = f_j,\ 1 \leq j \leq N. \tag{3.1}$$

What degree should P_N have? A polynomial of degree $N-1$,

$$P_N(x) = c_1 + c_2 x + c_3 x^2 + \cdots + c_N x^{N-1},$$

has N free parameters, the coefficients c_k. Since the polynomial must satisfy conditions at N points x_j, called *interpolating points* or *nodes*, we might expect to need this many parameters to satisfy the conditions. In an exercise you are asked to show by example that if the interpolating polynomial is allowed to be of degree N or higher, there are many polynomials satisfying the interpolation conditions. It is easy to show by example that if the degree is less than $N-1$, it may not be possible to satisfy all the conditions. Degree $N-1$ is just right; with this degree there is always a solution to the interpolation problem and only one.

Theorem 3.1. *Given N distinct points $\{x_j\}$ there is one and only one polynomial $P_N(x)$ of degree less than N that interpolates a given function $f(x)$ at these points.*

Proof. We first show that such a polynomial $P_N(x)$ exists and then show that it is unique. Write

---

the *Lagrange form* of the interpolating polynomial

$$P_N(x) = \sum_{k=1}^{N} f_k L_k(x), \tag{3.2}$$

---

where the functions $\{L_k(x)\}$ are at our disposal and are to be chosen independently of $f(x)$. If $P_N(x)$ is to be a polynomial of degree less than N for any choice of the data $f_1, \ldots, f_N$, then each $L_k(x)$ must also be a polynomial of degree less than N. Furthermore, in order to have $P_N(x_j) = f_j$ for $1 \leq j \leq N$, again for any choice of the data, then the $L_k(x)$ must satisfy

$$L_k(x_j) = \begin{cases} 0 & \text{if } j \neq k \\ 1 & \text{if } j = k. \end{cases}$$

This says that the polynomial $L_k(x)$ has zeros at each x_j with $j \neq k$, and so must have the form $L_k(x) = C \prod_{\substack{j=1 \\ j \neq k}}^{N} (x - x_j)$ for some constant C. The condition $L_k(x_k) = 1$

implies that $C = 1/\prod_{\substack{j=1\\j\neq k}}^{N}(x_k - x_j)$; hence

$$L_k(x) = \prod_{\substack{j=1\\j\neq k}}^{N} \frac{x - x_j}{x_k - x_j}. \tag{3.3}$$

To show that $P_N(x)$ is unique, let $Q_N(x)$ be another polynomial of degree less than N that satisfies $Q_N(x_j) = f_j$ for $1 \leq j \leq N$. The difference $D(x) = P_N(x) - Q_N(x)$ is also a polynomial of degree less than N, and $D(x_j) = P_N(x_j) - Q_N(x_j) = f_j - f_j = 0$ for $1 \leq j \leq N$. This says that D has N distinct zeros. But the Fundamental Theorem of Algebra states that any polynomial of degree less than N can have at most $N-1$ zeros unless it is identically zero. Accordingly $D \equiv 0$, which is to say that $P_N \equiv Q_N$ and the interpolating polynomial is unique. ■

The polynomial $L_k(x)$ given by (3.3) is called a *fundamental interpolating polynomial* or *shape function.* The fundamental interpolating polynomials are of exact degree $N-1$, but the interpolant can be of lower degree. Indeed, it is important to appreciate that when $f(x)$ is a polynomial of degree less than N, the interpolant must be $f(x)$ itself. After all, the polynomial $f(x)$ interpolates itself and its degree is less than N, so by uniqueness, $P_N(x) = f(x)$.

Example 3.1. Let $f(x) = \sin x$. Find the $P_3(x)$ that interpolates $f(x)$ at the three points $\{0, \pi/2, \pi\}$. The corresponding function values are $\{0, 1, 0\}$, so

$$\begin{aligned}
P_3(x) &= 0 \cdot L_1(x) + 1 \cdot L_2(x) + 0 \cdot L_3(x) \\
&= (x-0)(x-\pi)/\left[\left(\frac{\pi}{2}-0\right)\left(\frac{\pi}{2}-\pi\right)\right] \\
&= -\frac{4}{\pi^2}x(x-\pi).
\end{aligned}$$

■

Example 3.2. Let $P_4(x)$ interpolate the tabular data

x	1.82	2.50	3.65	4.03
y	0.00	1.30	3.10	2.52

Then the Lagrange form of $P_4(x)$ is

$$\begin{aligned}
P_4(x) &= 0 \cdot L_1(x) + 1.30L_2(x) + 3.10L_3(x) + 2.52L_4(x) \\
&= 1.30\frac{(x-1.82)(x-3.65)(x-4.03)}{(0.68)(-1.15)(-1.53)} + 3.10\frac{(x-1.82)(x-2.50)(x-4.03)}{(1.83)(1.15)(-0.38)} \\
&\quad + 2.52\frac{(x-1.82)(x-2.50)(x-3.65)}{(2.21)(1.53)(0.38)} \\
&= 1.09(x-1.82)(x-3.65)(x-4.03) - 3.88(x-1.82)(x-2.50)(x-4.03) \\
&\quad + 1.96(x-1.82)(x-2.50)(x-3.65).
\end{aligned}$$

■

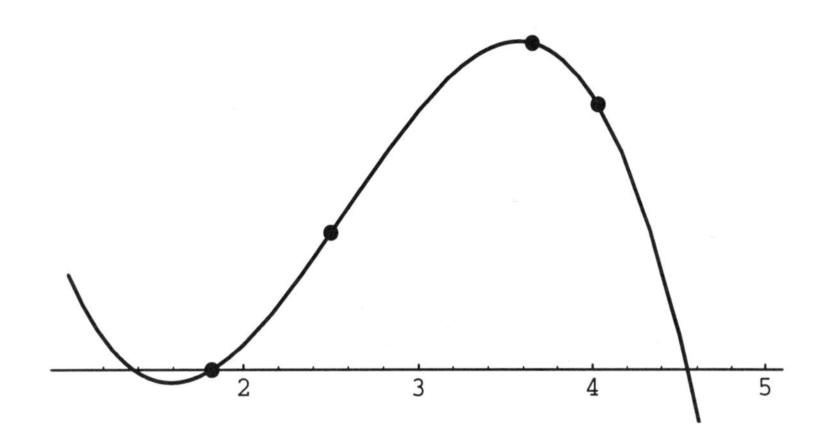

Figure 3.1 Plot of $P_4(x)$ for Example 3.2.

The polynomial $P_4(x)$ is plotted in Figure 3.1. Figure 3.2 shows plots of the four fundamental polynomials L_1, L_2, L_3, and L_4 associated with these data.

How well does P_N approximate f? Will interpolating at more points (increasing N) improve the approximation? The next theorem helps answer these questions.

Theorem 3.2. *Assume that $f(x)$ has N continuous derivatives on an interval I containing the interpolating points $\{x_j\}_{j=1}^{N}$. If $P_N(x)$ is the polynomial of degree less than N interpolating f on these data, then for any x in I there is a point ξ_x in I such that*

---

the error in polynomial interpolation is

$$f(x) - P_N(x) = \frac{1}{N!} f^{(N)}(\xi_x) w_N(x), \tag{3.4}$$

---

where

$$w_N(x) = \prod_{j=1}^{N} (x - x_j) \tag{3.5}$$

and

$$\min(x_1, \ldots, x_N, x) < \xi_x < \max(x_1, \ldots, x_N, x).$$

Proof. Clearly the equality is valid for $x = x_j$, $1 \leq j \leq N$. For x not equal to any of the interpolating points, define the new function

$$G(t) = f(t) - P_N(t) - \frac{f(x) - P_N(x)}{w_N(x)} w_N(t).$$

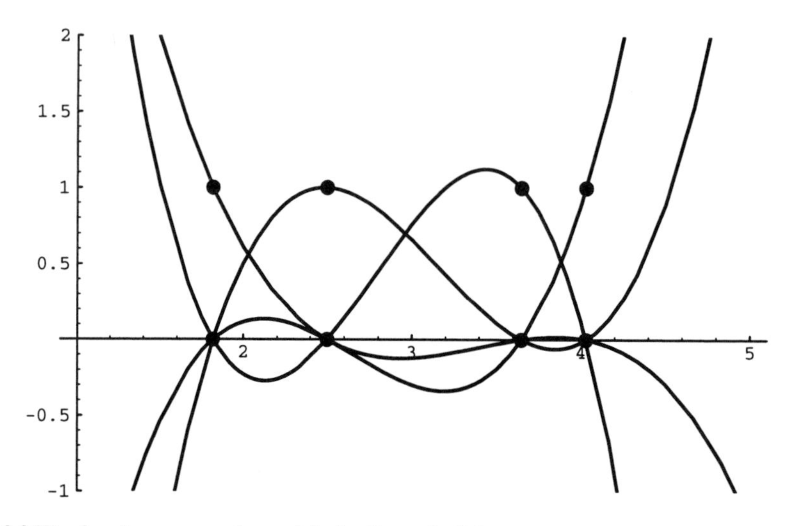

Figure 3.2 The four Lagrange polynomials for Example 3.2.

Now G has N continuous derivatives on I and

$$G(x_j) = f_j - f_j - 0 \cdot [f(x) - P_N(x)]/w_N(x) = 0,\ 1 \le j \le N.$$

Also, $G(x) = f(x) - P_N(x) - w_N(x)[f(x) - P_N(x)]/w_N(x) = 0$, so G has $N+1$ distinct zeros. By Rolle's theorem (see the appendix), G' has at least N distinct zeros in I. Repeating this argument, G'' has at least $N-1$ distinct zeros in I, and so on, and $G^{(N)}$ has at least one zero in I. Denoting a zero of $G^{(N)}$ by ξ_x, we find $0 = G^{(N)}(\xi_x) = f^{(N)}(\xi_x) - P_N^{(N)}(\xi_x) - w_N^{(N)}(\xi_x)[f(x) - P_N(x)]/w_N(x)$. The polynomial P_N is of degree less than N, so its Nth derivative is identically zero. The polynomial $w_N(t)$ is of degree N with leading term t^N, so its Nth derivative is $N!$. These observations show that

$$0 = f^{(N)}(\xi_x) - N![f(x) - P_N(x)]/w_N(x),$$

which is merely a rearrangement of (3.4). ∎

If the interval $I = [a,b]$ and we let

$$M_N = \max_{x \in I} |f^{(N)}(x)|,$$

then two upper bounds for the interpolation error are

$$|f(x) - P_N(x)| \le \frac{M_N}{N!} |w_N(x)| \tag{3.6}$$

$$\le \frac{M_N (b-a)^N}{N!} \quad \text{for } x \text{ in } (a,b). \tag{3.7}$$

A sharper version of this last bound is

$$\max_{a \le x \le b} |f(x) - P_N(x)| \le \frac{M_N}{N!} \max_{a \le x \le b} |w_N(x)|.$$

Theorem 3.2 is a classic result about the error of interpolation. Sometimes it can be used directly. For example, along with the tables of [13] are instructions about the degree of interpolating polynomial needed to get full tabular accuracy. As a concrete example, the scaled exponential integral $x\exp(x)E_1(x)$ is tabulated for $x = 2.0, 2.1, 2.2, \ldots, 10.0$ and it is said that degree 4 is appropriate. This is an example of interpolation for which we can choose the nodes from a set of possibilities. The assertion for the table assumes that the nodes have been chosen in an appropriate way, a matter to be taken up shortly. For a specific function that can be differentiated readily and for specific nodes, a bound on the error of interpolation might be obtained from the theorem. More often in practice we use guidelines that will be deduced below from the theorem.

Example 3.3. Again consider $f(x) = \sin x$ and suppose values are known at the five points $\{0.0, 0.2, 0.4, 0.6, 0.8\}$. Inequality (3.6) can be used to bound the error in approximating $\sin 0.28$ by $P_5(0.28)$. Since

$$M_5 = \max_{t\in[0,0.8]} |\sin^{(5)}(t)| = \max_{t\in[0,0.8]} |\cos t| = 1,$$

we have the bound

$$\begin{aligned} |\sin(0.28) - P_5(0.28)| &\leq |0.28(0.28-0.2)(0.28-0.4)(0.28-0.6)(0.28-0.8)|/5! \\ &= 3.7 \times 10^{-6}. \end{aligned}$$

An actual evaluation shows $P_5(0.28) = 0.2763591$, while $\sin 0.28 = 0.2763556$, so the exact error is -3.5×10^{-6}. ■

Theorem 3.2 and the bounds provide insight and guidelines for practical interpolation. The factor $w_N(x)$ in the error expression increases near the ends of the data interval and increases very rapidly as x is taken farther away from $[a,b]$; the higher the degree, the more this is true. Because of this the *bound* (3.6) increases rapidly, but the sharper *equality* (3.4) shows that this effect might be lessened for a given f and given x by a derivative factor that gets smaller. Approximating $f(x)$ by $P_N(x)$ outside the smallest interval I containing the points of interpolation (the nodes) is sometimes called *extrapolation.* In general, it is clearly dangerous to extrapolate very far outside the span of the data, especially when using polynomials of high degree. On the other hand, $w_N(x)$ is relatively small for x in the middle of the nodes. And, of course, because of continuity, the error must be small close to a node. These two observations suggest that when possible, it is best to interpolate at nodes centered about the x of interest and as close to x as possible. This is what was meant earlier when reference was made to choosing the nodes in an appropriate way.

The plot of $w_9(x) = (x+4)(x+3)\cdots(x-3)(x-4)$ on $[-4,4]$ seen in Figure 3.3 shows the qualitative behavior of this factor in the error expression. This function grows extremely fast outside the span of the nodes and is large toward the ends, but it is of modest size in the middle (as verified in Exercise 3.7). Figure 3.6 shows a polynomial interpolant of high degree. Figure 3.7 displays the same data and interpolant over a subinterval in the middle of the data. The physical origin of the data and

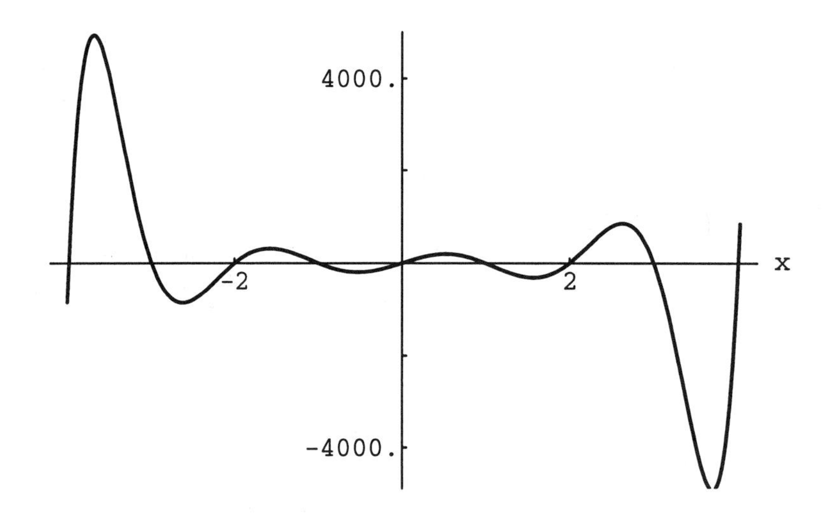

Figure 3.3 Plot of $w_9(x)$ on the interval $[-4,4]$.

the plot of the data suggest that the underlying $f(x)$ should vary slowly. Clearly the interpolant $P_{16}(x)$ does not provide a plausible approximation to $f(x)$ over the whole interval. Nevertheless, in the middle of the range the fit appears to be quite acceptable. The qualitative behavior seen here for a high degree polynomial interpolant might have been expected from consideration of the $w_N(x)$ factor in the error.

Sometimes we can evaluate a function anywhere we like in an interval, but wish to approximate it by a simpler function so as to approximate its derivative or integral or .... It is then natural to ask if there is a good choice of nodes for interpolating in the sense of making

$$\max_{a \le x \le b} |w_N(x)| \tag{3.8}$$

small. The answer is known (e.g., [12, pp. 227–228]). The points

$$x_j = \frac{(b+a)}{2} + \frac{(b-a)}{2} \cos \frac{(2j-1)\pi}{2N}, \; j = 1, \ldots, N, \tag{3.9}$$

called *Chebyshev points*, make (3.8) as small as possible. More details about the quality of this approximation are provided in the next section.

If derivative information is available, we can interpolate it along with the values of f. There are a great many possibilities and only one, called Hermite interpolation, will be described. Suppose that in addition to the values f_j given at x_j for $j = 1, \ldots, N$, we have the first derivative values f'_j. With $2N$ independent values it is reasonable to try to interpolate with a polynomial of degree $2N-1$. We leave as an exercise the fact that the fundamental polynomials $\phi_k(x), \Phi_k(x)$ of degree less than $2N$ that satisfy

$$\phi_k(x_j) = \begin{cases} 0 & \text{if } j \neq k \\ 1 & \text{if } j = k, \end{cases} \qquad \phi'_k(x_j) = 0 \quad \text{all } j$$

$$\Phi'_k(x_j) = \begin{cases} 0 & \text{if } j \neq k \\ 1 & \text{if } j = k, \end{cases} \qquad \Phi_k(x_j) = 0 \quad \text{all } j,$$

are given by

$$\phi_k(x) = [1 - 2L_k'(x_k)(x - x_k)]L_k^2(x)$$
$$\Phi_k(x) = (x - x_k)L_k^2(x). \tag{3.10}$$

Here the $L_k(x)$ are given by (3.3). It is now obvious that the polynomial

$$P(x) = \sum_{k=1}^{N} f_k \phi_k(x) + \sum_{k=1}^{N} f_k' \Phi_k(x)$$

satisfies

$$P(x_j) = f_j,\ j = 1, \ldots, N$$
$$P'(x_j) = f_j',\ j = 1, \ldots, N.$$

There is an expression like (3.4) for Hermite interpolation:

$$f(x) - P(x) = \frac{1}{2N!} f^{(2N)}(\xi_x) w_N^2(x),$$

where $w_N(x)$ is again given by (3.5).

Two cases of Hermite interpolation are of special interest in this book. In the solution of ordinary differential equations we use fifth degree polynomials (quintics) to interpolate f and f' at three points. In this chapter we use third degree polynomials (cubics) to interpolate f and f' at two points. It will be convenient later to express the Hermite cubic interpolant in a different form. If we write

$$H(x) = a + b(x - x_n) + c(x - x_n)^2 + d(x - x_n)^3 \tag{3.11}$$

and require that

$$H(x_n) = f_n,\ H'(x_n) = f_n'$$
$$H(x_{n+1}) = f_{n+1},\ H'(x_{n+1}) = f_{n+1}',$$

then it is easy to show that for $h = x_{n+1} - x_n$,

$$a = f_n,\ b = f_n' \tag{3.12}$$
$$c = [3(f_{n+1} - f_n)/h - 2f_n' - f_{n+1}']/h \tag{3.13}$$
$$d = [f_n' + f_{n+1}' - 2(f_{n+1} - f_n)/h]/h^2. \tag{3.14}$$

EXERCISES

3.1 Is the interpolating polynomial (3.2) always of exact degree $N - 1$? If not, illustrate by an example.

3.2 Suppose that $f(x)$ is a polynomial of degree $N - 1$ or less. Prove that if $P_N(x)$ interpolates $f(x)$ at N distinct points, then $P_N(x) \equiv f(x)$. Make up an example ($N \geq 3$) and verify by direct calculation.

3.3 For the data

x	1	2
$f(x)$	2	4

construct $P_2(x)$ using (3.2). Find a polynomial $Q(x)$ of degree 2 that also interpolates these data. Does this contradict the theory about uniqueness of the interpolating polynomial? Explain.

3.4 An alternate method for computing $P_N(x)$ is to write

$$P_N(x) = c_1 + c_2x + c_3x^2 + \cdots + c_Nx^{N-1}.$$

Then the interpolation conditions, $P_N(x_i) = f_i$ for $1 \le i \le N$, yield a system of N equations in the N unknowns $c_1, c_2, \ldots, c_N$ that can be solved using the codes Factor/Solve. Unfortunately, there are two difficulties with this method: (1) it is expensive ($N^3/3$ multiplications), and (2) the coefficient matrix can be very ill-conditioned.

(a) Implement this algorithm.

(b) Test it on the data in Example 3.5. Use the same six interpolating values and evaluate (see Exercise 3.5) at the remaining points. What is COND? How do the answers compare with those in the text for $P_6(x)$ computed by another method?

3.5 There are several ways to evaluate $P_N(x) = c_1 + c_2x + \cdots + c_Nx^{N-1}$. As a first algorithm we could use

```
P := c1
for i = 2,3,...,N begin
    P := P + ci * x^(i-1)
end i.
```

How many multiplications does this algorithm require? A better approach is based on the nested form of $P_N(x)$ used in Section 3.3:

$$c_1 + x\{c_2 + x[c_3 + x(c_4 + \cdots + c_Nx)\cdots]\}.$$

For example,

$$3 + 4x - 6x^2 + 5x^3 = 3 + x[4 + x(-6 + 5x)].$$

The new algorithm is

```
P := cN
for i = N-1, N-2, ..., 1 begin
    P := P * x + ci
end i.
```

Compare the number of multiplications for this algorithm to those for the first one.

3.6 Use the algorithm suggested by Exercise 3.4 to compute $P_{12}(w)$, which interpolates all the data in Example 3.5. Plot $P_{12}(w)$ for $5 \le w \le 100$. Does it look reasonable?

3.7 Verify the plot in Figure 3.3 of $w_9(x)$ to see that it has smallest magnitude near the middle of the data. Choose $x_i = -5 + i$ and evaluate $w_9(x)$ at $\{\pm 0.5, \pm 1.5, \ldots, \pm 4.5, \pm 5\}$.

3.8 Derivatives of $f(x)$ can be estimated by the corresponding derivative of $P_N(x)$ for some choice of N and $\{x_n\}_{n=1}^N$. The usual approach is to try

$$f^{(N-1)}(x) \approx P_N^{(N-1)}(x).$$

Since $P_N(x)$ has degree at most $N-1$, $P_N^{(N-1)}(x)$ must be a constant function (i.e., independent of x).

(a) Use (3.2) to show that

$$P_N^{(N-1)}(x) = (N-1)! \sum_{k=1}^{N} \frac{f_k}{\prod_{j \ne k}(x_k - x_j)}.$$

(b) What approximation to $f'(x)$ results when $N = 2$?

3.9 Verify that the functions given in (3.10) have the fundamental interpolation properties claimed.

3.10 Derive equations (3.12)–(3.14). Start with

$$H(x) = a + b(x - x_n) + c(x - x_n)^2 + d(x - x_n)^3,$$

and then derive and solve the four equations for a, b, c, and d resulting from the conditions

$$H(x_n) = f_n, \; H(x_{n+1}) = f_{n+1}$$
$$H'(x_n) = f'_n, \; H'(x_{n+1}) = f'_{n+1}.$$

3.2 MORE ERROR BOUNDS

Far more can be said about the error in polynomial interpolation. In this section some useful results are discussed and some results are given for the error made in approximating derivatives by derivatives of an interpolating polynomial.

One way to measure how well $P_N(x)$ approximates $f(x)$ on an interval $[a, b]$ is by the worst error:

$$\|f - P_N\| = \max_{a \le x \le b} |f(x) - P_N(x)|.$$

A fundamental theorem due to Weierstrass (see [4]) states that any function $f(x)$ continuous on a finite interval $[a,b]$ can be approximated arbitrarily well by a polynomial. Stated formally, given any $\varepsilon > 0$, there is a polynomial $P(x)$ such that $||f - P|| \leq \varepsilon$. It is plausible that interpolating f at more and more points in $[a,b]$ would lead to a better and better approximation. The bound (3.7) shows that if M_N does not grow too fast as $N \to \infty$, the interpolants P_N approximate f arbitrarily well. Unfortunately, this is not true for all continuous f. A result due to Faber says that for *any* given set of nodes $\{x_1^{(1)}\}, \{x_1^{(2)}, x_2^{(2)}\}, \ldots$ in $[a,b]$, there is a function $f(x)$ continuous on $[a,b]$ such that the interpolants $P_N(x)$ of degree less than N defined by

$$P_N(x_i^{(N)}) = f(x_i^{(N)}), \, i = 1, \ldots, N,$$

do not even have $||f - P_N||$ bounded as $N \to \infty$. Runge's function

$$f(x) = \frac{1}{1+x^2} \tag{3.15}$$

on $[-5,5]$ is a classic example. It seems obvious that interpolating such a smooth function at more and more equally spaced points should lead to convergence, but it is found that for even modest N, the interpolants are completely unacceptable (see Exercise 3.11).

It turns out that if one can interpolate at the good nodes (3.9), interpolation is about as good a way to approximate $f(x)$ by a low degree polynomial as possible. In fact, Runge's function can be quite accurately approximated by a polynomial that interpolates at the Chebyshev points (see Exercise 3.12). In general, there is a polynomial $P_n^*(x)$ of degree less than N that is the *best* approximation to $f(x)$ on $[a,b]$ in the sense that $||f - P_N^*||$ provides the smallest value of $||f - P||$ for all polynomials P of degree less than N. Let $P_N(x)$ interpolate $f(x)$ at the nodes $x_1, \ldots, x_N$ in $[a,b]$. For any x,

$$f(x) - P_N(x) = f(x) - P_N^*(x) + P_N^*(x) - P_N(x).$$

Now $P_N^*(x) - P_N(x)$ is a polynomial of degree less than N, so

$$P_N^*(x) - P_N(x) = \sum_{k=1}^{N} (P_N^*(x_k) - P_N(x_k)) L_k(x)$$

because Lagrangian interpolation at N points is exact for such polynomials. Using the fact that $P_N(x_k) = f_k$, we then find that

$$\begin{aligned} |f(x) - P_N(x)| &\leq |f(x) - P_N^*(x)| + \sum_{k=1}^{N} |P_N^*(x_k) - f_k| |L_k(x)| \\ &\leq ||f - P_N^*|| \left(1 + \max_{a \leq x \leq b} \sum_{k=1}^{N} |L_k(x)| \right), \end{aligned}$$

and then

$$||f - P_N|| = \max_{a \leq x \leq b} |f(x) - P_N(x)| \leq ||f - P_N^*|| \left(1 + \max_{a \leq x \leq b} \sum_{k=1}^{N} |L_k(x)| \right).$$

This inequality relates the error of P_N to the error of the best possible polynomial P_N^* by a factor

$$1+\max_{a\le x\le b}\sum_{k=1}^{N}|L_k(x)|,$$

which is given in terms of the points of interpolation alone. A simple analytical bound for the particular nodes (3.9) is found in [17]:

$$1+\max_{a\le x\le b}\sum_{k=1}^{N}|L_k(x)|\le 1+\frac{1}{N}\sum_{k=1}^{N}\tan\frac{(2k-1)\pi}{4N}.$$

The surprising thing is that the bound is so small for moderate degrees N. For $N\le 20$, it is less than 4. Thus

$$\|f-P_N^*\|\le\|f-P_N\|\le 4\|f-P_N^*\|$$

for all $N\le 20$. These easily constructed interpolating polynomials are, then, about as good as possible.

Interpolation is not so effective when the nodes cannot be chosen, and as Faber's theorem suggests, high order interpolation can be quite unsatisfactory. It is common in practice that a high order interpolating polynomial exhibits large amplitude oscillations even when the data appear to come from a smooth function $f(x)$. Examples are given in the next section; for extreme examples the reader need only try interpolating the data in Exercises 3.22, 3.23, and 3.26 by polynomials. For these reasons data are usually fit with relatively low degree polynomials, and a variety of devices are used to prevent oscillations. Some of these are studied in Section 3.5.

Polynomial interpolation is a basic tool in numerical analysis. As an example, derivatives of an interpolant $P_N(x)$ to $f(x)$ can be used to approximate derivatives of $f(x)$. An argument very similar to that of Theorem 3.2 (see [15, pp. 289–290] for details) can be used to show that for any $r<N$

$$f^{(r)}(x)-P_N^{(r)}(x)=\frac{f^{(N)}(\xi_x)}{(N-r)!}\prod_{k=1}^{N-r}(x-\zeta_k),$$

where the points $\{\zeta_k\}$ are known to be distinct and to satisfy

$$x_k<\zeta_k<x_{k+r},\ 1\le k\le N-r.$$

The point ξ_x depends on x and lies in the same interval I as the ξ_x in Theorem 3.2. It has been assumed here that $x_1<x_2<\cdots<x_N$. As a consequence,

$$|f^{(r)}(x)-P_N^{(r)}(x)|\le\frac{M_N(x_N-x_1)^{N-r}}{(N-r)!} \tag{3.16}$$

as long as $x_1\le x\le x_N$. The Lagrange form of the interpolating polynomial is convenient for deriving formulas for numerical differentiation. To approximate a derivative of $f(x)$ at a point z, given values f_k at points $\{x_1,\dots,x_N\}$, we simply form the interpolant, differentiate it, and evaluate it at $x=z$:

$$f^{(r)}(z)\approx P_N^{(r)}(z)=\sum_{k=1}^{N}f_kL_k^{(r)}(z).$$

Because the coefficients in this expression depend only on the nodes, we have here a formula that can be used for any $f(x)$. The programs provided with this chapter for the computation of cubic splines approximate the first derivative at the ends of the range of nodes in this way. At one end they define a cubic polynomial $C(x)$ interpolating at the nodes x_1, x_2, x_3, x_4 and then approximate $f'(x_1)$ by $C'(x_1)$, and similarly at the other end. For closely spaced data, this provides an accurate approximation.

Error bounds like (3.16) can be derived for the Hermite polynomials considered at the end of the preceding section (see [2] for details). Using the earlier notation, if f has four continuous derivatives for any x in the interval $[x_n, x_n + h]$, then with $M_4 = \max_{x_n \le x \le x_{n+h}} |f^{(4)}(x)|$,

$$|f(x) - H(x)| \le \frac{1}{384} M_4 h^4 \tag{3.17}$$

$$|f'(x) - H'(x)| \le \frac{\sqrt{3}}{216} M_4 h^3 \tag{3.18}$$

$$|f''(x) - H''(x)| \le \frac{1}{12} M_4 h^2 \tag{3.19}$$

$$|f'''(x) - H'''(x)| \le \frac{1}{2} M_4 h. \tag{3.20}$$

EXERCISES

3.11 Verify that using polynomials interpolating at the $N = 2m + 1$ equally spaced points $x_j = -5 + 5(j - 1)/m$ give poor approximations to Runge's function $f(x) = 1/(1 + x^2)$ on $[-5, 5]$.

(a) Compute the maximum value of $|f(x) - P_{2m+1}(x)|$ over a large set of x values (not interpolating points) in $[-5, 5]$ for $m = 7$, $m = 10$, and $m = 13$. Are the errors increasing or decreasing as m gets bigger?

(b) Repeat (a) but this time only compute error on $[-1, 1]$. Use the same $\{x_j\}$ and the same three m values as in (a). What happens this time as N increases?

3.12 Verify that using polynomials interpolating at the Chebyshev points (3.9) gives good approximations to Runge's function (3.15). As in the preceding exercise, compute the maximum value of $|f(x) - P_N(x)|$ over a large set of x values (not interpolating points) in $[-5, 5]$ for $N = 15$, $N = 21$, and $N = 27$. What is the behavior of the errors as N gets bigger?

3.13 Repeat Exercise 3.11b for the function $f(x) = |x|$ on $[-1, 1]$. The $\{x_j\}$ are now $x_j = -1 + (j - 1)/m$ for $j = 1, 2, \ldots, 2m + 1$.

3.14 Repeat Exercise 3.12 for the function $f(x) = |x|$ on $[-1, 1]$. Use $N = 21, 41$, and 61 this time.

3.3 NEWTON DIVIDED DIFFERENCE FORM

We have repeatedly used the fact that there is exactly one polynomial $P_N(x)$ of degree less than N that assumes given values f_j at N distinct points x_j. The Lagrange form (3.2) is just one way to represent this polynomial. As we have seen in the case of differentiation, it is well suited for many applications because of the simple dependence on the f_j. On the other hand, the nodes x_j do not appear in a simple way, and this is inconvenient for some tasks. In particular, it is not convenient when we do not know in advance what degree is appropriate. This is the most common situation when approximating data, so an alternative form due to Newton is preferred for practical interpolation by polynomials. Polynomial interpolation underlies two kinds of methods widely used for the numerical solution of ordinary differential equations, Adams methods and backward differentiation formulas (Gear's methods). At each step of the numerical solution of an initial value problem, the codes attempt to find the most appropriate degree for the underlying polynomial interpolant. For this reason such codes use either the Newton form of the polynomial or a closely related form. Although the machinery that must be developed for the Newton form may seem formidable at first, the calculations are easy to learn.

A basic tactic of numerical analysis is to estimate the error in a quantity by comparing it to a quantity believed to be more accurate. If $P_N(x)$ interpolates at the nodes $\{x_1,\ldots,x_N\}$ and $P_{N+1}(x)$ interpolates at the same nodes plus x_{N+1}, then in suitable circumstances the latter is a better approximation to $f(x)$ and $f(x) - P_N(x) \approx P_{N+1}(x) - P_N(x)$. If we do not know what degree is appropriate, this suggests a way to proceed. Start with the constant polynomial $P_1(x) = f_1$ interpolating at x_1. Having computed $P_N(x)$, compute $P_{N+1}(x)$ and use it to estimate the error of $P_N(x)$. If the estimated error is too big, increase the degree by interpolating at another node and repeat. This process is the basis of the Newton form of the interpolating polynomial.

For each n, the interpolant $P_n(x)$ is constructed as a "correction" to $P_{n-1}(x)$. Because $P_{n-1}(x)$ is of degree less than $n-1$ and $P_n(x)$ is of degree at most $n-1$, their difference must be a polynomial $Q_n(x)$ of degree at most $n-1$:

$$P_n(x) = P_{n-1}(x) + Q_n(x). \tag{3.21}$$

The polynomial $P_n(x)$ interpolates at $x_1,\ldots,x_{n-1}$ just as $P_{n-1}(x)$ does, so for $j = 1,\ldots,n-1$,

$$f_j = P_n(x_j) = P_{n-1}(x_j) + Q_n(x_j) = f_j + Q_n(x_j).$$

This implies that the $x_1,\ldots,x_{n-1}$ are roots of $Q_n(x)$. Because its degree is at most $n-1$, $Q_n(x)$ must have the form

$$Q_n(x) = c_n(x-x_1)(x-x_2)\cdots(x-x_{n-1})$$

for some constant c_n. The polynomial $P_n(x)$ also interpolates at x_n: $f_n = P_n(x_n) = P_{n-1}(x_n) + Q_n(x_n) = P_{n-1}(x_n) + c_n\prod_{j=1}^{n-1}(x_n - x_j)$. Because the nodes are distinct, none of the factors $(x_n - x_j)$ can vanish, and

$$c_n = \left(f_n - P_{n-1}(x_n)\right) \Big/ \prod_{j=1}^{n-1}(x_n - x_j). \tag{3.22}$$

The relations (3.21) and (3.22) along with $P_1(x_1) = f_1$ provide the Newton form of the interpolating polynomial. The coefficient c_n is called the $(n-1)$*st divided difference of f over the points* $x_1,\ldots,x_n$. A number of notations are seen. A common one is

$$c_n = f[x_1,\ldots,x_n].$$

In this notation

---

the *Newton divided difference form* is

$$P_N(x) = f[x_1] + f[x_1,x_2](x-x_1) + f[x_1,x_2,x_3](x-x_1)(x-x_2) + \cdots$$
$$+ f[x_1,x_2,\ldots,x_N] \prod_{j=1}^{N-1} (x-x_j). \quad (3.23)$$

It is clear from (3.23) that the leading coefficient (the coefficient of the highest degree term) of $P_N(x)$ is $f[x_1,\ldots,x_N]$. Some authors take this as the definition of the $(N-1)$st divided difference.

Before working some examples we present a theorem that relates an nth order divided difference to a pair of $(n-1)$st order divided differences. The relation leads to an algorithm for computing the c_n that is computationally more convenient than (3.22).

Theorem 3.3. *For distinct nodes $\{x_j\}$ and any $k > i$,*

$$f[x_i,\ldots,x_{k-1},x_k] = \frac{f[x_{i+1},\ldots,x_k] - f[x_i,\ldots,x_{k-1}]}{x_k - x_i} \quad (3.24)$$

and

$$f[x_i] = f_i.$$

Proof. Let $R_1(x)$ be the polynomial of degree less than $k-i$ that interpolates $f(x)$ on $x_{i+1},\ldots,x_k$ and let $R_2(x)$ be the polynomial of degree less than $k-i$ that interpolates on $x_i,\ldots,x_{k-1}$. The polynomial

$$S(x) = \frac{(x_k - x)R_2(x) + (x - x_i)R_1(x)}{x_k - x_i} \quad (3.25)$$

has a degree at most one higher than the degrees of $R_1(x)$ and $R_2(x)$. Accordingly, its degree is less than $k-i+1$. For $j = i+1,\ldots,k-1$,

$$S(x_j) = \frac{(x_k - x_j)f_j + (x_j - x_i)f_j}{x_k - x_i} = f_j,$$

so $S(x)$ interpolates $f(x)$ on $x_{i+1},\ldots,x_{k-1}$. Moreover, $S(x_i) = f_i$ and $S(x_k) = f_k$. By Theorem 3.1, $S(x)$ is *the* interpolating polynomial of degree less than $k-i+1$ that interpolates $f(x)$ on all the data. The result (3.24) simply expresses the fact that the leading coefficient of the left-hand side of (3.25) equals the leading coefficient of the right-hand side. ■

To illustrate the use of this theorem, we construct a divided difference table. Suppose that three rows and columns of differences have already been computed and written in a lower triangular array as follows:

$$\begin{array}{llll} x_1 & f[x_1] & & \\ x_2 & f[x_2] & f[x_1,x_2] & \\ x_3 & f[x_3] & f[x_2,x_3] & f[x_1,x_2,x_3]. \end{array}$$

To add another row corresponding to the node x_4, start with the data $f[x_4] = f_4$. Then

$$\begin{aligned} f[x_3,x_4] &= \frac{f[x_4]-f[x_3]}{x_4-x_3} \\ f[x_2,x_3,x_4] &= \frac{f[x_3,x_4]-f[x_2,x_3]}{x_4-x_2} \\ f[x_1,x_2,x_3,x_4] &= \frac{f[x_2,x_3,x_4]-f[x_1,x_2,x_3]}{x_4-x_1}. \end{aligned}$$

Notice the pattern in these calculations:

$$\begin{array}{llllllll} x_1 & f[x_1] & & & & & & \\ x_2 & f[x_2] & & f[x_1,x_2] & & & & \\ x_3 & f[x_3] & & f[x_2,x_3] & & f[x_1,x_2,x_3] & & \\ & & \searrow & & \searrow & & \searrow & \\ x_4 & f[x_4] & \rightarrow & f[x_3,x_4] & \rightarrow & f[x_2,x_3,x_4] & \rightarrow & f[x_1,x_2,x_3,x_4]. \end{array}$$

In general, the first column of the divided difference table is x_j, the second is f_j, the next is the first divided difference, and so on. The table provides a convenient device for constructing the required divided differences: the coefficients of the interpolating polynomial are the quantities along the diagonal.

Example 3.4. For the data from Example 3.2, first form the difference table:

x_j	f_j	$f[\,,\,]$	$f[\,,\,,\,]$	$f[\,,\,,\,,\,]$
1.82	0.00			
2.50	1.30	$\frac{1.30-0.00}{2.50-1.82} = 1.91$		
3.65	3.10	$\frac{3.10-1.30}{3.65-2.50} = 1.56$	$\frac{1.56-1.91}{3.65-1.82} = -0.19$	
4.03	2.52	$\frac{2.52-3.10}{4.03-3.65} = -1.53$	$\frac{-1.53-1.56}{4.03-2.50} = -2.02$	$\frac{-2.02+0.19}{4.03-1.82} = -.83$

Then according to (3.23),

$$\begin{aligned} P_4(x) = 0.0 + 1.91(x-1.82) - 0.19(x-1.82)(x-2.50) \\ -0.83(x-1.82)(x-2.50)(x-3.65). \end{aligned}$$

For computational efficiency this should be evaluated in the nested form

$$P_4(x) = (x-1.82)\{1.91 + (x-2.50)[-0.19 - 0.83(x-3.65)]\}.$$

Of course, if you expand this out, you should get the same (except for roundoff) as the Lagrange form. ■

There are two parts to an algorithm for calculating the Newton divided difference form of the interpolating polynomial. The first computes the divided differences needed for the coefficients of $P_N(x)$. It is not necessary to save the whole table as we can use a vector c_j to save the entry in the current row j as long as we compute one diagonal at a time (and do the calculations in the correct order):

```
c_N := f_N
for j = N - 1,...,1 begin
    c_j := f_j
    for k = j + 1,...,N begin
        c_k := (c_k - c_{k-1})/(x_k - x_j)
    end k
end j.
```

Once these coefficients are available, the second part of the algorithm is to evaluate $P_N(x)$ for a given x:

```
P_N := c_N
for k = N - 1,...,1 begin
    P_N := P_N * (x - x_k) + c_k
end k.
```

Divided differences can be related to the derivatives of $f(x)$ using Theorem 3.2. Application of the theorem to $P_{n-1}(x)$ with $x = x_n$ leads to

$$f(x_n) - P_{n-1}(x_n) = \frac{f^{(n-1)}(\xi_n)}{(n-1)!} \prod_{j=1}^{n-1} (x_n - x_j),$$

where

$$\min(x_1, \ldots, x_n) < \xi_n < \max(x_1, \ldots, x_n).$$

However, we also have

$$f(x_n) - P_{n-1}(x_n) = P_n(x_n) - P_{n-1}(x_n) = c_n \prod_{j=1}^{n-1} (x_n - x_j).$$

Equating the two expressions shows that

$$f[x_1, \ldots, x_n] = \frac{f^{(n-1)}(\xi_n)}{(n-1)!} \tag{3.26}$$

for a point ξ_n in the span of the data $x_1, \ldots, x_n$. With the efficient way of computing divided differences just presented and (3.26), we have a way to approximate derivatives of a function $f(x)$ known only from its values at certain points.

This last result gives us a better understanding of the error estimate we used to motivate the approach. According to Theorem 3.2,

$$f(x) - P_N(x) = \frac{1}{N!} f^{(N)}(\xi) \prod_{j=1}^{N} (x - x_j).$$

We have just seen that

$$P_{N+1}(x) - P_N(x) = f[x_1, \ldots, x_{N+1}] \prod_{j=1}^{N} (x - x_j)$$

$$= \frac{1}{N!} f^{(N)}(\eta) \prod_{j=1}^{N} (x - x_j).$$

Comparison of these two expressions shows that if $f^{(N)}$ does not vary much over the span of the data, the error of $P_N(x)$ can be estimated by comparing it to $P_{N+1}(x)$. It should be appreciated that the Newton form of the interpolating polynomial was used to obtain this result, but it is true regardless of the form used for computing the polynomial.

The Newton form (3.23) is closely related to the Taylor series of $f(x)$ about the point x_1:

$$f(x_1) + \frac{f^{(1)}(x_1)}{1!}(x - x_1) + \frac{f^{(2)}(x_1)}{2!}(x - x_1)^2 + \frac{f^{(3)}(x_1)}{3!}(x - x_1)^3 + \cdots + \frac{f^{(N-1)}(x_1)}{(N-1)!}(x - x_1)^{N-1} + \cdots.$$

As a consequence of (3.26), the Newton form of the interpolating $P_N(x)$ becomes the Taylor polynomial of degree $N - 1$ when the nodes $x_2, \ldots, x_n$ all tend to x_1.

EXERCISES

3.15 For the data

x	-1	0	1	2
f	2	2	2	5

calculate $P_4(x)$

(a) in the Lagrange form (3.2),

(b) using the matrix method discussed in Exercise 3.4 (the linear system here is small enough to be solved by hand), and

(c) in the Newton divided difference form (3.23).

3.16 For the data

x	-2	-1	0	1	2
f	-4	1	1	2	10

calculate $P_5(x)$

(a) in the Lagrange form (3.2),

(b) using the matrix method discussed in Exercise 3.4 (the linear system here is small enough to be solved by hand), and

(c) in the Newton divided difference form (3.23).

3.17 Compute the divided difference table and $P_3(x)$ for the data from Example 3.1. Verify that this polynomial is the same as the one in Lagrange form.

3.18 What is the operation count for the evaluation of the coefficients in the Newton divided difference form of the interpolating polynomial? What is the operation count for each evaluation of $P_N(x)$? How does this compare to the Lagrange form?

3.19 Implement the algorithms for the Newton divided difference form. See if you can reproduce the graphs in Figures 3.4 and 3.5. Try your algorithm on some of the other data sets in the exercises to test the conjecture that high degree polynomial interpolation often results in approximations whose graphs show undesirable behavior.

3.4 ASSESSING ACCURACY

How do we know when we have a good approximation? We have already seen a couple of possibilities. One is to use (3.4), that is, $f(x) - P_N(x) = f^{(N)}(\xi_x) w_N(x)/N!$. Since w_N is a polynomial, it is easily evaluated at any x. The derivative factor presents problems, however, since we certainly do not know ξ_x and probably do not even know

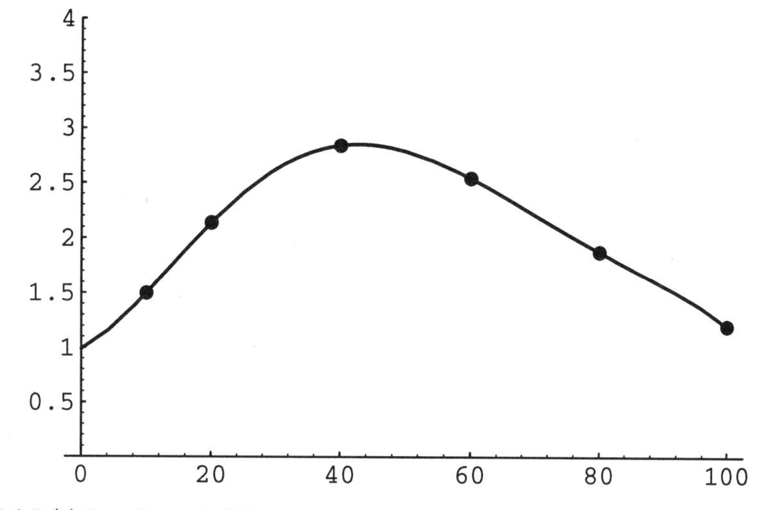

Figure 3.4 $P_6(x)$ from Example 3.5.

$f^{(N)}$. Another possibility is to compare the result of interpolating on one set of nodes to that of a result of higher degree obtained by interpolating on the same set plus one more node. A variation on this is to compare results of the same degree obtained by interpolating on different sets of nodes. Often the best approach is to reserve some data and evaluate the exact error $f(x) - P_N(x)$ at these nodes. A realistic appraisal may require that a lot of data be held in reserve, and it is far from clear how to decide which nodes to use for interpolation and which to keep in reserve. Usually we have some idea about the behavior of the underlying function. A graph of the data and the interpolant is then a great help in deciding whether the interpolant reproduces this behavior adequately.

It is illuminating to see some examples of polynomial interpolation. The programs used to calculate the interpolating polynomials below are straightforward implementations of the Newton divided difference form. We do not provide the codes because they are easy to write and, as will be seen, interpolating with high degree polynomials is generally not a good idea.

Example 3.5. The following table of the relative viscosity V of ethanol as function of the percent of anhydrous solute weight w is taken from [12, p. D-236]:

w	5	10	15	20	30	40
$V(w)$	1.226	1.498	1.822	2.138	2.662	2.840
	50	60	70	80	90	100
	2.807	2.542	2.210	1.877	1.539	1.201

To see how good or bad $P(w)$ is, some data points will be held in reserve. Specifically, we define $P_6(w)$ as the polynomial interpolating at $\{10, 20, 40, 60, 80, 100\}$. The error of this interpolant is assessed by evaluating it at the remaining nodes where we know

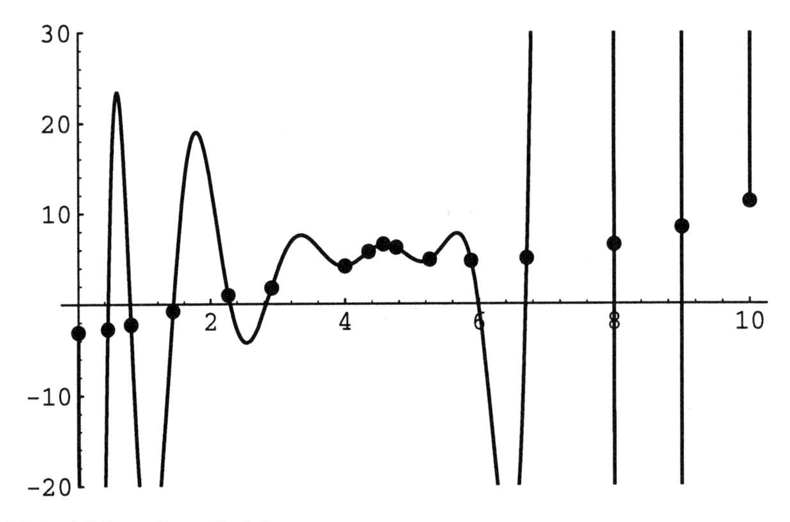

Figure 3.5 $P_{16}(\nu)$ from Example 3.6.

the value of the function:

w	5	15	30	50	70	90
$P_6(w)$	1.201	1.824	2.624	2.787	2.210	1.569
$V(w) - P_6(w)$	0.025	−0.002	0.038	0.020	0.000	−0.030

This is probably sufficiently accurate for most purposes. Figure 3.4 shows that $P_6(x)$ provides a fit that looks reasonable. However, if all 12 data points are interpolated, the resulting $P_{12}(x)$ is not nearly so nice (see Exercise 3.6). ∎

Example 3.6. As a second example, we consider some data taken from [19, p. 84]. Here ν is the reciprocal of the wavelength of light and the function $E(\nu)$ measures the relative extinction of light at this frequency due to observation and scattering by interstellar materials.

ν	$E(\nu)$	ν	$E(\nu)$	ν	$E(\nu)$
>0.00	−3.10	3.65	3.10	>5.88	4.77
0.29	−2.94	>4.00	4.19	6.25	5.02
>0.45	−2.72	4.17	4.90	>6.71	5.05
>0.80	−2.23	>4.35	5.77	7.18	5.39
1.11	−1.60	>4.57	6.57	>8.00	6.55
>1.43	−0.78	>4.76	6.23	8.50	7.45
1.82	0.00	5.00	5.52	>9.00	8.45
>2.27	1.00	>5.26	4.90	9.50	9.80
2.50	1.30	5.56	4.65	>10.00	11.30
>2.91	1.80				

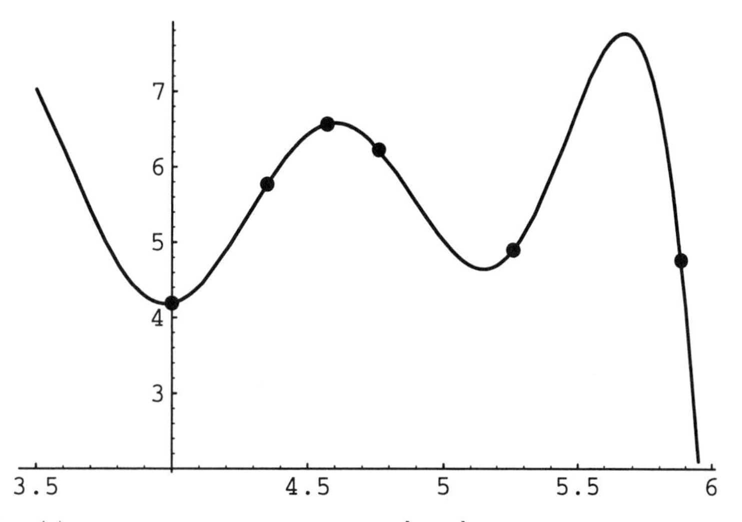

Figure 3.6 $P_{16}(\nu)$ from Example 3.6 over the interval $[3.5, 6]$.

The data points marked (>) were interpolated by $P_{16}(\nu)$, which is graphed in Figure 3.5. This is not a good idea! Still, Figure 3.6 shows that the fit is acceptable in the middle of the span of data. As discussed earlier, we might have expected this from the form of the error expression. Several values of P_{16} at points held in reserve are $P_{16}(0.29) = -108.8$, $P_{16}(1.11) = -28.3$, $P_{16}(5) = 5.01$, $P_{16}(8.5) = -5035$, and $P_{16}(9.5) = 60,749$. ■

3.5 SPLINE INTERPOLATION

The error expression of Theorem 3.2 suggests that raising the degree of the interpolating polynomial will provide a more accurate approximation. Unfortunately, other factors play a role and this tactic often does not succeed in practice. The expression suggests another tactic that will succeed. The error depends strongly on the length of the interval containing the nodes. If we can somehow reduce this length, the theorem says that we will then get a better approximation. The basic idea of this section is to approximate $f(x)$ by a polynomial only on a piece of the interval. The approximations over all the pieces form an interpolant called a *spline*. (In this book the word spline is a synonym for piecewise polynomial function.) More specifically, the function $f(x)$ is to be approximated on $[x_1, x_N]$. The interval $[x_1, x_N]$ is split into subintervals $[x_n, x_{n+1}]$, where $x_1 < x_2 < \cdots < x_N$. A spline is a polynomial on each interval $[x_n, x_{n+1}]$. In this context the $\{x_i\}$ are called the *breakpoints* or *knots*. In the subsections that follow, a selection of splines that arise from interpolation with constraints are studied. A key issue is how smoothly the polynomials connect at the knots, and this governs the order in which they are taken up.

DISCONTINUOUS AND CONTINUOUS SPLINES

The simplest splines are those arising from interpolation done independently on each subinterval $[x_n, x_{n+1}]$. The bound (3.7) can be applied to each subinterval. For example, suppose that any four nodes are chosen in each subinterval $[x_n, x_{n+1}]$. Let the spline interpolant $S(x)$ consist of the cubic polynomial interpolants on the subintervals. If $h = \max(x_{n+1} - x_n)$ and

$$M_4 = \max_{x_1 \le x \le x_N} |f^{(4)}(x)|,$$

then

$$|f(x) - S(x)| \le \frac{M_4}{4!} h^4 \quad \text{for} \quad x_1 \le x \le x_N.$$

As $h \to 0$, a good approximation is obtained over the whole interval. Evidently the tactic of fixing the degree and approximating the function on pieces is more promising for practical interpolation than approximating the function over the whole interval by means of increasing the degree.

Generally the polynomial on $[x_n, x_{n+1}]$ does not agree at x_n with the polynomial on $[x_{n-1}, x_n]$, so this spline is generally discontinuous at the knots. When approximating a continuous function $f(x)$ this may not be acceptable. It is easy to modify this construction to obtain a continuous spline. All we need do is include the ends of each subinterval among the points where $f(x)$ is interpolated. The polynomial on $[x_n, x_{n+1}]$ then has the value $f(x_n)$ at x_n and so does the polynomial on $[x_{n-1}, x_n]$.

Only data from $[x_{n-1}, x_n]$ are used in constructing the spline on this subinterval, so the error depends only on the behavior of $f(x)$ on this subinterval. This will not be the case for splines taken up later. In some contexts the spline is to be constructed before all the data are available and this property of the construction is essential.

The simplest continuous spline is one that is piecewise linear, that is, $S(x)$ is a broken-line function (see Figure 3.7). If $S(x)$ is required to interpolate $f(x)$ at the knots, then on $[x_n, x_{n+1}]$ for $1 \le n \le N-1$ the Lagrange form is

$$S(x) = f_n \frac{x - x_{n+1}}{x_n - x_{n+1}} + f_{n+1} \frac{x - x_n}{x_{n+1} - x_n},$$

which can be rewritten as

$$S(x) = f_n + \frac{f_{n+1} - f_n}{x_{n+1} - x_n}(x - x_n). \tag{3.27}$$

Example 3.7. Given the data (5, 1.226) (30, 2.662) (60, 2.542) (100, 1.201) taken from Example 3.5, (3.17) yields

$$S(x) = \begin{cases} 1.266 + 0.05744(x-5), & 5 \le x \le 30 \\ 2.662 - 0.00400(x-30), & 30 \le x \le 60 \\ 2.542 - 0.03352(x-60), & 60 \le x \le 100 \end{cases}.$$

■

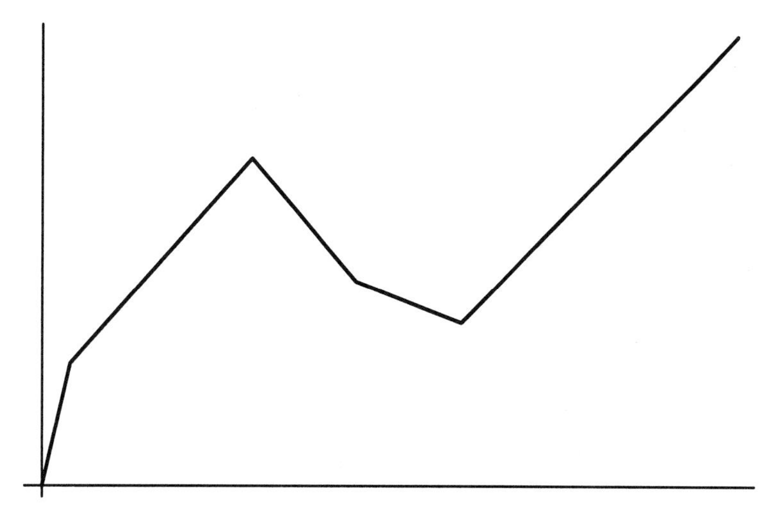

Figure 3.7 A typical linear spline function.

The linear interpolating spline (3.27) is very easy to evaluate once the proper subinterval has been located. All spline evaluation routines must contain an algorithm for finding the right subinterval. Often this takes longer than the actual evaluation of the polynomial there. For linear interpolation, an error bound is

$$|f(x) - S(x)| \leq \frac{1}{8} M_2 h^2 \text{ for } x_1 \leq x \leq x_N, \tag{3.28}$$

where $M_2 = \max_{x_1 \leq x \leq x_N]} |f''(x)|$. Convergence is guaranteed as $h \to 0$ if $|f''|$ is bounded. A similar argument using (3.16) yields

$$|f'(x) - S'(x)| \leq M_2 h, \; x_n < x < x_{n+1}, \; 1 \leq n < N-1. \tag{3.29}$$

Thus, $S'(x)$ can be used as an estimate for $f'(x)$ that gets better as $h \to 0$.

Continuous splines are used in the solution by finite elements of boundary value problems for second order differential equations. The simple program CODE1 in [1] allows the user to specify the degree of the interpolating polynomial on each element—the subinterval $[x_n, x_{n+1}]$—in the range 1 to 3. The higher the degree, the more accurate the approximation, but the greater the possibility of unwanted oscillations. This may not matter when using the spline for finite elements, but it is to be avoided when representing data. For the latter purpose a good compromise seems to be the use of cubic polynomials, so in the rest of this section we concentrate on them.

The error of a continuous cubic spline constructed by interpolation independently on each subinterval can be analyzed using the error expressions developed for polynomial interpolation. On each subinterval

$$|f^{(k)}(x) - P_4^{(k)}(x)| \leq C_k h^{4-k}$$

for $k = 0, 1, \ldots, 3$ and suitable constants C_k. It is not so easy to prove and the powers of h differ, but similar results can be established for all the cubic splines we take up.

The point, though, is that on each subinterval $P_4'(x) \approx f'(x)$. This implies that for sufficiently small h, $P_4'(x)$ has the same sign as $f'(x)$ as long as $f'(x) \neq 0$. Put differently, except near the extrema of $f(x)$, for small h the spline is monotonely increasing (decreasing) wherever $f(x)$ is. The same argument applies to the second derivative and leads to the conclusion that except near inflection points of $f(x)$, for small h the spline is concave (convex) wherever $f(x)$ is. We conclude that for small h, the spline will reproduce the shape of the function it interpolates. The same will be true of all the cubic splines we take up. This is one reason why spline interpolation is much more satisfactory than interpolation by high degree polynomials. But what if h is not "small"? When the data are sparse, it is necessary to impose conditions on the spline to preserve the shape of the function, one of the matters we take up in the next subsection.

CONTINUOUS FIRST DERIVATIVE

If we have derivative data available, it is easy to extend the approach of the preceding subsection to obtain an interpolant with a continuous derivative. For example, we could interpolate $f(x_n), f'(x_n), f(x_{n+1}), f'(x_{n+1})$ by the cubic Hermite interpolating polynomial on $[x_n, x_{n+1}]$. Doing this on each subinterval produces a spline $H(x)$ with a continuous first derivative. Each interval is treated independently, so the bounds (3.17)–(3.20) hold and show that a good approximation is obtained. In the chapter on differential equations, we produce, for successive n, approximations to a function $y(x)$ and its derivative at the points x_n, $x_n + h/2$, and $x_n + h$. By forming the quintic (degree 5) Hermite interpolant to these data, a spline with a continuous derivative is formed that approximates $y(x)$ and $y'(x)$ for all x. It is especially important in this context that the interval $[x_n, x_n + h]$ is handled independently because generally it is only the data on this interval that are available when interpolation is done.

Let us now consider the representation of data when only $f(x_i)$ values are known and there are not many of them. It has been found that a cubic spline $H(x)$ yields a plot pleasing to the eye if it has a continuous derivative and if it preserves monotonicity. By the latter is meant that if $f_n < f_{n+1}$, then $H(x)$ increases on (x_n, x_{n+1}) and if $f_n > f_{n+1}$, then $H(x)$ decreases. The point is to avoid oscillations that do not appear in the data. A moment's thought shows that linear splines preserve monotonicity. The problem with them is that their graphs have "corners." By going to cubics and a continuous first derivative, we avoid the corners. Such a "shape-preserving" interpolant can be constructed along the lines of the cubic Hermite interpolant. The cubics on $[x_{n-1}, x_n]$ and $[x_n, x_{n+1}]$ both interpolate to f_n at x_n. If the first derivative is to be continuous, the first derivatives of the two cubics must have the same value at x_n, but now the value of this derivative is an unknown parameter that we choose to achieve monotonicity.

As in (3.11) the cubic is written in the form

$$H(x) = a_n + b_n(x - x_n) + c_n(x - x_n)^2 + d_n(x - x_n)^3$$

for $x_n \leq x \leq x_{n+1}$, $1 \leq n \leq N - 1$. Note that the parameter b_n is just the slope of $H(x)$ at the point x_n. Proceeding as in the derivation of (3.12)–(3.14) with the notation $h_n = x_{n+1} - x_n$ and $\Delta_n = (f_{n+1} - f_n)/h_n$ yields

$$a_n = f_n$$

$$c_n = (3\Delta_n - 2b_n - b_{n+1})/h_n \qquad (3.30)$$
$$d_n = (b_n + b_{n+1} - 2\Delta_n)/h_n^2.$$

These equations result from solving the three interpolation conditions $H(x_n) = f_n$, $H(x_{n+1}) = f_{n+1}$, and $H'(x_{n+1}) = b_{n+1}$ for the three unknowns a_n, c_n, and d_n.

The quantity Δ_n is the slope of the line through (x_n, f_n) and (x_{n+1}, f_{n+1}). If $\Delta_n = 0$, it seems reasonable to force $H(x)$ to be constant on $[x_n, x_{n+1}]$, that is, to make the slopes $b_n = b_{n+1} = 0$. If $\Delta_n \neq 0$, let us define $\alpha_n = b_n/\Delta_n$ and $\beta_n = b_{n+1}/\Delta_n$. To preserve monotonicity it is necessary that the sign of the slope of $H(x)$ at x_n and x_{n+1} be the same as that of Δ_n. Mathematically this is $\alpha_n \geq 0, \beta_n \geq 0$.

A sufficient condition on α and β to preserve monotonicity was found by Ferguson and Miller [7]. This was independently discovered by Fritsch and Carlson and published in the more accessible reference [10]. The argument involves studying $H'(x)$ as a function of α_n and β_n. This is not too complicated since $H'(x)$ is a quadratic on (x_n, x_{n+1}). A simple condition that guarantees monotonicity is preserved is that α_n and β_n lie in the interval $[0,3]$. There are many formulas for α_n and β_n that satisfy this restriction. One given in [9] that works pretty well is to use

$$b_n = \frac{\Delta_{n-1}\Delta_n}{r_n\Delta_n + (1 - r_n)\Delta_{n-1}} \qquad (3.31)$$

with

$$r_n = \frac{h_{n-1} + 2h_n}{3(h_{n-1} + h_n)} \qquad (3.32)$$

for $n = 2, 3, \ldots, N-1$. If $\Delta_{n-1}\Delta_n < 0$, then the slopes change sign at x_n. In such a case we probably should not impose any requirements on the slope of $H(x)$ at x_n. Some people suggest setting $b_n = 0$ when this happens. Others say to go ahead and use (3.31) as long as there is no division by zero. The heuristic actually chosen can have a significant impact on the behavior of the shape-preserving cubic spline near those regions where $\Delta_{n-1}\Delta_n < 0$. At the ends the simplest rule is to use $b_1 = \Delta_1$ and $b_N = \Delta_{N-1}$. A better choice is to use the end slope of the quadratic interpolating the three closest data points (assuming it satisfies the constraint on α and β); other possibilities are given in [9]. With (3.31) and the simple choice for b_1 and b_N it is easy to show that the sufficient conditions on α_n and β_n are satisfied. At the ends $\alpha_1 = 1$ and $\beta_{N-1} = 1$, which are certainly in $[0,3]$. For $n = 2, 3, \ldots, N-1$, clearly $\frac{1}{3} \leq r_n \leq \frac{2}{3}$, so $\alpha_n = \Delta_{n-1}/[r_n\Delta_n + (1 - r_n)\Delta_{n-1}] \leq 1/(1 - r_n) \leq 3$ and $\beta_{n-1} = \Delta_n/[r_n\Delta_n + (1 - r_n)\Delta_{n-1}] \leq 1/r_n \leq 3$ as desired.

The algorithm for $H(x)$ is very simple. Compute b_1 by whatever formula you choose; for $n = 2, 3, \ldots, N-1$ take $b_n = 0$ if $\Delta_{n-1}\Delta_n \leq 0$, otherwise compute b_n from (3.31), (3.32). Compute b_N. The values c_n and d_n can be computed from (3.30) for $n = 1, \ldots, N-1$ either at the same time the b_n are computed or in a separate pass over the data. Later in this chapter we provide a routine SVALUE/Spline_value for the evaluation of $H(x)$.

Examination of the algorithm shows that on the subinterval $[x_n, x_{n+1}]$, the spline depends on the data (x_{n-1}, f_{n-1}), (x_n, f_n), (x_{n+1}, f_{n+1}), and (x_{n+2}, f_{n+2}). It should be no surprise that it depends on data from adjacent subintervals because the first

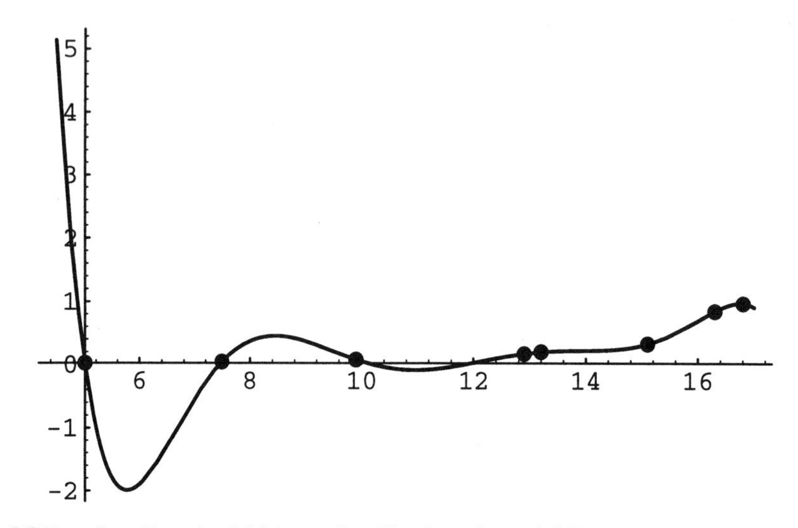

Figure 3.8 Data from Exercise 3.30 interpolated by the polynomial P_8.

derivatives of the polynomials in adjacent subintervals have to match at the nodes. Although not as local as the cubic splines of the preceding subsection, the construction of the shape-preserving spline on a subinterval requires only data from the subinterval itself and the two adjacent subintervals. As might be expected this $H(x)$ works very well on data that are always monotone but is less successful on oscillatory data. See Huynh [14] for some alternatives.

This spline is not very accurate as $h = \max(x_{n+1} - x_n)$ tends to zero, but that is not its purpose. It should be used when the data are "sparse" and qualitative properties of the data are to be reproduced. It is a simple and effective automatic French curve.

Example 3.8. Exercise 3.30 describes a situation in which a shape-preserving spline is particularly appropriate for the approximation of a function $f(C)$. There are only eight data points and it is necessary to approximate the derivative $f'(C)$. The concentration C and diffusion coefficient $D(C)$ that make up the function $f(C) = CD(C)$ to be approximated are nonnegative. As can be seen in Figure 3.8, a polynomial interpolant to $f(C)$ is unsatisfactory because it fails to reproduce this fundamental property. Also, it does not reproduce the monotonicity of the data, casting doubt on its use for approximating $f'(C)$. The shape-preserving spline requires the monotonicity of the interpolant to match that of the data; Figure 3.9 shows that the result is a much more plausible fit. ■

CONTINUOUS SECOND DERIVATIVE

The last spline considered has a historical origin. To draw smooth curves through data points, drafters once used thin flexible strips of plastic or wood called splines. The

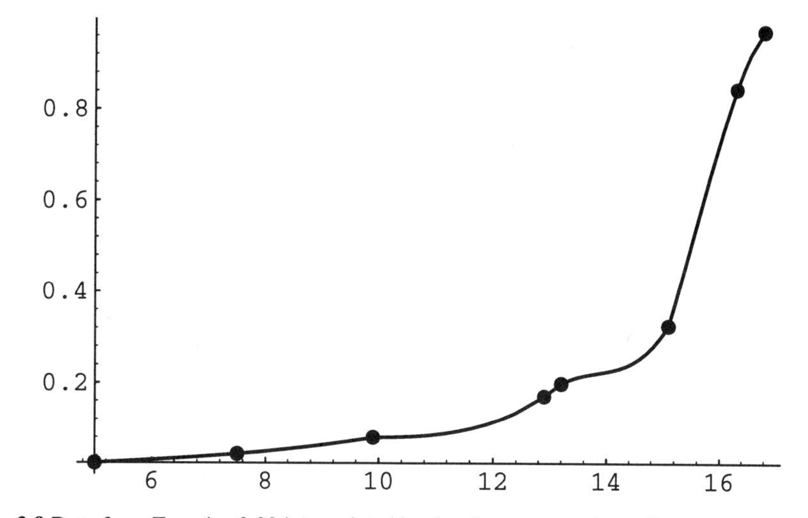

Figure 3.9 Data from Exercise 3.30 interpolated by the shape-preserving spline.

data were plotted on graph paper and a spline was held on the paper with weights so that it went over the data points. The weights were constructed so that the spline was free to slip. The flexible spline straightened out as much as it could subject to the constraint that it pass over the data points. The drafter then traced along the spline to get the interpolating curve. The smooth cubic spline presented here is the solution of a linearized model of the physical spline. The physical analogy already points out something very different about this spline—its value at any point depends on all the data.

To construct the smooth cubic spline, we write once again

$$S(x) = a_n + b_n(x - x_n) + c_n(x - x_n)^2 + d_n(x - x_n)^3 \tag{3.33}$$

on each $[x_n, x_{n+1}]$, $1 \leq n \leq N-1$. There are $4(N-1)$ free parameters to be determined. The interpolation conditions require that for $1 \leq n \leq N-1$

$$S(x_n^+) = f_n, \text{ and } S(x_{n+1}^-) = f_{n+1}, \tag{3.34}$$

giving $2N-2$ conditions. There remain $2N-2$ degrees of freedom that can be used to make $S(x)$ smooth on all of $[x_1, x_N]$. Note that (3.34) automatically ensures that S is continuous on $[x_1, x_N]$. For S' to be continuous at interior knots,

$$S'(x_n^-) = S'(x_n^+), 2 \leq n \leq N-1. \tag{3.35}$$

This provides $N-2$ conditions, so N degrees of freedom remain. For S'' to be continuous at interior knots,

$$S''(x_n^-) = S''(x_n^+), 2 \leq n \leq N-1 \tag{3.36}$$

for another $N-2$ conditions. Exactly 2 degrees of freedom are left. This is not enough to achieve a continuous S''' (this is undesirable anyway since the resulting S would be a cubic polynomial rather than a *piecewise* cubic polynomial). There are many possibilities for the two additional constraints.

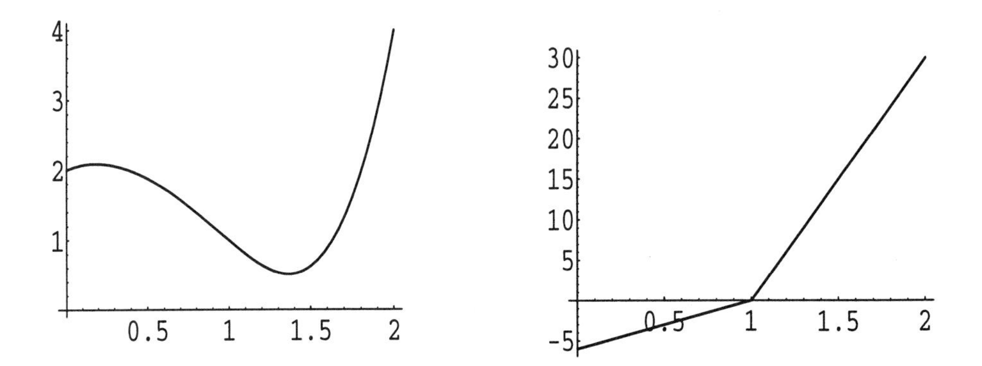

Figure 3.10 Graphs of $S(x)$ and $S''(x)$ from Example 3.9.

Type 1. $S'(x_1) = f'(x_1), S'(x_N) = f'(x_N)$.

Type 2. $S''(x_1) = S''(x_N) = 0$.

Type 3. $S''(x_1) = f''(x_1), S''(x_N) = f''(x_N)$.

Type 4. $S'(x_1) = S'(x_N), S''(x_1) = S''(x_N)$.

For obvious reasons these conditions are known as *end conditions*. The second condition is the one leading to a spline that approximates the physical spline. The physical spline straightens out as much as possible past the last data point on each end, so it becomes a straight line with zero second derivative. In the form stated, the first and third conditions are useful only if extra information is available about f. However, the exact slopes or curvatures needed here are often replaced by polynomial approximations in practice. The last end condition is appropriate when f is periodic with period $x_N - x_1$ because then $f(x)$ and all its derivatives have the same values at x_1 and x_N.

Example 3.9. Let

$$S(x) = \begin{cases} 2 + x - 3x^2 + x^3, & 0 \le x \le 1 \\ 1 - 2(x-1) + 5(x-1)^3, & 1 \le x \le 2. \end{cases}$$

It is easily verified that S is in $C^2[0,2]$, and satisfies the interpolation conditions $S(0) = 2$, $S(1) = 1$, $S(2) = 4$ and the end conditions $S'(0) = 1$, $S'(2) = 13$. Graphs of S and S'' are shown in Figure 3.10. Note that the graph of S appears very smooth, while that for S'' has an obvious corner at the knot $x = 1$. ■

Returning to the earlier characterization of $S(x)$, we had $4N - 4$ conditions on the $4N - 4$ unknowns given by (3.33). A matrix method is in order, but some preliminary manipulations will simplify the task considerably. On each interval $[x_n, x_{n+1}]$

$$S'(x) = b_n + 2c_n(x - x_n) + 3d_n(x - x_n)^2 \tag{3.37}$$

$$S''(x) = 2c_n + 6d_n(x - x_n). \tag{3.38}$$

The interpolation conditions immediately yield, from (3.33),

$$a_n = f_n,\ 1 \leq n \leq N-1, \tag{3.39}$$

and also $f_{n+1} = a_n + b_n h_n + c_n h_n^2 + d_n h_n^3$ which can be rewritten as

$$b_n = (f_{n+1} - f_n)/h_n - c_n h_n - d_n h_n^2,\ 1 \leq n \leq N-1. \tag{3.40}$$

This eliminates half of the unknowns. The continuity condition (3.36) on S'' says that $2c_n = 2c_{n-1} + 6d_{n-1}h_{n-1}$ or [with $c_N = S''(x_N)/2$]

$$d_{n-1} = \frac{c_n - c_{n-1}}{3h_{n-1}},\ 2 \leq n \leq N. \tag{3.41}$$

Only formulas for $c_1, \ldots, c_N$ remain. They are provided by the two end conditions and the global continuity of S'. From (3.35) and (3.37) it follows that $b_n = b_{n-1} + 2c_{n-1}h_{n-1} + 3d_{n-1}h_{n-1}^2$ for $2 \leq n \leq N-1$. Substitution in (3.40) and (3.41) gives

$$\frac{f_{n+1} - f_n}{h_n} - c_n h_n - \frac{1}{3}h_n(c_{n+1} - c_n) = \frac{f_n - f_{n-1}}{h_{n-1}} + c_{n-1}h_{n-1} + \frac{2}{3}h_{n-1}(c_n - c_{n-1})$$

for $2 \leq n \leq N-1$ and a rearrangement yields

$$h_{n-1}c_{n-1} + 2(h_{n-1} + h_n)c_n + h_n c_{n+1} = 3\left(\frac{f_{n+1} - f_n}{h_n} - \frac{f_n - f_{n-1}}{h_{n-1}}\right). \tag{3.42}$$

Only the first type of end conditions (prescribed slopes) is taken up here. From (3.33), (3.40), and (3.41),

$$\begin{aligned} f'(x_1) = S'(x_1) = b_1 &= (f_2 - f_1)/h_1 - c_1 h_1 - d_1 h_1^2 \\ &= (f_2 - f_1)/h_1 - c_1 h_1 - \frac{1}{3}h_1(c_2 - c_1), \end{aligned}$$

so

$$2h_1 c_1 + h_1 c_2 = 3\{(f_2 - f_1)/h_1 - f'(x_1)\}. \tag{3.43}$$

Similarly, $f'(x_N) = S'(x_N)$ leads to

$$h_{N-1}c_{N-1} + 2h_{N-1}c_N = 3\left(f'(x_N) - \frac{f_N - f_{N-1}}{h_{N-1}}\right). \tag{3.44}$$

Equations (3.41)–(3.44) provide a set of N equations in the N unknowns $c_1, c_2, \ldots, c_N$. The coefficient matrix has the very special structure

$$\begin{pmatrix} 2h_1 & h_1 & & & & \\ h_1 & 2(h_1 + h_2) & h_2 & & & \\ & \ddots & \ddots & \ddots & & \\ & & & h_{N-2} & 2(h_{N-2} + h_{N-1}) & h_{N-1} \\ & & & & h_{N-1} & 2h_{N-1} \end{pmatrix}.$$

Such a matrix is called tridiagonal (all the nonzero entries lie in three diagonal bands), symmetric (the entry in row i of column j equals the one in row j of column i), and diagonally dominant (in each column the magnitude of the entry on the main diagonal exceeds the sum of the magnitudes of the other entries in the column). We saw in Section 2.5.2 that for such matrices the system has a unique solution for each right-hand side and the solution can be found accurately using Gaussian elimination *without any row interchanges.*

Theorem 3.4. *Given the knots $x_1 < x_2 < \cdots < x_N$ and $f_n = f(x_n)$, $1 \le n \le N$, there exists one and only one function $S(x)$ that satisfies each of the following:*

1. *$S(x)$ is a cubic polynomial in each $[x_n, x_{n+1}]$, $1 \le n \le N-1$.*
2. *$S(x)$ is in $C^2[x_1, x_N]$.*
3. *$S(x_n) = f_n$, $1 \le n \le N$.*
4. *$S'(x_1) = f'(x_1), S'(x_N) = f'(x_N)$.*

For this choice of end conditions, $S(x)$ is called the *complete cubic spline.* The coefficient matrix has the same structure for end conditions of types (2) and (3) and similar results are true for them. With the choice of type (2), $S(x)$ is called the *natural cubic spline.* The matrix has a somewhat different form for the periodic end conditions, type (4), but similar results are true and the spline can be computed conveniently.

Because the smooth cubic spline depends on all the data, a system of linear equations must be solved to construct it. Often large data sets are to be fit, and if the solution of the linear equations were anything like as expensive as for a general system, the approach would be impractical. Fortunately, the system is very special and it is practical to interpolate data involving thousands of nodes. First the tridiagonal system (3.42)–(3.44) must be solved. Since it is not necessary to do row interchanges in this case, the elimination formulas are very simple. For reinforcement, let us work through the details.

$$\begin{aligned} \alpha_1 c_1 + \beta_1 c_2 &= \gamma_1 \\ \beta_1 c_1 + \alpha_2 c_2 + \beta_2 c_3 &= \gamma_2 \\ &\vdots \\ \beta_{N-1} c_{N-1} + \alpha_N c_N &= \gamma_N. \end{aligned}$$

To eliminate the first entry in row 2, multiply row 1 by β_1/α_1 and subtract. The remaining equations have the same pattern, so at the kth stage multiply row k by the current β_k/α_k and subtract from row $k+1$. The algorithm for elimination and modification of the right-hand side is

for $k = 2, 3, \ldots, N$ begin

$p := \beta_{k-1}/\alpha_{k-1}$
$\alpha_k := \alpha_k - p * \beta_{k-1}$
$\gamma_k := \gamma_k - p * \gamma_{k-1}$

end k.

Back substitution is also easy:

$c_N := \gamma_N/\alpha_N$
for $k = N-1, N-2, \ldots, 1$ begin
$\quad c_k := (\gamma_k - \beta_k * c_{k+1})/\alpha_k$
end k.

The whole computation costs only $3N-3$ multiplications and $2N-1$ divisions. Once the **c** vector is known, vector **d** can be computed from (3.41), and vector **b** from (3.40). The storage required is a small multiple of N rather than the N^2 needed for a general system of equations.

We finish this section by discussing some of the mathematical properties of the complete cubic interpolatory spline $S(x)$. The physical spline used by drafters can be modeled using the theory of thin beams. In general, the curvature $\kappa(x)$ of a function $f(x)$ is

$$\kappa(x) = \frac{|f''(x)|}{(1+(f'(x))^2)^{3/2}}$$

and in this theory the expression is linearized to $\kappa(x) \approx |f''(x)|$. When $(S'(x))^2 \ll 1$, the quantity $\int_{x_1}^{x_N} (S'')^2\,dx$ can be regarded as a measure of the curvature of the spline $S(x)$. We prove now that in this measure, any smooth interpolating function satisfying the type (1) end conditions must have a curvature at least as large as that of $S(x)$. This is sometimes referred to as the *minimum curvature* property of the complete cubic spline. The same result is true for the natural cubic spline when the requirement that the interpolant satisfy the type (1) end conditions is dropped.

Theorem 3.5. *If g is any $C^2[x_1, x_N]$ function that interpolates f over $\{x_1, \ldots, x_N\}$ and satisfies the type (1) end conditions, then*

$$\int_{x_1}^{x_N} (S'')^2\,dx \le \int_{x_1}^{x_N} (g'')^2\,dx,$$

where $S(x)$ is the complete cubic interpolatory spline. The same inequality holds for g that do not necessarily satisfy the type (1) end conditions when $S(x)$ is the natural cubic interpolatory spline.

Proof. First observe that

$$\int_{x_1}^{x_N} (g''-S'')^2\,dt = \int_{x_1}^{x_N} (g'')^2\,dt - 2\int_{x_1}^{x_N} (g''-S'')S''\,dt - \int_{x_1}^{x_N} (S'')^2\,dt.$$

If it can be shown that the second integral on the right is zero, then

$$\int_{x_1}^{x_N} (S'')^2\,dt = \int_{x_1}^{x_N} (g'')^2\,dt - \int_{x_1}^{x_N} (g''-S'')^2\,dt \le \int_{x_1}^{x_N} (g'')^2\,dt,$$

since the integral of a nonnegative function is always nonnegative, and we are finished. To establish that the desired integral is zero, note that

$$\int_{x_1}^{x_N} (g''-S'')S''\,dt = \sum_{n=1}^{N-1} \int_{x_n}^{x_{n+1}} (g''-S'')S''\,dt,$$

and two integrations by parts give

$$\int_{x_n}^{x_{n+1}} (g'' - S'')S''\,dt = (g' - S')S''\Big|_{x_n}^{x_{n+1}} - \int_{x_n}^{x_{n+1}} (g' - S')S'''\,dt$$

$$= [(g' - S')S'' - (g - S)S''']\Big|_{x_n}^{x_{n+1}} + \int_{x_n}^{x_{n+1}} (g - S)S^{(4)}\,dt.$$

Since S is a cubic on each $[x_n, x_{n+1}]$, it follows that $S^{(4)} \equiv 0$, so the last integral is zero. Also, $(g - S)|_{x_n}^{x_{n+1}} = (f_{n+1} - f_{n+1}) - (f_n - f_n) = 0$ since both g and S interpolate f. Thus,

$$\int_{x_1}^{x_N} (g'' - S'')S''\,dt = \sum_{n=1}^{N-1} [(g' - S')S''|_{x_{n+1}} - (g' - S')S''|_{x_n}],$$

which telescopes to $(g' - S')S''|_{x_N} - (g' - S')S''|_{x_1}$, and the type (1) end conditions force these terms to vanish. The terms vanish without assuming that g satisfies type (1) end conditions when $S(x)$ is the natural cubic spline because it satisfies the end conditions $S''(x_1) = S''(x_N) = 0$. ■

While the minimum curvature property is nearly always desirable, there are circumstances in which it is a disadvantage. Note that f certainly interpolates itself, so Theorem 3.5 implies $\int_{x_1}^{x_N} (S'')^2\,dx \le \int_{x_1}^{x_N} (f'')^2\,dx$. In examples where f has very large curvature, there can be a considerable discrepancy between S and f unless there are enough knots (data points) in the region of large curvature that S can turn sufficiently fast. Several illustrative examples are given in the next section.

Convergence rates analogous to (3.17)–(3.20) for the Hermite cubic spline can be established in the complete cubic case. However, proofs are more difficult because $S(x)$ is determined by all the data and it is not possible to treat each subinterval independently. The following result is from [11].

Theorem 3.6. *If f is in $C^4[x_1, x_N]$, and $S(x)$ is the complete cubic interpolatory spline for f with knots $\{x_1 < \cdots < x_N\}$, then for any x in $[x_1, x_N]$*

$$|f(x) - S(x)| \le \frac{5}{384} h^4 M_4$$

$$|f'(x) - S'(x)| \le \frac{1}{24} h^3 M_4$$

$$|f''(x) - S''(x)| \le \frac{3}{8} h^2 M_4,$$

where $M_4 = \max |f^{(4)}(x)|$ for x in $[x_1, x_N]$.

In contrast to polynomial interpolation, $S(x)$ does converge to $f(x)$ as $N \to \infty$ as long as $h \to 0$. The first and second derivatives of the spline also converge to the corresponding derivatives of f. Because of this, the spline inherits the shape of f when h is small. For example, at a point t where $f'(t) > 0$, convergence implies that for all sufficiently small h, $S'(t) > 0$. Accordingly, the smooth spline inherits the monotonicity of f for small h, except possibly near the extrema of f. The shape-preserving spline is required only when the data are so sparse that we must impose directly the property of monotonicity on the spline. The same argument shows that for small h, the smooth spline is convex where f is, except possibly near inflection points.

The complete cubic spline will converge when f has fewer than four continuous derivatives on $[a,b]$, just not as fast. Experimentation with a physical spline shows that the farther a node x_k is from a given point t, the less the effect of the value of f_k on $S(t)$. This is also true of the mathematical spline, and a careful analysis, see [16], of convergence reveals that the rate of convergence at t depends only on how smooth $f(x)$ is near t. In particular, the convergence rates of the theorem hold on subintervals of $[a,b]$ where f has four continuous derivatives.

In practice, it is usually impossible to use the conclusions of Theorem 3.6 to estimate errors, given only discrete data, since M_4 is not available. As was suggested for polynomial interpolation, it is wise to reserve some data as a check on the approximation. A graph of S can help in making judgments about the quality of the fit.

ROUTINES FOR CUBIC SPLINE INTERPOLATION

Two routines are provided for the calculation of the complete cubic interpolatory spline S. One, SPCOEF in FORTRAN, Spline_coeff in C, sets up the tridiagonal system (3.42)–(3.44) for $\{c_i\}$, solves it, and computes $\{d_i\}$ and $\{b_i\}$ from (3.41) and (3.40). This routine should be called only *once* for a given set of data. The coefficients output from this routine are then used in the evaluation routine, SVALUE in FORTRAN, Spline_value in C. It is called once for each point t where $S(t)$ is to be evaluated. A routine to compute the coefficients defining the shape-preserving interpolant is quite useful. It can be written easily by modifying SPCOEF or Spline_coeff so as to use the formulas of Section 3.5.2. Proceeding in this way, SVALUE or Spline_value can be used for the evaluation of both kinds of spline.

Instead of using the slopes $f'(x_1)$ and $f'(x_N)$ needed for the end conditions of the complete cubic spline, the routines provided interpolate the four data points nearest each end with cubics, and the slopes of these approximations are used in (3.43) and (3.44). As $h \to 0$, the resulting spline converges to the complete cubic spline. In practice this approximation works well enough if N is not too small. The approximation is not plausible for Example 3.8 because there are only eight data points, all of which are used to approximate the derivatives at the end as well as the function throughout the interval. When the data are this sparse, the shape-preserving spline is more appropriate.

A typical call in FORTRAN is

```
CALL SPCOEF(N,X,F,B,C,D,FLAG)
```

and

```
flag = Spline_coeff(n, x, f, b, c, d);
```

in the C and C++ versions. The input vectors X and F hold the data points (x_i, f_i) to be interpolated and N is the number of such points. The output vectors B, C, and

D contain the coefficients of the cubics. In normal circumstances the output variable FLAG is set to zero. However, if the input $N < 2$, then no calculations are performed and FLAG := −1. If the entries of X are not correctly ordered (so that some $h_i \leq 0$), then FLAG := −2.

To evaluate the spline the FORTRAN version SVALUE first finds an index i such that $x_i \leq t < x_{i+1}$ and then the ith cubic is evaluated to get $S(t)$. A typical call in FORTRAN is

```
CALL SVALUE(N,X,F,B,C,D,T,INTERV,S,FLAG)
```

and

```
flag = Spline_value(n, x, f, b, c, d, t, interval, s);
```

in the C++ version. The last two parameters are output, so their addresses must explicitly be passed in C:

```
flag = Spline_value(n, x, f, b, c, d, t, &interval, &s);
```

As usual, arrays in the C and C++ versions are indexed starting at 0 rather than 1 as is typical of FORTRAN. The parameters N, X, F, B, C, and D have the same meaning for SPCOEF and Spline_coeff. The last three are input quantities that must have been set by a prior call to SPCOEF or Spline_coeff. The variable T holds the point where the evaluation is to be made and the answer comes back in S. If the index i satisfying $x_i \leq T < x_{i+1}$ is known, this can be input using the variable INTERV or interval, as the case may be. However, it is not necessary to do this since the code will calculate the correct value and assign it to INTERV or interval. The normal value of FLAG (the return value in the C version) is zero. When N < 2, FLAG is returned with the value −1. If T $< x_1$, then FLAG is set to 1, and the cubic for $[x_1, x_2]$ is used for S. If T $> x_N$, then FLAG is set to 2, and the cubic for $[x_{N-1}, x_N]$ is used for S.

Example 3.10. A sample driver is provided to interpolate $\sin x$ over { 0, 0.2, 0.4, 0.6, 0.8 } ; the resulting $S(x)$ is then tabulated at { 0.1, 0.3, 0.5, 0.7, 0.9 } to yield the following.

```
T = 1.000000000000000E-001 S = 9.984780844446299E-002 FLAG = 0
T = 3.000000000000000E-001 S = 2.955172204918649E-001 FLAG = 0
T = 5.000000000000000E-001 S = 4.794156789356284E-001 FLAG = 0
T = 7.000000000000001E-001 S = 6.442480866255412E-001 FLAG = 0
T = 9.000000000000000E-001 S = 7.830836113227835E-001 FLAG = 2
```

Note that only one call is made to SPCOEF or Spline_coeff even though the spline is evaluated at five points. Why is FLAG = 2 in the last line? ■

Example 3.11. A graph of the spline $S(v)$ interpolating the 16 indicated data points from Example 3.6 appears in Figure 3.11. It is a dramatically better approximation than the polynomial of high degree appearing in Figure 3.2. Values of S at some of the reserved data points are $S(0.29) = -2.88$, $S(1.11) = -1.57$, $S(5) = 5.50$, $S(8.5) =$

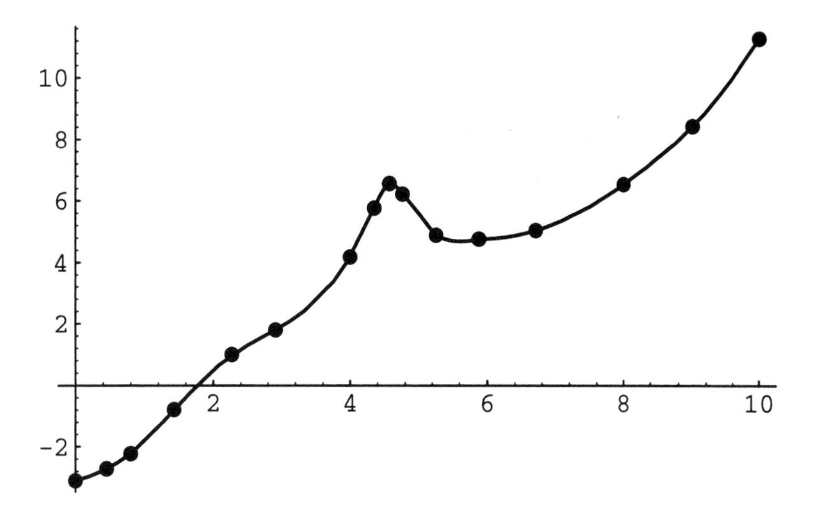

Figure 3.11 $S(\mathrm{v})$ from Example 3.11.

7.41, and $S(9.5) = 9.74$. They are in good agreement with the actual values. ■

Example 3.12. Observed values for the thrust (T) versus time (t) curve of a model rocket are

t	>0.00	>0.05	>0.10	>0.15	>0.20	0.25	>0.30
T	0.0	1.0	5.0	15.0	33.5	38.0	33.0
t	0.35	>0.40	0.45	>0.50	0.55	>0.60	0.65
T	21.5	16.5	16.0	16.0	16.0	16.0	16.0
t	>0.70	0.75	>0.80	>0.85	>0.90	>0.95	>1.00
T	16.0	16.0	16.0	16.0	6.0	2.0	0.0

The 15 values indicated by $(>)$ were used as data. The resulting complete cubic spline $S(x)$ is graphed in Figure 3.12. Note that the large curvatures near $t = 0.40$ and $t = 0.85$ are difficult to handle. Values of S at some reserved data points are $S(0.25) = 39.1$, $S(0.35) = 23.6$, and $S(0.65) = 16.1$. ■

EXERCISES

3.20 Derive equation (3.44) from the end condition $f'(x_N) = S'(x_N)$.

3.21 If the end conditions $S''(x_1) = f''(x_1)$ and $S''(x_N) = f''(x_N)$ are used, what equations should replace (3.43) and (3.44)?

3.22 The vapor pressure P of water (in bars) as a function of temperature T (° C) is

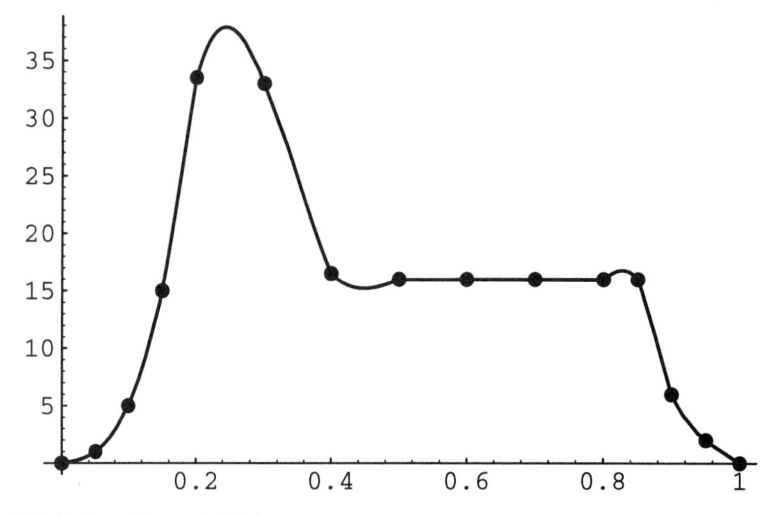

Figure 3.12 $S(t)$ from Example 3.12.

T	0	10	20	30
$P(T)$	0.006107	0.012277	0.023378	0.042433
T	40	50	60	70
$P(T)$	0.073774	0.12338	0.19924	0.31166
T	80	90	100	
$P(T)$	0.47364	0.70112	1.01325	

Interpolate these data with the cubic spline $S(x)$. It is also known that $P(5) = 0.008721$, $P(45) = 0.095848$, and $P(95) = 0.84528$. How well does $S(x)$ do at these points?

3.23 The following data give the absorbance of light (A) as a function of wavelength (λ) for vanadyl D-tartrate dimer.

λ	> 3125	> 3250	3375	> 3500	3625	> 3750
$A(\lambda)$	0.700	0.572	0.400	0.382	0.449	0.560
λ	3875	> 4000	4125	> 4250	4375	
$A(\lambda)$	0.769	0.836	0.750	0.530	0.315	
λ	> 4500	4625	> 4750	4875	> 5000	
$A(\lambda)$	0.170	0.144	0.183	0.252	0.350	

Use the cubic spline $S(x)$ to interpolate the nine indicated (>) data points. Explore the effects of scaling and shifting the independent variable (x = wavelength) with each of the following.

(a) The data as is.

(b) Replace x by $x/1000$ for all inputs.

(c) Replace x by $(x - 4000)/1000$ for all inputs.

For each case evaluate $S(x)$ at the remaining noninterpolated wavelengths. How well do these values compare with the known absorbances? Does shifting and/or scaling affect the accuracy of $S(x)$?

3.24 Repeat Exercise 3.23 except use $P_9(x)$ instead of $S(x)$. Use the method suggested in Exercise 3.4. What effect does the scaling have on COND?

3.25 The absorption of sound (at 20°C, 40% humidity) as a function of frequency, f, is

f	>20	>40	63	>100	200
$A(f)$	0.008	0.030	0.070	0.151	0.359
f	>400	800	>1250	2000	>4000
$A(f)$	0.592	0.935	1.477	2.870	9.618
f	10,000	$>16{,}000$	$>40{,}000$	$>80{,}000$	
$A(f)$	53.478	122.278	429.310	850.536	

Use the cubic spline $S(x)$ to interpolate the nine indicated ($>$) points in the following two ways.

(a) The data as is.

(b) $\log f$ versus $\log A(f)$.

Which seems to be better?

3.26 The following table gives values for a property of titanium as a function of temperature T.

T	605	645	685	725	765	795	825
$C(T)$	0.622	0.639	0.655	0.668	0.679	0.694	0.730
T	845	855	865	875	885	895	905
$C(T)$	0.812	0.907	1.044	1.336	1.881	2.169	2.075
T	915	925	935	955	975	1015	1065
$C(T)$	1.598	1.211	0.916	0.672	0.615	0.603	0.601

Compute and plot the cubic spline $S(T)$ for these data (use about 15 interpolating points). How well does it do?

3.27 In performing potentiometric titrations one obtains a potential difference curve plotted against volume of titrant added. The following table gives the measurements for the potentiometric titration of Fe^{2+} solution with 0.1095N Ce^{4+} solution using platinum and calomel electrodes.

Added sol. (ml)	1.0	5.0	10.0	15.0	20.0	21.0	22.0
E (mV)	373	415	438	459	491	503	523
Added sol. (ml)	22.5	22.6	22.7	22.8	22.9	23.0	23.1
E (mV)	543	550	557	565	575	590	620
Added sol. (ml)	23.2	23.3	23.4	23.5	24.0	26.0	30.0
E (mV)	860	915	944	958	986	1067	1125

Compute the cubic spline $S(x)$ for these data (use about 15 interpolating points). Graph $S(x)$ for x in [20, 24]. How well does it behave? The physical problem has exactly one inflection point. Is this true for $S(x)$?

3.28 The potential energy of two or more interacting molecules is called van der Waal's interaction energy. A theoretical calculation for two interacting helium atoms has the set of energies $V(r)$ for various values of the internuclear distance r given below. The energy exhibits repulsion ($V > 0$) for small r and attraction ($V < 0$)

for larger values of r.

r (bohrs)	4.6	4.8	5.0	5.1	5.2
$V(r)$	32.11	9.00	−3.52	−7.11	−9.22
r	5.3	5.4	5.5	5.6	5.7
$V(r)$	−10.74	−11.57	−11.95	−12.00	−11.73
r	5.8	5.9	6.0	6.5	7.0
$V(r)$	−11.23	−10.71	−10.13	−7.15	−4.77
r	7.5	8.0	9.0	10.0	
$V(r)$	−3.17	−2.14	−1.03	−0.54	

Compute the cubic spline $S(x)$ using about 12 interpolating points. How well does it work?

3.29 Modify the routine SVALUE or Spline_value to return $S'(x)$ and $S''(x)$ as well as $S(x)$.

3.30 In [5] a method is given for deducing the diffusion coefficient D for chloroform in polystyrene from uptake measurements. Using several assumptions, they arrive at the quantity

$$\widehat{D}(C_0) = \frac{1}{C_0}\int_0^{C_0} D(C)\,dC,$$

which can be measured for a number of C_0 values. A differentiation with respect to C_0 gives an expression for D in terms of the quantity

$$\frac{d}{dC_0}\left[C_0\widehat{D}(C_0)\right].$$

Using the data

C_0	5.0	7.5	9.9	12.9
$\widehat{D}(C_0)$	0.0240	0.0437	0.0797	0.1710
C_0	13.2	15.1	16.3	16.8
$\widehat{D}(C_0)$	0.1990	0.3260	0.8460	0.9720

approximate D for each C_0 value by differentiating the appropriate spline fit.

3.31 Show that the cubic spline $S(x)$ has a critical point z in $[x_n, x_{n+1}]$, that is, $S'(z) = 0$, if and only if the following are true:

(i) $z = x_n + (-c_n \pm \sqrt{c_n^2 - 3b_n d_n})/(3d_n)$
(ii) $x_n \le z$
(iii) $z \le x_{n+1}$.

Why is it not sufficient merely to use (i) and the test $b_n b_{n+1} = S'(x_n)S'(x_{n+1}) < 0$?

3.32 Show that the cubic spline $S(x)$ has an inflection point z in (x_n, x_{n+1}), that is, $S''(z) = 0$, if and only if $c_n c_{n+1} < 0$, in which case $z = x_n - c_n/(3d_n)$.

3.33 Use the formula in Exercise 3.31 to find all local minima for the data in Exercise 3.23.

3.34 Use the formula in Exercise 3.31 to find the local maximum (near $T = 905$) for the data in Exercise 3.26.

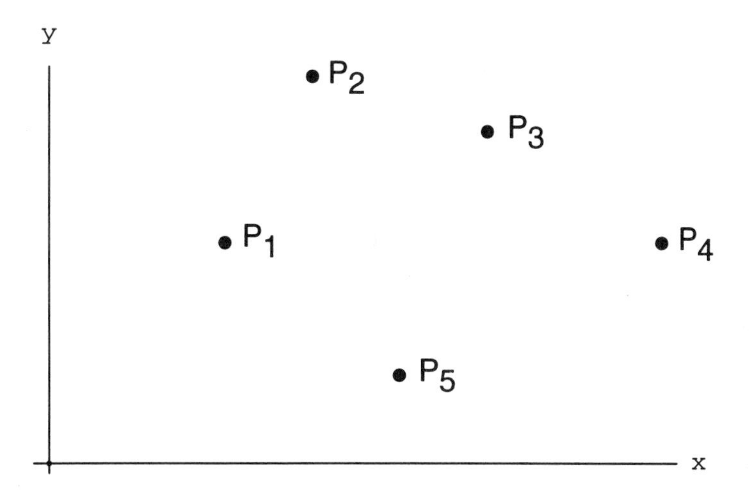

Figure 3.13 Scattered data in the plane.

3.35 For the data in Exercise 3.28 the global minimum at $r = r_e$ corresponds to stable equilibrium $(V' = 0)$. There is also an inflection point (where $V'' = 0$) at $r = r_i$. What does $S(x)$ yield for r_e and r_i? Are the answers reasonable?

3.36 Use the formulas in Exercises 3.31 and 3.32 to find the local maximum (near $\nu = 4.5$) and all inflection points for the data in Example 3.5.

3.6 INTERPOLATION IN THE PLANE

In this section a few of the ideas involved in interpolating functions of several variables are taken up. Although the ideas of the case of one variable generalize, there are new difficulties arising from geometrical considerations. To be specific, only the case of two independent variables will be considered.

Suppose we have values f_i given at distinct points p_i for $i = 1, \ldots, N$ in a region Ω in the $x-y$ plane (see Figure 3.13), and we seek a polynomial in the two variables x and y that interpolates the data. This is easily accomplished in a way similar to Lagrangian interpolation. If $p = (x, y)$ is a general point in Ω, an interpolating polynomial is given by

$$Q(x,y) = \sum_{i=1}^{N} f_i \phi_i(p) = \sum_{i=1}^{N} f_i \phi_i(x,y)$$

provided that

$$\phi_i(p_j) = \begin{cases} 1 & \text{if } i = j \\ 0 & \text{if } i \neq j \end{cases}$$

and each ϕ_i is a polynomial in x and y. It is easy to verify that

$$\phi_i(x,y) = \prod_{\substack{j=1 \\ j \neq i}}^{N} \left[\frac{(x-x_j)^2 + (y-y_j)^2}{(x_i-x_j)^2 + (y_i-y_j)^2} \right], \; i = 1, \ldots, N,$$

satisfies the requirements. Thus, it is easy to construct a polynomial in two variables,

$$Q(x,y) = \sum_{m,n=0}^{M} a_{mn} x^m y^n,$$

which interpolates given values at any set of distinct points in the plane.

The interpolating polynomial given is not closely analogous to that for one variable because the degree is much higher than the number of nodes. Unfortunately, the facts are simply different when there is more than one independent variable. This can be seen by considering a general quadratic polynomial in two variables (x,y):

$$P(x,y) = a_0 + a_1 x + a_2 y + a_3 x^2 + a_4 xy + a_5 y^2.$$

There are six parameters a_i. The analog of the result for one variable would be that there is a unique polynomial interpolating at six distinct points (x_i, y_i) in the plane. For each node the interpolation condition is

$$f_i = P(x_i, y_i) = a_0 + a_1 x_i + a_2 y_i + a_3 x_i^2 + a_4 x_i y_i + a_5 y_i^2.$$

Interpolation at six nodes provides six linear equations for the six parameters. Suppose that five of the nodes are $\{(0,0),(1,0),(0,-1),(-1,0),(0,1)\}$ and the sixth is (α, β). In the equation corresponding to each of the first five nodes, the coefficient of a_4 is zero. It is also zero in the sixth equation if $\alpha = 0$ or if $\beta = 0$. For any node (α, β) of this kind, the system amounts to six equations in only five unknowns. The system may or may not have a solution, but if there is a solution, it cannot be unique because a_4 can have any value. In several dimensions the placement of the nodes has a role in questions of existence and uniqueness that we did not see in one dimension.

If we had a choice about the points of interpolation, we might like to work on a rectangular grid. By this is meant there are n x-coordinates, $x_1 < x_2 < \cdots < x_n$, and m y-coordinates, $y_1 < y_2 < \cdots < y_m$, and the $n \times m$ points of interpolation consist of the pairs (x_i, y_j) for $1 \leq i \leq n$, $1 \leq j \leq m$; see Figure 3.14. In this special, but quite useful, case, the fundamental polynomials or shape functions are easily constructed from the functions in one variable. Recalling (3.3), let

$$L_i(x) = \prod_{\substack{k=1 \\ k \neq i}}^{n} \frac{x - x_k}{x_i - x_k}, \; i = 1, \ldots, n$$

$$l_j(y) = \prod_{\substack{k=1 \\ k \neq j}}^{m} \frac{y - y_k}{y_j - y_k}, \; j = 1, \ldots, m.$$

Then an interpolant $Q(x,y)$ such that

$$Q(x_i, y_j) = f_{ij}, \; 1 \leq i \leq n, \; 1 \leq j \leq m, \tag{3.45}$$

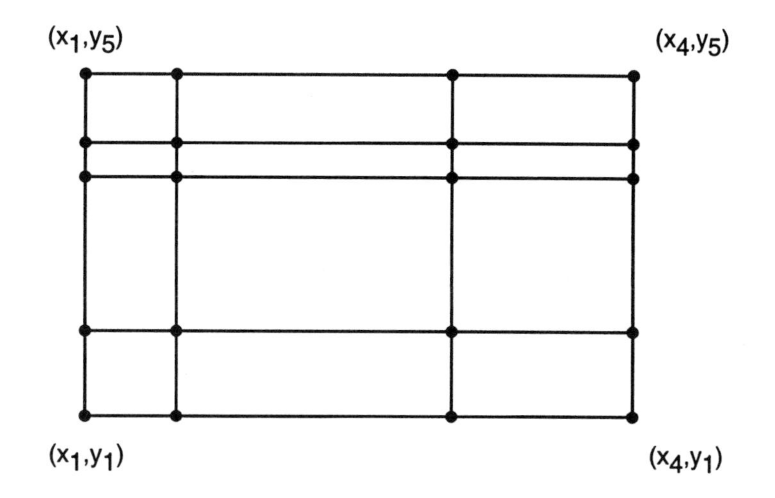

Figure 3.14 Data on a rectangular grid ($n = 4, m = 5$).

is given by

$$Q(x,y) = \sum_{i=1}^{n} \sum_{j=1}^{m} L_i(x) l_j(y) f_{ij}. \tag{3.46}$$

If $Q(x,y)$ is multiplied out, it clearly has the form

$$Q(x,y) = \sum_{s=0}^{n} \sum_{t=0}^{m} a_{st} x^s y^t. \tag{3.47}$$

We now show that the coefficients a_{st} are uniquely determined by the interpolation conditions (3.45). Choose any i with $1 \leq i \leq n$ and consider the polynomial in the one variable y:

$$Q(x_i, y) = \sum_{t=0}^{m} \left(\sum_{s=0}^{n} a_{st} x_i^s \right) y^t.$$

We know that there is exactly one polynomial

$$\sum_{t=0}^{m} b_{it} y^t$$

that interpolates the values f_{ij} at y_j for $j = 1, \ldots, m$. Because $Q(x_i, y)$ does this, it must be that

$$b_{it} = \sum_{s=0}^{n} a_{st} x_i^s \quad \text{for} \quad t = 0, \ldots, m.$$

This equation holds for each i. Now choose a t with $0 \le t \le m$. There is exactly one polynomial

$$R(x) = \sum_{s=0}^{n} c_{st} x^s$$

such that

$$R(x_i) = b_{it} \quad \text{for} \quad i = 1, \ldots, n.$$

Because the polynomial

$$\sum_{s=0}^{n} a_{st} x^s$$

does this, it must be the case that $c_{st} = a_{st}$ for $0 \le s \le n$ and for any $0 \le t \le m$. Thus the coefficients a_{st} are uniquely determined by the interpolation coefficients as we wanted to show.

As a simple example, let us consider interpolation at the four corners of a rectangle: (x_1, y_1), (x_1, y_2), (x_2, y_1), (x_2, y_2).

$$L_1(x) = \frac{x - x_2}{x_1 - x_2}, \; L_2(x) = \frac{x - x_1}{x_2 - x_1}$$

$$\ell_1(y) = \frac{y - y_2}{y_1 - y_2}, \; l_2(y) = \frac{y - y_1}{y_2 - y_1}$$

$$Q(x,y) = f_{11}\left(\frac{x - x_2}{x_1 - x_2}\right)\left(\frac{y - y_2}{y_1 - y_2}\right) + f_{12}\left(\frac{x - x_2}{x_1 - x_2}\right)\left(\frac{y - y_1}{y_2 - y_1}\right) \tag{3.48}$$

$$+ f_{21}\left(\frac{x - x_1}{x_2 - x_1}\right)\left(\frac{y - y_2}{y_1 - y_2}\right) + f_{22}\left(\frac{x - x_1}{x_2 - x_1}\right)\left(\frac{y - y_1}{y_2 - y_1}\right).$$

Figure 3.15 displays a typical interpolant for $n = 4$ and $m = 3$.

Interpolants constructed in this way are called tensor product interpolants. The example (3.48) is said to be a bilinear interpolant over a rectangle because it has the form

$$a_{00} + a_{10}x + a_{01}y + a_{11}xy, \tag{3.49}$$

that is, it is linear in each variable when the other is held fixed. The general first degree polynomial has the form (3.49) with $a_{11} = 0$. A biquadratic has the form

$$a_{00} + a_{10}x + a_{01}y + a_{20}x^2 + a_{11}xy + a_{02}y^2 + a_{21}x^2y + a_{22}x^2y^2 + a_{12}xy^2,$$

while the general second degree polynomial has here $a_{21} = a_{22} = a_{12} = 0$. Generalizations to bicubic versus cubic and higher degrees should be clear.

In studying how well a function of two variables is approximated by a particular kind of interpolating polynomial, a critical matter is the highest degree for which the approximation is exact. For example, in the rectangle of Figure 3.16, we can interpolate at the nine indicated points using a biquadratic. However, it is exact only for second degree polynomials in spite of the presence of the higher degree terms x^2y, yx^2, and x^2y^2. In fact, only six interpolating points are needed to construct a quadratic interpolating polynomial that is exact to second degree. It is not at all clear how to choose

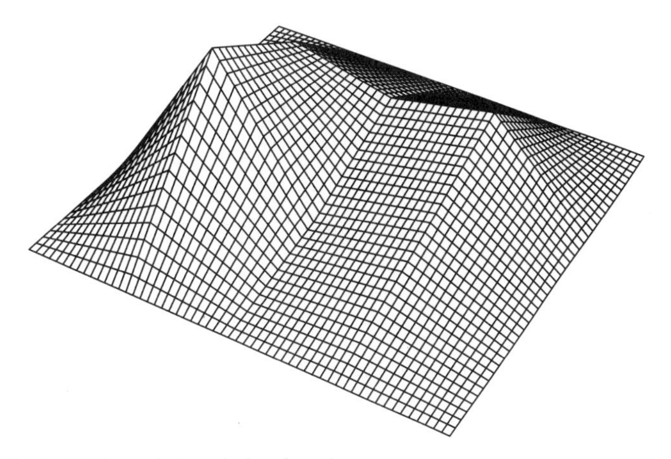

Figure 3.15 A typical bilinear interpolating function.

the six points symmetrically from the rectangular grid. Because of this, biquadratics or bicubics are generally used for interpolation on rectangular grids.

Just as in the case of one variable, piecewise polynomial interpolation may provide more satisfactory interpolants than a polynomial interpolant over the whole region. If the region can be broken up into rectangles, tensor product interpolants can be used on each piece. In the case of one variable, two pieces connect at a single point, but in the plane they connect along a line, and more than one piece can touch at a point. In contrast to the ease with which the polynomial pieces could be connected smoothly in the case of one variable, it is hard to get much smoothness where polynomials in several variables are joined.

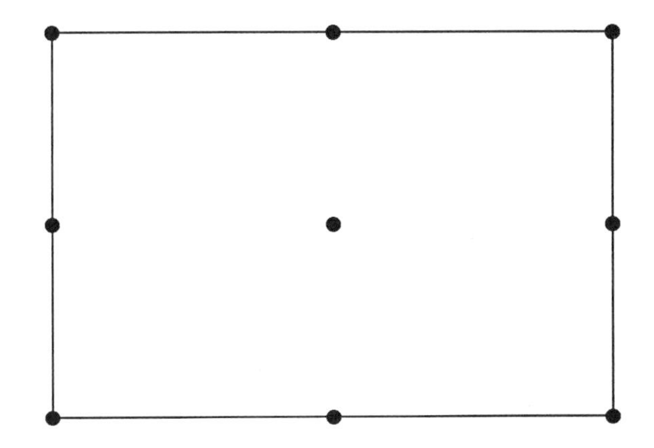

Figure 3.16 Interpolation points for biquadratics.

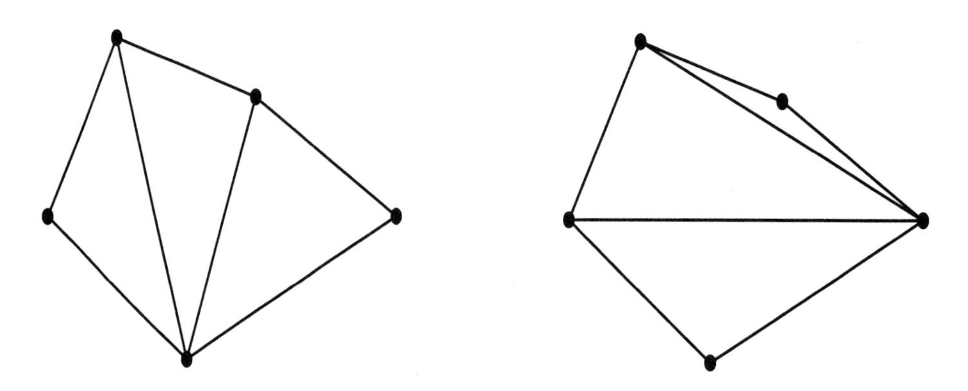

Figure 3.17 Two triangulations for the data in Figure 3.13.

In piecewise polynomial interpolation the idea is to work with regions for which interpolants are readily constructed and to decompose the region of interest into regions of this kind. A popular alternative to rectangles is triangles. For example, one might triangulate the region of Figure 3.13 in the two ways sketched in Figure 3.17. As a rule it is best to avoid the "skinny" triangles of the second possibility, and routines for triangulating regions generally try to avoid nearly degenerate triangles. It is not hard to write down the shape functions for linear interpolation on the general triangle of Figure 3.18. They are

$$\phi_1(x,y) = \frac{1}{2A}[(x_2y_3 - x_3y_2) + (y_2 - y_3)x + (x_3 - x_2)y]$$
$$\phi_2(x,y) = \frac{1}{2A}[(x_3y_1 - x_1y_3) + (y_3 - y_1)x + (x_1 - x_3)y]$$
$$\phi_3(x,y) = \frac{1}{2A}[(x_1y_2 - x_2y_1) + (y_1 - y_2)x + (x_2 - x_1)y],$$

where

$$2A = x_2y_3 + x_1y_2 + x_3y_1 - y_1x_2 - y_2x_3 - y_3x_1,$$

and A is the area of the triangle. Then the linear function that has given values at the corners

$$f_i \text{ given at } (x_i, y_i),\ i = 1,2,3,$$

is

$$Q(x,y) = f_1\phi_1(x,y) + f_2\phi_2(x,y) + f_3\phi_3(x,y).$$

Note that on this triangle Q has the form

$$Q(x,y) = a + bx + cy \tag{3.50}$$

for some a, b, and c. See Figure 3.19 for an illustration of piecewise linear interpolation on a triangular grid.

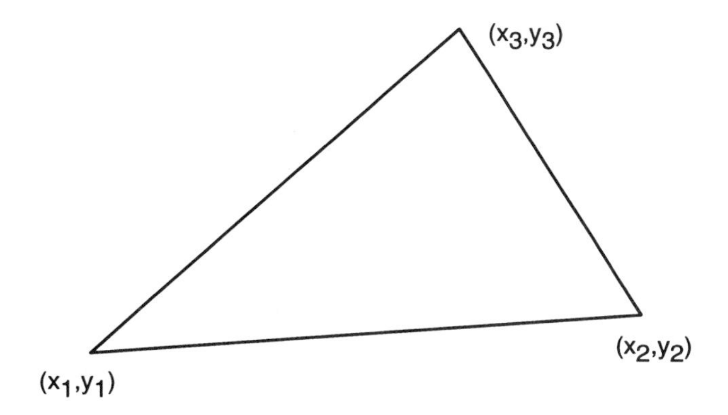

Figure 3.18 A general triangle.

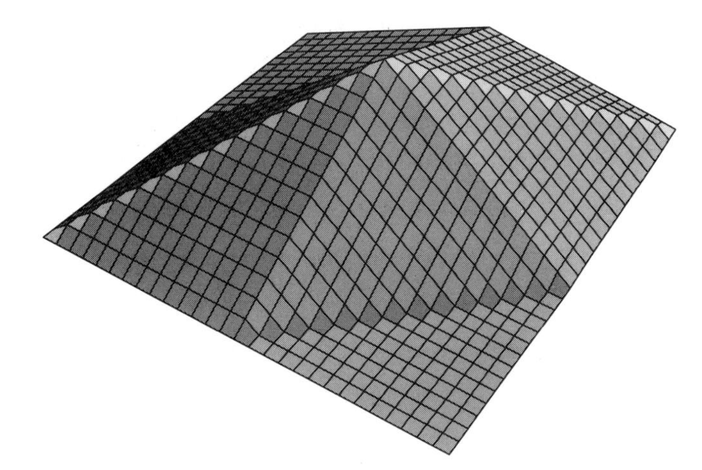

Figure 3.19 Linear interpolation on a triangular grid.

In finite element analysis one tries to determine a piecewise polynomial approximation to the solution of an ordinary or partial differential equation. Polynomial approximations over subregions prove to be very convenient in computation. Also, representing solutions in the Lagrangian, or nodal, form

$$\sum f_i \phi_i(x,y)$$

is convenient because the f_i are approximate solution values at the nodes (x_i, y_i). One difficulty is that rectangles or triangles are too restrictive. A way to accommodate subregions with curved sides is to transform the region to a "standard," "reference," or "master" region. For example, we might choose to work with a standard triangle such as the one given in Figure 3.20. Suppose we want to represent a function $f(x,y)$ on a region Ω. If it is known how to map the region onto the standard triangle, then

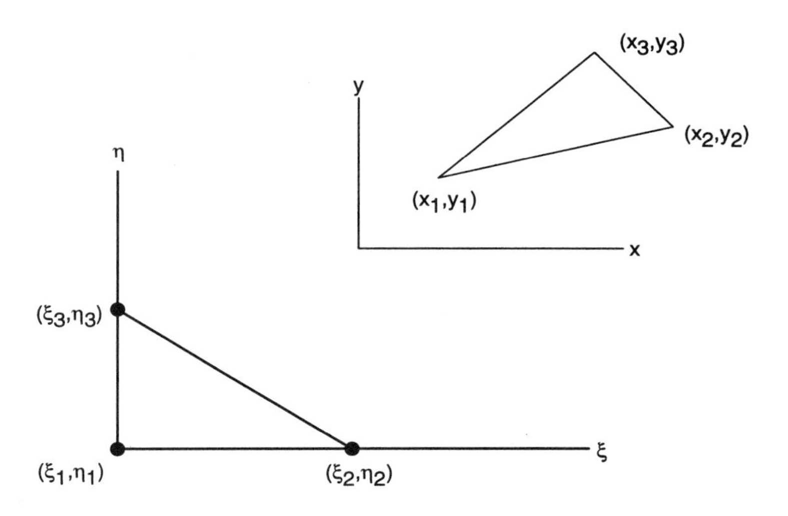

Figure 3.20 Mapping from the standard triangle.

interpolation can be done. Let T be a mapping such that

$$T: \begin{array}{l} x = x(\xi,\eta) \\ y = y(\xi,\eta), \end{array}$$

and as (ξ,η) ranges over the standard triangle, the values (x,y) range over Ω. Then $f(x,y)$ for (x,y) in Ω has

$$f(x,y) = f(x(\xi,\eta), y(\xi,\eta)) = \hat{f}(\xi,\eta),$$

and we can simply interpolate $\hat{f}$ over the triangle. Of course, the mapping must be a proper one, meaning that as (ξ,η) ranges over the triangle, the (x,y) cover all of Ω, and there is no overlap [different (ξ,η) go into different (x,y)]. A nice idea used for finite elements is to construct the mapping by interpolation, too. As an example, let us construct the mapping from the standard triangle to the general triangle of Figure 3.20. The shape functions for the standard triangle are obviously

$$\begin{aligned} \phi_1(\xi,\eta) &= \xi \\ \phi_2(\xi,\eta) &= \eta \\ \phi_3(\xi,\eta) &= 1-\xi-\eta \end{aligned}$$

when we let

node 1 be (1, 0)
node 2 be (0, 1)
node 3 be (0, 0)

because, for example,

$$\phi_1(1,0) = 1, \phi_1(0,1) = 0, \phi_1(0,0) = 0.$$

Interpolating x we have

$$\begin{aligned} x &= x_1\phi_1(\xi,\eta) + x_2\phi_2(\xi,\eta) + x_3\phi_3(\xi,\eta) \\ &= x_1\xi + x_2\eta + x_3(1-\xi-\eta) \end{aligned}$$

and, similarly,

$$y = y_1\xi + y_2\eta + y_3(1-\xi-\eta).$$

In this particular case the mapping carries straight lines into straight lines and there is no difficulty about a proper mapping. If higher degree interpolation were used, the triangle would be mapped into a region with curved sides. Roughly speaking, if the region is not too different from a triangle, the mapping will be proper. Continuing with the example, suppose we interpolate the f_i by the same basis functions used to construct the mapping. The interpolant $Q(x,y)$ is

$$\begin{aligned} Q(x(\xi,\eta),y(\xi,\eta)) &= f_1\phi_1(\xi,\eta) + f_2\phi_2(\xi,\eta) + f_3\phi_3(\xi,\eta) \\ &= f_1\xi + f_2\eta + f_3(1-\xi-\eta). \end{aligned}$$

If we were to solve the relation $x(\xi,\eta), y(\xi,\eta)$ to get the inverse mapping

$$T^{-1}: \begin{array}{l} \xi = \xi(x,y) \\ \eta = \eta(x,y) \end{array}$$

and eliminate ξ and η in this interpolant, we would get the expression earlier for the bilinear interpolant. This seems a complicated way to get a simple result. The virtue of the procedure is that interpolation is done on regions with curved boundaries by transforming them to a simple, standard region for which interpolation is comparatively easy. In finite element computations it is found that the process is easily programmed and very powerful. All we aim to do here is sketch the process. For details the reader may consult one of the great many books devoted to finite elements. The books span a great range of mathematical sophistication; a good introduction is [1].

EXERCISES

3.37 The formula (3.48) could be called a Lagrange form of the bilinear interpolating polynomial; consider the "Newton form"

$$Q(x,y) = a + b(x-x_1) + c(y-y_1) + d(x-x_1)(y-y_1).$$

Solve for a, b, c, and d so that Q interpolates a function $f(x,y)$ at the four corners. As in (3.48) let $f_{ij} = f(x_i,y_j)$.

3.38 Show that $Q(x,y)$, which is a quadratic polynomial in x and y [generalizing (3.50)], has six coefficients. On a triangle, its interpolating points are usually chosen to be the three triangle vertices and the three edge midpoints. For the triangle with vertices $(0,0)$, $(1,0)$, and $(0,1)$ compute the shape function that is one at $(0,0)$ and zero at the remaining five interpolating points.

3.7 CASE STUDY 3

This case study has two parts, one applying continuous splines and the other, smooth splines. Integrals of the form

$$\int_a^b f(x)\cos(\omega x)\,dx \text{ and } \int_a^b f(x)\sin(\omega x)\,dx$$

with finite a and b are called finite Fourier integrals. For large ω such integrals present special difficulties for numerical methods because of the rapid oscillation of the integrand. Filon's method [8] for approximating finite Fourier integrals will be developed here by means of a continuous spline. Accurate evaluation of the coefficients of the method was discussed in Chapter 1. Other aspects of the task will be discussed in Chapter 5. The second part of the case study takes up the use of smooth splines for fitting data with curves instead of functions.

Broadly speaking, Filon approximates finite Fourier integrals in a manner like that used in Chapter 5 for integrals of the form

$$\int_a^b f(x)w(x)\,dx$$

when $w(x)$ presents some difficulty. Namely, first approximate $f(x)$ with a convenient function $S(x)$ and then compute analytically

$$\int_a^b S(x)w(x)\,dx$$

as an approximation to the desired integral. In detail there are important differences because here the weight function $w(x)$ does not have one sign and oscillates rapidly for large frequencies ω. Also, the approach of Chapter 5 would apply to particular ω, and we would like a method that can be applied conveniently for any ω. Insight is provided by a classic technique of applied mathematics for approximating Fourier integrals when ω is "large." If derivatives of $f(x)$ are available, asymptotic approximations can be obtained by means of integration by parts. For example, integrating by parts twice gives

$$\int_a^b f(x)\cos(\omega x)\,dx = \frac{1}{\omega}\left[f(b)\sin(\omega b) - f(a)\sin(\omega a)\right] + R(\omega),$$

where

$$R(\omega) = -\frac{1}{\omega^2}\left[f'(b)\cos(\omega b) - f'(a)\cos(\omega a)\right] + \frac{1}{\omega^2}\int_a^b f'(x)\cos(\omega x)\,dx.$$

If M_1 is a bound on the magnitude of $f'(x)$, then

$$|R(k)| \le \omega^{-2}\left(2 + (b-a)\right)M_1,$$

that is, $R(\omega)$ is $O(\omega^{-2})$ as $\omega \to \infty$. Accordingly, the asymptotic approximation

$$\int_a^b f(x)\cos(\omega x)\,dx \approx \frac{1}{\omega}\left[f(b)\sin(\omega b) - f(a)\sin(\omega a)\right]$$

is accurate to $O(\omega^{-2})$. However, the integral itself ordinarily goes to zero like ω^{-1}, so the relative error is ordinarily only $O(\omega^{-1})$. The situation is typical of classical asymptotic approximation of integrals. The bigger ω is, the better the asymptotic approximation and the more difficult the integral is for conventional numerical methods. On the other hand, for a given ω, the asymptotic approximation may not be sufficiently accurate and there is no easy way to improve it. When ω is "small," finite Fourier integrals are easy for conventional numerical methods, and when ω is "large," asymptotic approximations are satisfactory. Filon's method provides a way to compute accurate integrals when ω is of moderate size.

Filon divides the interval $[a,b]$ into $2N$ subintervals of equal length h. Let us define $x_j = a + jh$ for $j = 0,\ldots,2N$. The function $f(x)$ is approximated by a continuous spline $S(x)$ that is a quadratic polynomial on each $[x_{2m}, x_{2m+2}]$ defined there by interpolation to $f(x_j)$ for $j = 2m, 2m+1, 2m+2$. Each of the integrals in the approximation

$$\int_a^b f(x)\cos(\omega x)\,dx \approx \int_a^b S(x)\cos(\omega x)\,dx = \sum_{m=0}^{N} \int_{x_{2m}}^{x_{2(m+1)}} S(x)\cos(\omega x)\,dx$$

can be evaluated analytically by integration by parts. A lengthy calculation results in Filon's method:

$$\int_a^b f(x)\cos(\omega x)\,dx \approx h\left[\alpha\left(f(b)\sin(\omega b) - f(a)\sin(\omega a)\right) + \beta C_e + \gamma C_o\right].$$

Here $\theta = \omega h$ and

$$\alpha = \left(\theta^2 + \theta\sin(\theta)\cos(\theta) - 2\sin^2(\theta)\right)/\theta^3$$

$$\beta = 2\left(\theta\left(1 + \cos^2(\theta)\right) - 2\sin(\theta)\cos(\theta)\right)/\theta^3$$

$$\gamma = 4\left(\sin(\theta) - \theta\cos(\theta)\right)/\theta^3.$$

Also,

$$C_e = 0.5f(a)\cos(\omega a) + \sum_{m=1}^{N-1} f(x_{2m})\cos(\omega x_{2m}) + 0.5f(b)\cos(\omega b)$$

$$C_o = \sum_{m=0}^{N-1} f(x_{2m+1})\cos(\omega x_{2m+1}).$$

There is a similar approximation when cosine is replaced by sine. The formula is

$$\int_a^b f(x)\sin(\omega x)\,dx \approx h\left[-\alpha\left(f(b)\cos(\omega b) - f(a)\cos(\omega a)\right) + \beta S_e + \gamma S_o\right].$$

The coefficients α, β, γ are the same, and S_e and S_o are like C_e and C_o with the cosines replaced by sines.

Using the results developed in this chapter for the error of interpolation, it is easy to bound the error of Filon's method. On each subinterval $[x_{2m}, x_{2m+2}]$, $S(x)$ is a quadratic interpolant to $f(x)$. If M_3 is a bound on the third derivative of $f(x)$ on all of $[a,b]$, we found that

$$|f(x) - S(x)| \le \frac{M_3(2h)^3}{3!} = \frac{4}{3}M_3h^3$$

uniformly for $a \leq x \leq b$. Then

$$\left|\int_a^b f(x)\cos(\omega x)\,dx - \int_a^b S(x)\cos(\omega x)\,dx\right| \leq \int_a^b |f(x) - S(x)|\,dx \leq \frac{4}{3}(b-a)M_3h^3.$$

This is a bound on the absolute error that depends on $f(x)$ and the sampling interval h, but not on ω. If we want a meaningful result for "large" ω, we have to take into account that the integral is $O(\omega^{-1})$. This leads to a bound on the relative error that is $O(\theta^{-1}h^4)$.

In a subsection of Chapter 5 about applying a general-purpose code to problems with oscillatory integrands, the example

$$\int_0^\pi \frac{\sin(20x)}{1+x^2}\,dx$$

is discussed. The function $f(x) = 1/(1+x^2)$ ought to be approximated well by a quadratic interpolatory spline with a relatively crude mesh, so Filon's method ought to be quite effective. A matter not usually discussed with Filon's method is how to estimate the accuracy of the result. One way to proceed is to compute a result that we believe to be more accurate and estimate the error of the less accurate result by comparison. If the accuracy is acceptable, often the more accurate result is the one taken as the answer. Inspection of the formulas shows that if h, or equivalently θ, is halved, we can reuse all the evaluations of f, $\sin(x)$, and $\cos(x)$ made in the first approximation to keep down the cost. According to the bounds, the error should be reduced enough by halving h to get a good estimate of the error by comparison. Using Filon's method with $\theta = 0.4$, we obtained an approximate integral of 0.04566373122996. Halving θ resulted in the approximation 0.04566373690838. Estimating the accuracy by comparison suggests that we have an answer with an error smaller than 6×10^{-9}. Reuse of the function evaluations by virtue of halving θ holds the cost to 315 evaluations of the integrand. The quadrature code Adapt developed in Chapter 5 asks the user to specify the accuracy desired. Using the code in the manner outlined in that chapter and experimenting some with the error tolerances, we obtained an approximation 0.04566373866324 with an estimated error of about -1.6×10^{-9}. This cost 1022 evaluations of the integrand.

Let us now change the subject from applying a continuous spline to applying a smooth spline. In this chapter we have been looking at the approximation of functions $y = f(x)$, but sometimes we want to approximate curves. This will be discussed in the plane for it will then be clear how to deal with a curve in three dimensions. The basic idea is to use a parametric representation $(x(s), y(s))$ for the curve and approximate independently the coordinate functions $x(s)$ and $y(s)$. The parameter s can be anything, but often in the theory it is taken to be arc length. Having chosen somehow nodes s_i, $i = 1, \ldots, N$, we can interpolate the data $x_i = x(s_i)$ by a spline $S_x(s)$ and likewise the data $y_i = y(s_i)$ by a spline $S_y(s)$. The curve $(x(s), y(s))$ is then approximated by $(S_x(s), S_y(s))$. This yields a curve in the plane that passes through all the points $(x(s_i), y(s_i))$ in order. It is natural to use the smooth cubic spline of SPCOEF because it leads to a curve with continuous curvature, but if the data are sparse, we might have to resort to the shape-preserving spline to get a curve of the expected shape. All these computations are familiar except for the selection of the nodes s_i. One way to proceed

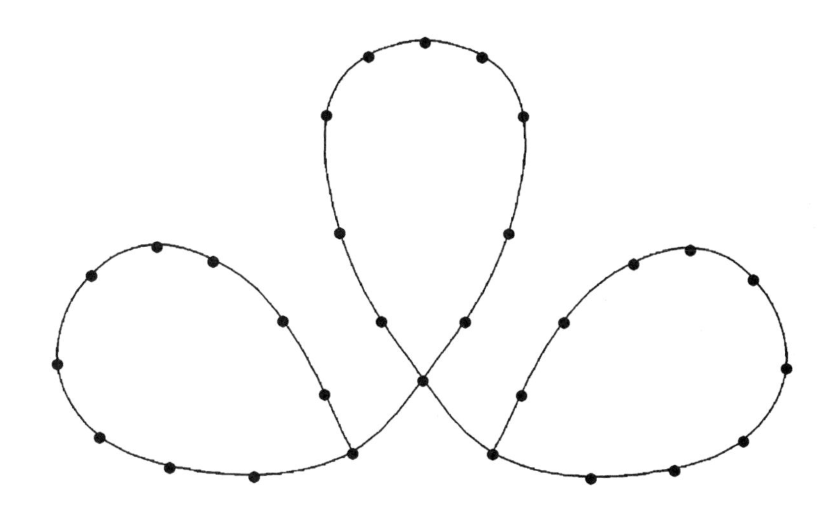

Figure 3.21 Curve fit for a sewing machine pattern.

is to choose them so that the parameter s approximates the arc length of the curve. This is done by taking $s_1 = 0$ and defining the difference between s_i and s_{i+1} as the distance between the points (x_i, y_i) and (x_{i+1}, y_{i+1}), namely

$$s_{i+1} = s_i + \sqrt{(x_{i+1} - x_i)^2 + (y_{i+1} - y_i)^2}.$$

Exercise 3.39 suggests an alternative. See Farin [6] for many other approaches.

An interesting example of the technique is furnished by a need to approximate curves for automatic control of sewing machines. Arc length is the natural parameter because a constant increment in arc length corresponds to a constant stitch length. An example taken from [18] fits the data (2.5, −2.5), (3.5, −0.5), (5, 2), (7.5, 4), (9.5, 4.5), (11.8, 3.5), (13, 0.5), (11.5, −2), (9, −3), (6, −3.3), (2.5, −2.5), (0, 0), (−1.5, 2), (−3, 5), (−3.5, 9), (−2, 11), (0, 11.5), (2, 11), (3.5, 9), (3, 5), (1.5, 2), (0, 0), (−2.5, −2.5), (−6, −3.3), (−9, −3), (−11.5, −2), (−13, 0.5), (−11.8, 3.5), (−9.5, 4.5), (−7.5, 4), (−5, 2), (−3.5, −0.5), (−2.5, −2.5). The resulting spline curve is seen in Figure 3.21. ∎

REFERENCES

1. E. Becker, G. Carey and J. T. Oden, *Finite Elements: An Introduction*, Vol. I, Prentice Hall, Englewood Cliffs, N.J., 1981.
2. G. Birkhoff and A. Priver, "Hermite interpolation errors for derivatives," *J. Math. and Physics*, 46 (1967), pp. 440–447.

3. C. de Boor, *A Practical Guide to Splines*, Springer-Verlag, New York, 1978.
4. E. W. Cheney, *Introduction to Approximation Theory*, McGraw-Hill, New York, 1966.
5. J. Crank and G. Park, "Evaluation of the diffusion coefficient for $CHCl_3$ in polystyrene from simple absorption experiments," *Trans. Faraday Soc.*, 45 (1949), pp. 240–249.
6. G. Farin, *Curves and Surfaces for Computer Aided Geometric Design*, Academic Press, San Diego, 1988.
7. J. Ferguson and K. Miller, "Characterization of shape in a class of third degree algebraic curves," TRW Report 5322-3-5, 1969.
8. L. N. G. Filon, "On a quadrature formula for trigonometric integrals," *Proc. Roy. Soc. Edinburgh*, 49 (1928–1929), pp. 38–47.
9. F. Fritsch and J. Butland, "A method for constructing local monotone piecewise cubic interpolants," *SIAM J. Sci. Stat. Comp.*, 5 (1984), pp. 300–304.
10. F. Fritsch and R. Carlson, "Monotone piecewise cubic interpolation," *SIAM J. Numer. Anal.*, 17 (1980), pp. 238–246.
11. C. Hall and W. Meyer, "Optimal error bounds for cubic spline interpolation," *J. Approx. Theory*, 16 (1976), pp. 105–122.
12. *Handbook of Chemistry and Physics*, 63rd ed., CRC Press, Cleveland, 1982–1983.
13. *Handbook of Mathematical Functions*, M. Abramowitz and I. Stegun, eds., Dover, Mineola, N.Y., 1964.
14. H. Huynh, "Accurate monotone cubic interpolation," *SIAM J. Numer. Anal.*, 30 (1993), pp. 57–100.
15. E. Isaacson and H. Keller, *Analysis of Numerical Methods*, Dover, Mineola, N.Y., 1994.
16. W. J. Kammerer and G. W. Reddien, "Local convergence of smooth cubic spline interpolates," *SIAM J. Numer. Anal.*, 9 (1972), pp. 687–694.
17. M. J. D. Powell, "On the maximum errors of polynomial approximation defined by interpolation and by least squares criteria," *Comp. J.*, 9 (1967), pp. 404–407.
18. P. Rentrop and W. Wever, "Interpolation algorithms for the control of a sewing machine," *Proceedings of ECMI II in Oberwolfach*, H. Neunzert, ed., Teubner-Kluwer, Dortdrecht, Netherlands, 1988, pp. 251–268.
19. B. Savage, and J. Mathis, "Observed properties of interstellar dust," in *Annual Review of Astronomy and Astrophysics*, 17 (1979).

MISCELLANEOUS EXERCISES FOR CHAPTER 3

3.39 As mentioned in the case study, there are other ways to select the nodes when fitting data with curves. A simple one is

$$t_1 = 0$$
$$t_{i+1} = t_i + |x_{i+1} - x_i| + |y_{i+1} - y_i|, \; 1 \leq i < N.$$

Using this scheme and SPCOEF or Spline_coeff, fit the data

x	1.00	0.34	−0.59	−0.58	0.04	0.38
y	0.00	0.88	0.54	−0.29	−0.51	−0.11
x	0.19	−0.13	−0.20	−0.03	0.12	
y	0.25	0.22	−0.04	−0.17	−0.08	

Using SVALUE or Spline_value, evaluate at sufficiently many points to sketch a smooth curve in the xy plane.

3.40 Find a technique that gives a good fit to the model rocket data from Example 3.11. Interpolate the indicated ($>$) data and avoid undesirable oscillations.

3.41 Implement the shape-preserving cubic spline described in Section 3.5. Test it out on some of the data sets from this chapter, for example, Exercises 3.22, 3.25, and 3.27. Sketch the graph of $H(x)$. Also try it on the nonmonotone data in Example 3.11, Exercise 3.26, and Exercise 3.28. How well does it do?

3.42 Implement the bilinear interpolating function $Q(x,y)$ given in (3.48). Test it on several different functions and several different grids.

3.43 Implement the linear interpolating function $Q(x,y)$ given in (3.50). Test it on several different functions and several different triangulations.

CHAPTER 4

ROOTS OF NONLINEAR EQUATIONS

Finding solutions of a system of nonlinear equations

$$f(x) = 0 \tag{4.1}$$

is a computational task that occurs frequently both on its own and as a part of a more complicated problem. Most of this chapter is devoted to the case of a continuous real function $f(x)$ of a single real variable x because it is important and can be discussed in elementary terms. The general case of n nonlinear equations in n unknowns is much more difficult both in theory and practice. Although the theory is too involved to be developed here, some of the basic methods are discussed briefly at the end of the chapter.

A *root* of (4.1), or a *zero* of $f(x)$, is a number α such that $f(\alpha) = 0$. A root is described more fully by its multiplicity m. This means that for x near α, $f(x)$ can be written in the form

$$f(x) = (x-\alpha)^m g(x), \tag{4.2}$$

where $g(x)$ is continuous near α and $g(\alpha) \neq 0$. If $m = 1$, the root is said to be *simple* and otherwise, *multiple*. The basic definition permits m to be a fraction. For example, with the function

$$f(x) = x\sqrt{x-1},$$

equation (4.1) has $\alpha = 1$ as a root of multiplicity 1/2 (and $\alpha = 0$ as a simple root). However, if $f(x)$ is sufficiently smooth, then m must be a positive integer. Indeed, if $f(x)$ has its first m derivatives continuous on an interval that includes α and

$$\begin{cases} f(\alpha) = 0 \\ f'(\alpha) = f''(\alpha) = \cdots = f^{(m-1)}(\alpha) = 0 \\ f^{(m)}(\alpha) \neq 0, \end{cases} \tag{4.3}$$

then α is a root of multiplicity m. This is seen by expanding $f(x)$ in a Taylor series about α to obtain

$$f(x) = f(\alpha) + (x-\alpha)f'(\alpha) + \frac{(x-\alpha)^2}{2}f''(\alpha) + \cdots + \frac{(x-\alpha)^{m-1}}{(m-1)!}f^{(m-1)}(\alpha) + \frac{(x-\alpha)^m}{m!}f^{(m)}(\xi_x),$$

where ξ_x lies between x and α. Using (4.3), this simplifies to

$$f(x) = \frac{(x-\alpha)^m}{m!} f^{(m)}(\xi_x). \tag{4.4}$$

If we take $g(x) = f^{(m)}(\xi_x)/m!$, then $g(\alpha) = f^{(m)}(\alpha)/m! \neq 0$. We shall always assume that $f(x)$ is sufficiently smooth near α that we can use (4.4) instead of the basic definition (4.2) and in particular, that roots are of integer multiplicity.

According to the definition of a root α, the graph of $f(x)$ touches the x axis at α (Figure 4.1). For a root of multiplicity m, the function $f^{(m)}(x)$ does not change sign near α because it is continuous and $f^{(m)}(\alpha) \neq 0$. This observation and the expression (4.4) show that if m is even, $f(x)$ is tangent to the x axis at α but does not cross there and that if m is odd, $f(x)$ crosses the axis at α.

A family of problems taken up in Case Study 4 has the form $0 = f(x) = F(x) - \gamma$ for a parameter $\gamma > 0$. Specifically, $F(x) = x\exp(-x)$ and a representative value of γ is 0.07. Curve sketching as in an introductory calculus course is often used to locate roots and determine their multiplicity. For this family, as $x \to -\infty$, $f(x) \to -\infty$, and as $x \to +\infty$, $f(x) \to -\gamma$. It is seen from the first derivative, $f'(x) = (1-x)\exp(-x)$, that f is strictly increasing for $x < 1$ and strictly decreasing for $x > 1$. At the extremum, $f(1) = e^{-1} - \gamma$ is positive for $\gamma = 0.07$. Also, $f(0) = -\gamma$ is negative. These facts and the continuity of f tell us that when $\gamma = 0.07$, there are exactly two roots that are simple. One is in $(0, 1)$ and the other is greater than 1. In general, for a root of $f(x)$ to be multiple, $f'(x)$ must vanish at the root. So, wherever the function is strictly increasing or strictly decreasing, any root it might have must be simple. For the family of functions, the fact that $f'(x) = 0$ only at $x = 1$ means that this is the only point where the function might have a multiple root. It is easily seen that there is a multiple root only when $\gamma = e^{-1}$ and the root then is of multiplicity 2 (a double root).

An approximate root z that results in a *computed* value $f(z) = 0$ is not unusual, especially when it approximates a multiple root α. After all, the aim is to find a z that makes $f(z)$ vanish. When the root is of multiplicity m, $f(z) \approx (z-\alpha)^m g(\alpha)$. Some numbers will help us to understand this. For a root of high multiplicity like $m = 10$, an approximation of modest accuracy like $z = \alpha + 10^{-4}$ leads to $f(z) = 10^{-40} g(\alpha)$. Then if $|g(\alpha)| \leq 1$, the function $f(z)$ underflows in IEEE single precision arithmetic.

As we shall see, standard methods are not as effective for multiple roots as they are for simple roots. To understand the performance of codes based on these methods, it is necessary to appreciate that roots that are close together "look" like multiple roots. Suppose that $f(x)$ has the two simple roots $\alpha_1 \neq \alpha_2$. The basic definition and a little argument show that $f(x) = (x-\alpha_1)(x-\alpha_2)G(x)$ for a $G(x)$ that does not vanish at either root. This expression can be rewritten as

$$f(x) = (x-\alpha_1)[(x-\alpha_1) - (\alpha_2 - \alpha_1)]G(x)\,.$$

For x far from the roots in the sense that $|x-\alpha_1| \gg |\alpha_2 - \alpha_1|$, the pair of simple roots "looks" like a double root because

$$f(x) \approx (x-\alpha_1)^2 G(x).$$

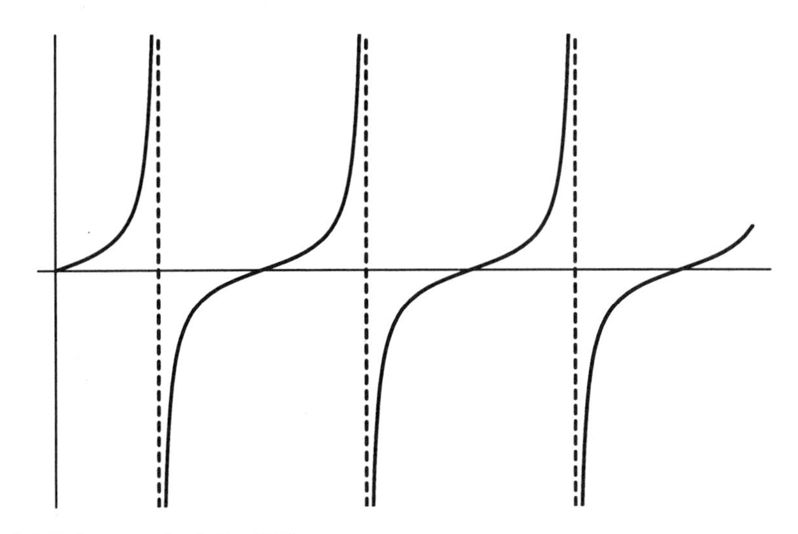

Figure 4.1 $f(x) = \tan x$ for $0 \leq x \leq 10$.

A concept from the theory of complex variables related to that of a root of multiplicity m is a *pole* of multiplicity m. If we can write

$$F(x) = (x - \alpha)^{-m} G(x),$$

where $G(\alpha) \neq 0$, then we say that α is a pole of $F(x)$ of multiplicity m. It is easy to see that if α is a root of $f(x)$ of multiplicity m, then it is a pole of $F(x) = 1/f(x)$ of the same multiplicity, and vice versa. A familiar example is $\tan(x) = \sin(x)/\cos(x)$, plotted in Figure 4.1. This function has a root where $\sin(x)$ vanishes and a pole where $\cos(x)$ vanishes. Functions change sign at poles of odd multiplicity, just as they do at roots of odd multiplicity.

One difficulty in computing a root of $f(x) = 0$ is deciding when an approximation z is good enough. The *residual* $f(z)$ seems an obvious way to assess the quality of an approximate root. MATHCAD does exactly this. It accepts z as a root when $|f(z)| <$ TOL, with TOL $= 10^{-3}$ by default. The trouble with a residual test is that there is no obvious measure of scale. Multiple roots present difficulties because the function is nearly flat in a considerable interval about the root. The issue is related to the conditioning of a root, but also to the way we set up the equation.

When we formulate a problem, we select a scaling. This may be no more than choosing a system of units, but often we use the fact that any zero of $f(x)$ is a zero of $F(x) = g(x)f(x)$. Introducing a scaling $g(x)$ can make quite a difference. For instance, the two problems $\sin(x) = 0$ and $F(x) = 10^{-38}\sin(x) = 0$ are mathematically equivalent, but the second is badly scaled because forming $F(z)$ for even a moderately good approximate root z will result in underflow in single precision IEEE arithmetic. Often we scale problems without giving any special thought to the matter, but a good scale can be quite helpful. It is quite a useful device for dealing with real or apparent singularities. The function $f(x) = \sin(x)/x$ is perfectly well behaved at $x = 0$ (it is analytic), but it has an apparent singularity there and some care is needed in its evaluation. This

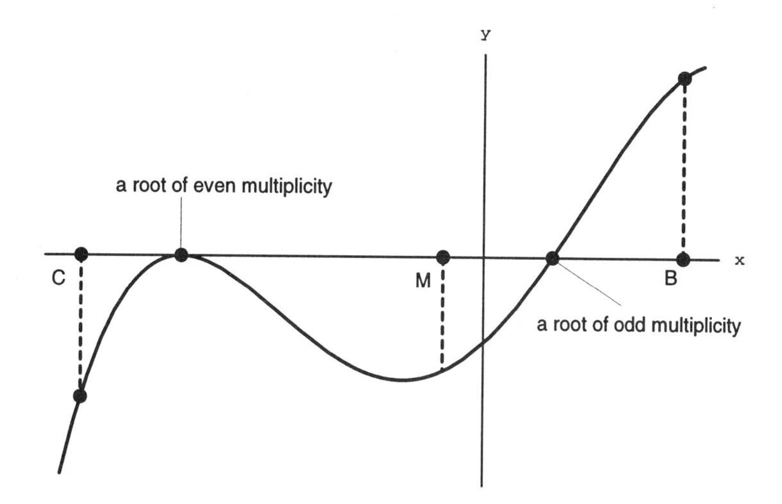

Figure 4.2 Graphical interpretation of roots.

can be circumvented by calculating the roots of the scaled function $F(x) = xf(x)$. It must be kept in mind that as with this example, $F(x)$ has all the roots of $f(x)$, but it might pick up additional roots from $g(x)$. A more substantial example is furnished by an equation to be solved in an exercise:

$$f(x) = \frac{1}{180} - \left(\frac{1-\cos(\pi/10)}{\cos(\pi/10)-\cos(x)}\right)\frac{\sin(x)}{x}.$$

This function has a simple pole at all the points where $\cos(x) = \cos(\pi/10)$ and an apparent singularity at $x = 0$. Scaling this function with $g(x) = x(\cos(\pi/10) - \cos(x))$ makes computing the roots more straightforward.

Sometimes a natural measure of scale is supplied by a coefficient in the equation. An example is provided by the family of problems $f(x) = F(x) - \gamma$ with $\gamma > 0$. Just as when solving linear equations, the residual $r = f(z) = F(z) - \gamma$ can be used in a backward error analysis. Obviously z is the exact solution of the problem $0 = F(x) - \gamma'$, where $\gamma' = \gamma + r$. If $|r|$ is small compared to $|\gamma|$, then z is the exact solution of a problem close to the given problem. For such problems we have a reasonable way to specify how small the residual ought to be.

4.1 BISECTION, NEWTON'S METHOD, AND THE SECANT RULE

If a continuous function $f(x)$ has opposite signs at points $x = B$ and $x = C$, then it has at least one zero in the interval with endpoints B and C. The method of bisection (or binary search) is based on this fact. If $f(B)f(C) < 0$, the function $f(x)$ is evaluated at the midpoint $M = (B+C)/2$ of the interval. If $f(M) = 0$, a zero has been found. Otherwise, either $f(B)f(M) < 0$ or $f(M)f(C) < 0$. In the first case there is at least one zero between M and B, as in Figure 4.2, and in the second case there is at least one zero between C and M. In this way an interval containing a root is found that has half the length of the original interval. The procedure is repeated until a root is located to whatever accuracy is desired.

In algorithmic form we have the

---

bisection method:

until $|B-C|$ is sufficiently small or $f(M)=0$ begin
  $M := (B+C)/2$
  if $f(B)f(M)<0$ then
    $C := M$
  else
    $B := M$
end until.

Example 4.1. When $f(x)=x^2-2$, the equation (4.1) has the simple root $\alpha=\sqrt{2}$. For $B=0$, $C=6$, the bisection method produces [note: 0.16 (+01) means 0.16×10^1]

B	C	$\|\alpha - M\|$
0.0	6.0	0.16(+01)
0.0	3.0	0.86(−01)
0.0	1.5	0.66(+00)
0.75	1.5	0.29(+00)
1.125	1.5	0.10(+00)
1.3125	1.5	0.80(−02)
1.40625	1.5	0.39(−01)
1.40625	1.453125	0.15(−01)

Note the erratic behavior of the error, although the interval width $|B-C|$ is halved at each step. ■

Bisection is often presented in programming books in this manner because it is a numerical algorithm that is both simple and useful. A more penetrating study of the method will make some points important to understanding many methods for computing zeros, points that we require as we develop an algorithm that attempts to get the best from several methods.

An interval $[B,C]$ with $f(B)f(C)\le 0$ is called a *bracket*. A graphical interpretation tells us somewhat more than just that $f(x)$ has a root in the interval. Zeros of even multiplicity between B and C do not cause a sign change and zeros of odd multiplicity do. If there were an even number of zeros of odd multiplicity between B and C, the sign changes would cancel out and f would have the same sign at both ends. Thus, if $f(B)f(C)<0$, there must be an odd number of zeros of odd multiplicity and possibly some zeros of even multiplicity between B and C. If we agree to count the number of zeros according to their multiplicity (i.e., a zero of multiplicity m counts as m zeros), then we see that there are an odd number of zeros between B and C.

A careful implementation of bisection takes into account a number of matters raised in Chapter 1. There is a test for values of f that are exactly zero; the test for a change of sign is not programmed as a test of $f(B)f(C) < 0$ because of the potential for underflow of the product; and the midpoint is computed as $M = B + (B - C)/2$ because it is just as easy to compute and more accurate than $M = (B + C)/2$.

We often try to find an approximate root z for which $f(z)$ is as small as possible. In attempting this, the finite word length of the computer must be taken into account and so must the details of the procedure for evaluating f. Eventually even the sign of the computed value may be incorrect. This is what is meant by *limiting precision.* Figure 1.2 shows the erratic size and sign of function values when the values are so small that the discrete nature of the floating point number system becomes important. If a computed function value has the wrong sign because the argument is very close to a root, it may happen that the bracket selected in bisection does not contain a root. Even so, the approximations computed thereafter will stay in the neighborhood of the root. It is usually said that a bisection code will produce an interval $[B, C]$ of specified length that contains a root because $f(B)f(C) < 0$. This is superficial. It should be qualified by saying that either this is true, or a root has been found that is as accurate as the precision allows. The qualification "as accurate as the precision allows" means here that either the computed $f(z)$ vanishes, or that one of the computed values $f(B)$, $f(C)$ has the wrong sign.

A basic assumption of the bisection method is that $f(x)$ is continuous. It should be no surprise that the method can fail when this is not the case. Because a bisection code pays no attention to the values of the function, it cannot tell the difference between a pole of odd multiplicity and a root of odd multiplicity [unless it attempts to evaluate $f(x)$ exactly at a pole and there is an overflow]. So, for example, if a bisection code is given the function $\tan(x)$ and asked to find the root in $[5, 7]$, it will have no difficulty. If asked to find the root in $[4, 7]$, it will not realize there is a root in the interval because the sign change due to the simple pole cancels out the sign change due to the simple root. And, what is worse, if asked to find a root in $[4, 5]$, it will locate a pole or cause an overflow. We see here another reason for scaling: removing odd order poles by scaling removes the sign changes that might cause bisection to locate a pole rather than a zero. Here this is done by $F(x) = \cos(x)\tan(x) = \sin(x)$. One of the examples of scaling given earlier is a less trivial illustration of the point. Because of the very real possibility of computing a pole of odd multiplicity, it is prudent when using a bisection code to inspect the residual $f(z)$ of an alleged root z—it would be highly embarrassing to claim that z results in a very *small* value of $f(z)$ when it actually results in a very *large* value!

A bisection code can converge to a pole because it makes no use of the value $f(M)$, just its sign. Because of this its rate of convergence is the same whether the root is simple or not and whether the function is smooth or not. Other methods converge much faster when the root is simple and the function is smooth, but they do not work so well when this is not the case.

Bisection has a number of virtues. Provided an initial bracket can be found, it will converge no matter how large the initial interval known to contain a root. It is easy to decide reliably when the approximation is good enough. It converges reasonably fast and the rate of convergence is independent of the multiplicity of the root and the smoothness of the function. The method deals well with limiting precision.

Bisection also has some drawbacks. If there are an even number of zeros between B and C, it will not realize that there are any zeros at all because there is no sign change. In particular, it is not possible to find a zero of even multiplicity except by accident. It can be fooled by poles. A major disadvantage is that for simple zeros, which seem to be the most common by far, there are methods that converge much more rapidly. There is no way to be confident of calculating a particular root nor of getting all the roots. This is troublesome with all the methods, but some are (much) better at computing the root closest to a guessed value. Bisection does not generalize to functions of a complex variable nor easily to functions of several variables.

Let us now take up two methods that are superior to bisection in some, although not all, of these respects. Both approximate $f(x)$ by a straight line $L(x)$ and then approximate a root of $f(x) = 0$ by a root of $L(x) = 0$.

Newton's method (Figure 4.3) will be familiar from calculus. It takes $L(x)$ as the line tangent to $f(x)$ at the latest approximation x_i and the next approximation (iterate) is the root x_{i+1} of $L(x) = 0$. Equivalently, approximating $f(x)$ by the linear terms of a Taylor's series about x_i,

$$f(x) \approx f(x_i) + (x - x_i) f'(x_i),$$

suggests solving

$$f(x_i) + (x - x_i) f'(x_i) = 0$$

for its root x_{i+1} to approximate α [assuming that $f'(x_i) \neq 0$]. The resulting method is known as

---

Newton's method:

$$x_{i+1} = x_i - \frac{f(x_i)}{f'(x_i)}. \tag{4.5}$$

When it is inconvenient or expensive to evaluate $f'(x)$, a related procedure called the secant rule is preferred because it uses only values of $f(x)$. Let $L(x)$ be the secant line that interpolates $f(x)$ at the two approximations x_{i-1}, x_i:

$$L(x) = \frac{x - x_i}{x_{i-1} - x_i} f_{i-1} + \frac{x - x_{i-1}}{x_i - x_{i-1}} f_i.$$

The next approximation x_{i+1} is taken to be the root of $L(x) = 0$. Hence, assuming that $f(x_i) \neq f(x_{i-1})$, we have the

---

secant rule:

$$x_{i+1} = x_i - f(x_i) \frac{x_i - x_{i-1}}{f(x_i) - f(x_{i-1})}. \tag{4.6}$$

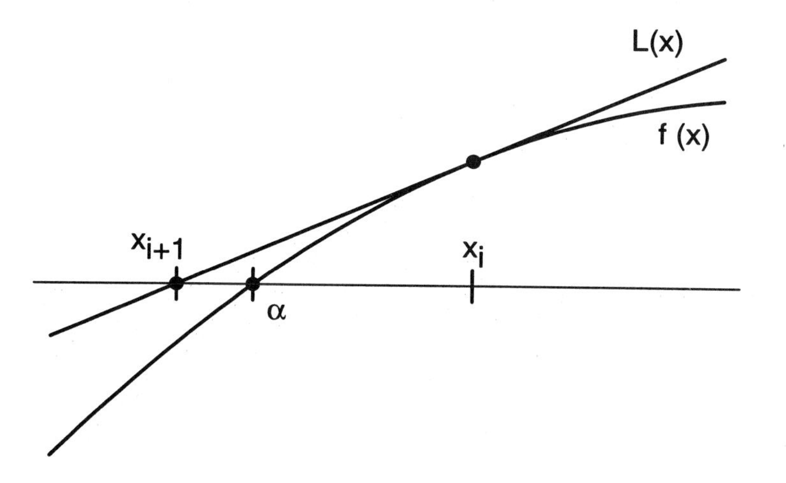

Figure 4.3 Newton's method.

The method is illustrated graphically in Figure 4.4. Although a picture furnishes a natural motivation for the method, an alternative approach is to approximate the derivative in Newton's method (4.5) by a difference quotient to get (4.6).

A little analysis shows that Newton's method and the secant rule converge much faster than bisection for a simple root of (4.1). Considering first Newton's method, we have from (4.5)

$$\alpha - x_{i+1} = \alpha - x_i + \frac{f(x_i)}{f'(x_i)}.$$

If x_i is near α, then

$$f(x_i) \approx f(\alpha) + (x_i - \alpha) f'(\alpha) + \frac{(x_i - \alpha)^2}{2} f''(\alpha)$$
$$f'(x_i) \approx f'(\alpha) + (x_i - \alpha) f''(\alpha).$$

Now $f(\alpha) = 0$ and $f'(\alpha) \neq 0$ for a simple root, so

$$\alpha - x_{i+1} \approx \alpha - x_i + \frac{(x_i - \alpha) f'(\alpha) + \frac{(x_i - \alpha)^2}{2} f''(\alpha)}{f'(\alpha) + (x_i - \alpha) f''(\alpha)}$$
$$\approx -(x_i - \alpha)^2 \frac{f''(\alpha)}{2f'(\alpha)}.$$

It is seen that if x_i is near a simple root, the error in x_{i+1} is roughly a constant multiple of the square of the error in x_i. This is called *quadratic* convergence.

A similar look at the secant rule (4.6) leads to

$$\alpha - x_{i+1} \approx -(x_i - \alpha)(x_{i-1} - \alpha) \frac{f''(\alpha)}{2f'(\alpha)}. \tag{4.7}$$

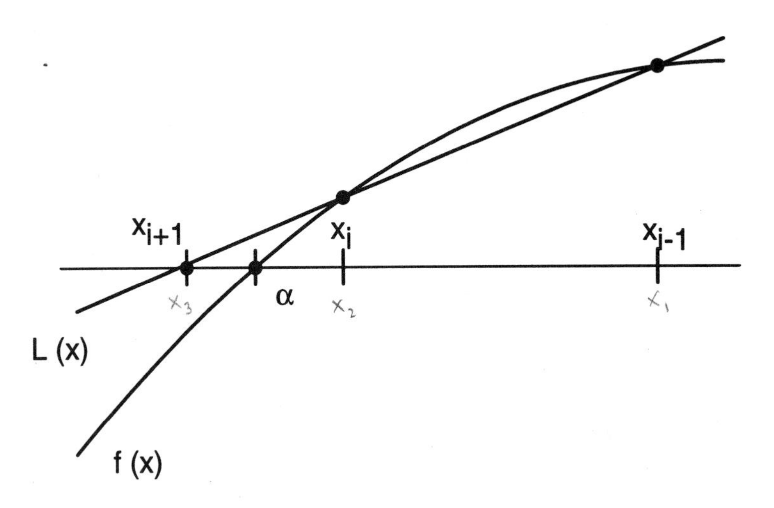

Figure 4.4 Secant rule.

This method does not converge as fast as Newton's method, but it is much faster than bisection. For both methods it can be shown that if the starting values are sufficiently close to a simple root and $f(x)$ is sufficiently smooth, the iterates will converge to that root. For Newton's method,

$$\lim_{x_i \to \alpha} \frac{(x_{i+1} - \alpha)}{(x_i - \alpha)^2} = C \neq 0$$

and for the secant rule,

$$\lim_{x_i \to \alpha} \frac{(x_{i+1} - \alpha)}{(x_i - \alpha)(x_{i-1} - \alpha)} = c \neq 0.$$

A careful treatment of the secant rule even shows that

$$\lim_{x_i \to \alpha} \frac{|x_{i+1} - \alpha|}{|x_i - \alpha|^p} = \gamma \neq 0,$$

where $p = (1 + \sqrt{5})/2 \approx 1.618$.

Example 4.2. As in Example 4.1, let $f(x) = x^2 - 2$. An easy calculation shows that for the secant rule started with $x_1 = 3$ and $x_2 = 2$,

i	x_i	$\lvert\alpha - x_i\rvert$
1	3.0000000000000000	0.16(+01)
2	2.0000000000000000	0.59(+00)
3	1.5999999999999999	0.19(+00)
4	1.4444444444444444	0.30(−01)
5	1.4160583941605840	0.18(−02)
6	1.4142330592571590	0.19(−04)
7	1.4142135750814935	0.13(−07)
8	1.4142135623731826	0.87(−13)

and for Newton's method started with $x_1 = 3$,

i	x_i	$\lvert\alpha - x_i\rvert$
1	3.0000000000000000	0.16(+01)
2	1.8333333333333335	0.42(+00)
3	1.4621212121212122	0.48(−01)
4	1.4149984298948028	0.78(−03)
5	1.4142137800471977	0.22(−06)
6	1.4142135623731118	0.17(−13)

Both methods converge quite rapidly and the quadratic convergence of Newton's method is apparent. Comparison with the bisection method of Example 4.1 shows the superiority of the secant rule and Newton's method (for this problem). ■

If an iteration is such that

$$\lim_{x_i \to \alpha} \frac{|x_{i+1} - \alpha|}{|x_i - \alpha|^r} = \gamma \neq 0,$$

the method is said to converge at rate r with constant γ. It has been argued that for a simple root, Newton's method converges at the rate $r = 2$ and it has been stated that the secant rule converges at the rate $r = p \approx 1.618$. Bisection does not fit into this framework; the width of the bracketing intervals are being halved at every step, but nothing can be said about

$$\lim_{x_i \to \alpha} \frac{|x_{i+1} - \alpha|}{|x_i - \alpha|}$$

(see Example 4.1).

The secant rule is a principal part of the code Zero developed in this chapter, so we now state conditions that guarantee its convergence to a simple root of (4.1) and study how fast it converges. As a first step we derive an expression that relates the function values at three successive iterates x_{i-1}, x_i, x_{i+1}. Let $L(x)$ be the polynomial of degree 1 interpolating $f(x)$ on the set $\{x_i, x_{i-1}\}$. The iterate x_{i+1} is the zero of $L(x)$. In Chapter 3 we developed an expression for the error of interpolation [see (3.4)], which in this case is

$$f(x_{i+1}) - L(x_{i+1}) = (x_{i+1} - x_{i-1})(x_{i+1} - x_i)\frac{f''(\xi)}{2}$$

or, since $L(x_{i+1}) = 0$,

$$f(x_{i+1}) = (x_{i+1} - x_{i-1})(x_{i+1} - x_i)\frac{f''(\xi)}{2} \tag{4.8}$$

for a suitable (unknown) point ξ. Some manipulation of equation (4.6) gives the two relations

$$x_{i+1} - x_i = -\frac{(x_i - x_{i-1})f(x_i)}{f(x_i) - f(x_{i-1})} \tag{4.9}$$

$$x_{i+1} - x_{i-1} = -\frac{(x_i - x_{i-1})f(x_{i-1})}{f(x_i) - f(x_{i-1})}. \tag{4.10}$$

A third relation is obtained from the mean value theorem for derivatives:

$$\frac{f(x_i) - f(x_{i-1})}{x_i - x_{i-1}} = f'(\zeta), \tag{4.11}$$

where ζ, a point between x_i and x_{i-1}, is unknown. Combining equations (4.8)–(4.11), we arrive at

$$f(x_{i+1}) = f(x_i)f(x_{i-1})\frac{f''(\xi)}{2[f'(\zeta)]^2}.$$

Let us assume that on an appropriate interval we have

$$|f''(x)| < M_2, 0 < m_1 \le |f'(x)| \le m_2 \tag{4.12}$$

and that we are computing a simple zero α. (Why *must* it be simple with these hypotheses?) Then these bounds and the expression above for $f(x_{i+1})$ imply that

$$|f(x_{i+1})| \le |f(x_i)||f(x_{i-1})|\frac{M_2}{2m_1^2}.$$

If we let

$$\varepsilon_i = |f(x_i)|\frac{M_2}{2m_1^2},$$

this inequality leads to

$$\varepsilon_{i+1} \le \varepsilon_i \varepsilon_{i-1}.$$

Supposing that

$$\varepsilon = \max(\varepsilon_0, \varepsilon_1) < 1,$$

it is easy to argue by induction that

$$\begin{aligned}
\varepsilon_2 &\leq \varepsilon^2 \\
\varepsilon_3 &\leq \varepsilon^2 \cdot \varepsilon = \varepsilon^3 \\
\varepsilon_4 &\leq \varepsilon^3 \cdot \varepsilon^2 = \varepsilon^5 \\
&\vdots \\
\varepsilon_i &\leq \varepsilon^{\delta_i},
\end{aligned}$$

where

$$\delta_i = \frac{1}{\sqrt{5}}\left[\left(\frac{1+\sqrt{5}}{2}\right)^{i+1} - \left(\frac{1-\sqrt{5}}{2}\right)^{i+1}\right].$$

The formal proof is left as an exercise. Since

$$\left|\frac{1+\sqrt{5}}{2}\right| > 1 > \left|\frac{1-\sqrt{5}}{2}\right|,$$

we see that for i large,

$$\delta_i \approx \frac{1}{\sqrt{5}}\left(\frac{1+\sqrt{5}}{2}\right)^{i+1}.$$

In any event, $\delta_i \to \infty$ as $i \to \infty$, and since $0 \leq \varepsilon < 1$, we must have $\varepsilon_i \to 0$, which is what we wanted to prove. Let us now state a formal theorem and complete the details of its proof.

Theorem 4.1. *The secant rule defined by (4.6) with initial guesses x_0, x_1 converges to a simple zero α of $f(x)$ if x_0, x_1 lie in a sufficiently small closed interval containing α on which $f'(x)$, $f''(x)$ exist and are continuous and $f'(x)$ does not vanish.*

Proof. Without loss of generality we assume that M_2 defined by (4.10) is positive. Otherwise $f''(x) \equiv 0$ near α, implying that $f(x)$ is a linear function and the secant rule converges in one step. With the assumptions on f' and f'', the bounds m_1, m_2, and M_2 are well defined. Using the mean value theorem for derivatives, we see that

$$\begin{aligned}
|f(x_0)| &= |f(\alpha) + (x_0 - \alpha)f'(\zeta_1)| \leq |x_0 - \alpha| m_2 \\
|f(x_1)| &= |f(\alpha) + (x_1 - \alpha)f'(\zeta_2)| \leq |x_1 - \alpha| m_2.
\end{aligned}$$

This implies that the quantity ε defined above is less than 1 if x_0 and x_1 are sufficiently close to α. The argument above shows that

$$|f(x_i)| \frac{M_2}{2m_1^2} \leq \varepsilon^{\delta_i}.$$

But

$$\begin{aligned}
|f(x_i)| &= |f(x_i) - f(\alpha)| = |(x_i - \alpha)f'(\eta)| \\
&\geq |x_i - \alpha| m_1.
\end{aligned}$$

Hence,

$$|x_i - \alpha| \leq \frac{2m_1}{M_2}\varepsilon^{\delta_i}.$$

This says that $x_i \to \alpha$. The argument suggests that the rate of convergence is the golden mean $(1+\sqrt{5})/2$ stated earlier. ■

Methods that converge at a rate $r > 1$ are said to be superlinearly convergent. We have seen that this is the case for Newton's method and the secant rule when computing a simple root. Unfortunately it is not the case when computing a multiple root. It is easy enough to see this for Newton's method. If x_i is near a root α of multiplicity $m > 1$, then

$$f(x) \approx \frac{(x-\alpha)^m}{m!} f^{(m)}(\alpha)$$
$$f'(x) \approx \frac{(x-\alpha)^{m-1}}{(m-1)!} f^{(m)}(\alpha).$$

This implies that

$$x_{i+1} - \alpha = x_i - \alpha - \frac{f(x_i)}{f'(x_i)} \approx \left(\frac{m-1}{m}\right)(x_i - \alpha).$$

This expression shows that for a root of multiplicity m, Newton's method is only linearly convergent with constant $(m-1)/m$.

An example from Wilkinson [11] illustrates several difficulties that can arise in the practical application of Newton's method.

Example 4.3. Consider the problem

$$x^{20} - 1 = 0.$$

In attempting to compute the simple root $\alpha = 1$ using Newton's method, suppose we start with $x_0 = 1/2$. Then from (4.5)

$$x_1 = \frac{1}{2} - \frac{\left(\frac{1}{2}\right)^{20} - 1}{20\left(\frac{1}{2}\right)^{19}} = 26214.875$$

because the tangent is nearly horizontal. Thus, a reasonably good guess for a root leads to a *much* worse approximation. Also, notice that if x_i is much larger than 1, then

$$x_{i+1} = x_i - \frac{x_i^{20} - 1}{20x_i^{19}} \approx x_i - \frac{x_i^{20}}{20x_i^{19}} = \frac{19}{20}x_i.$$

To the same degree of approximation,

$$\frac{x_{i+1} - 1}{x_i - 1} \approx \frac{x_{i+1}}{x_i} \approx \frac{19}{20},$$

which says that we creep back to the root at 1 at a rate considerably *slower* than bisection. What is happening here is that the roots of this equation are the roots of unity. The 20 simple roots lie on a circle of radius 1 in the complex plane. The roots are well separated, but when "seen" from as far away as 26000, they appear to form a root of multiplicity 20, as argued earlier in this chapter. Newton's method converges linearly with constant $19/20$ to a root of multiplicity 20, and that is exactly what is observed when the iterates are far from the roots.

Much is made of the quadratic convergence of Newton's method, but it is quadratically convergent only for simple roots. Even for simple roots, this example shows that quadratic convergence is observed only when "sufficiently" close to a root. And, of course, when "too" close to a root, finite precision arithmetic affects the rate of convergence. ■

Let us now consider the behavior of Newton's method and the secant rule at limiting precision. Figure 1.2 shows an interval of machine-representable numbers about α on which computed values of the function vary erratically in sign and magnitude. These represent the smallest values the computed $f(x)$ can assume when formed in the working precision, and quite frequently they have no digits in agreement with the true $f(x)$. For a simple root, $|f'(\alpha)|$ is not zero, and if the root is not ill-conditioned, the derivative is not small. As a consequence, the computed value of the first derivative ordinarily has a few digits that are correct. It then follows that the correction to x_i computed by Newton's method is very small at limiting precision and the next iterate stays near the root even if it moves away because $f(x_i)$ has the wrong sign. This is like bisection and is what is meant by the term "stable at limiting precision." The secant rule behaves differently. The correction to the current iterate,

$$\frac{x_i - x_{i-1}}{f(x_i) - f(x_{i-1})},$$

has unpredictable values at limiting precision. Clearly it is possible that the next iterate lie far outside the interval of limiting precision.

There is another way to look at the secant rule that is illuminating. One approach to finding a root α of $f(x)$ is to interpolate several values $y_i = f(x_i)$ by a polynomial $P(x)$ and then approximate α by a root of this polynomial. The secant rule is the case of linear interpolation. Higher order interpolation provides a more accurate approximation to $f(x)$, so it is plausible that it would lead to a scheme with a higher rate of convergence. This turns out to be true, although only the increase from linear to quadratic interpolation might be thought worth the trouble. The scheme based on quadratic interpolation is called Muller's method. Muller's method is a little more trouble than the secant rule, because it involves computing the roots of a quadratic, and it converges somewhat faster. There are some practical differences. For all methods based on interpolation by a polynomial of degree higher than 1, there is a question of which root to take as the next iterate x_{i+1}. To get convergence, the root closest to x_i should be used. An important difference between Muller's method and the secant rule is due to the possibility of a real quadratic polynomial having complex roots. Even with real iterates and a real function, Muller's method might produce a complex iterate. If complex roots are interesting, this is a virtue, but if only real roots are desired, it is a defect. The secant rule can be used to compute complex roots, but it will not leave the real line spontaneously like Muller's method. The MATHCAD documentation points out that its code, which is based on the secant rule, can be used to compute the roots of a complex-valued function by starting with a guess that is complex.

An alternative to direct interpolation is *inverse interpolation*. This approach is based on interpolating the inverse function $f^{-1}(y)$ of $y = f(x)$. To prevent confusion with the reciprocal of $f(x)$, the inverse function will be denoted here by $G(y)$. We assume that we have at our disposal only a procedure for evaluating $f(x)$. However, each value $f(x_i) = y_i$ provides a value of the inverse function because by definition $x_i = G(y_i)$. Finding a root of $f(x)$ corresponds to *evaluating* the inverse function: a root α satisfies $f(\alpha) = 0$, hence $\alpha = G(0)$. This is a familiar task that we solve in a familiar way. We are able to evaluate a function $G(y)$ at certain points y_i and we wish to approximate the value $G(0)$. This is done by approximating $G(y)$ with a polynomial interpolant $P(y)$ and then evaluating $P(0) \approx \alpha$. Of course, it is easy to interpolate $G(y)$ by whatever degree polynomial we want. However, as with direct interpolation, most of the improvement to the rate of convergence is gained on going to quadratic interpolation. An interesting fact left to an exercise is that the method derived from linear inverse interpolation is the same as that derived from linear direct interpolation, namely the secant rule. Examination of Figure 4.4 helps in understanding this. On the other hand, quadratic direct and inverse interpolation are quite different. For one thing, quadratic inverse interpolation cannot produce a complex iterate when the function and the previous iterates are real.

Inverse interpolation is attractive because of its simplicity. Unfortunately, there is a fundamental difficulty—f might not have an inverse on the interval of interest. This is familiar from the trigonometric functions. For instance, the function $y = \sin(x)$ does not have an inverse for all x. To invert the relationship with $x = \arcsin(y)$, the argument x is restricted to an interval on which $\sin(x)$ is monotone. In a plot like Figure 4.2, the inverse of f is found by "turning the picture on its side." Only on an interval where $f(x)$ is monotone does the inverse function exist as a single-valued function. At a simple root α, $f'(\alpha) \neq 0$, so there is an interval containing α on which $f'(x) \neq 0$ [$f(x)$ is monotone] and $G(y)$ exists. So, the usual kind of result is obtained. If we can start close enough to a simple root, there is an inverse function and we can compute the root with inverse interpolation. When some distance from a root or when the root is multiple, there may be serious difficulties with inverse interpolation because then the function does not have an inverse on the relevant interval.

With the exception of bisection, the methods we have studied are guaranteed to converge only when sufficiently close to a zero of a function that is sufficiently smooth. This is rather unsatisfactory when we have no idea about where the roots are. On the other hand, often we are interested in the root closest to a specific value. It is by no means certain that the methods will converge from this value to the nearest root since that depends on just how close the value is to the root, but it is a useful characteristic of the methods. In contrast, if the initial bracket given a bisection code contains several roots, the code might locate any one of them. The technique of *continuation* is useful when it is hard to find a starting guess good enough to get convergence to a particular root, or to any root. Many problems depend on a parameter λ and it may be that zeros can be computed easily for some values of the parameter. The family $x\exp(-x) - \gamma = 0$ is an example. Solutions are desired for values $\gamma > 0$, but it is obvious that $\alpha = 0$ is a root when $\gamma = 0$. It is generally the case, although not always, that the roots

$\alpha(\lambda)$ depend continuously on λ. The idea of continuation is to solve a sequence of problems for values of λ ranging from one for which the problem is solved easily to the desired value of the parameter. This may not be just an artifice; you may actually want solutions for a range of parameter values. Roots obtained with a value λ' are used as guesses for the next value λ''. If the next value is not *too* different, the guesses will be good enough to obtain convergence. In the case of the example, the smallest positive root is desired, so starting with $\alpha(0) = 0$ should result in the desired root $\alpha(\gamma)$. When there is no obvious parameter in $f(x) = 0$, one can be introduced artificially. A common embedding of the problem in a family is $0 = F(x,\lambda) = f(x) + (\lambda - 1)f(x_0)$. By construction, x_0 is a root of this equation for $\lambda = 0$ and the original equation is obtained for $\lambda = 1$. Another embedding is $0 = F(x,\lambda) = \lambda f(x) + (1-\lambda)(x - x_0)$, which is also to be started with the root x_0 for $\lambda = 0$ and a sequence of problems solved for λ increasing to 1.

A virtue of bisection is that it is easy to decide when an approximate root is good enough. The convergence of Newton's method, the secant rule, quadratic inverse interpolation, and the like cannot be judged in this way. Many codes use the size of the residual for this purpose, but this is hazardous for reasons already studied. Superlinear convergence provides another way to decide convergence. When the iterate x_i is sufficiently close to a root α, superlinear convergence implies that the next iterate is much closer, $|\alpha - x_{i+1}| \ll |\alpha - x_i|$. Because of this, the error of x_i can be approximated by

$$\alpha - x_i = (x_{i+1} - x_i) + (\alpha - x_{i+1}) \approx x_{i+1} - x_i.$$

This is computationally convenient for if the estimated error is too large, x_{i+1} is available for another iteration. In the case of Newton's method, this estimate of the error is

$$x_{i+1} - x_i = -\frac{f(x_i)}{f'(x_i)}.$$

This estimate might be described as a natural scaling of the residual $f(x_i)$. If x_i passes a convergence test based on superlinear convergence, it is assumed that x_{i+1} is a rather better approximation to α, so why not use it as the answer? Reporting x_{i+1} as the answer is called *local extrapolation.*

EXERCISES

4.1 The residual of an alleged root r of $F(x) = 0$ is $F(r)$. One often sees the statement that a residual is "small," so the approximate root must be "good." Is this reliable? What role does scaling play?

4.2 How are simple and multiple roots distinguished graphically? Interpret graphically how well the roots are determined. Compare with Exercise 4.1.

4.3 Geometrically estimate the root of the function $F(x)$ whose graph is given below.

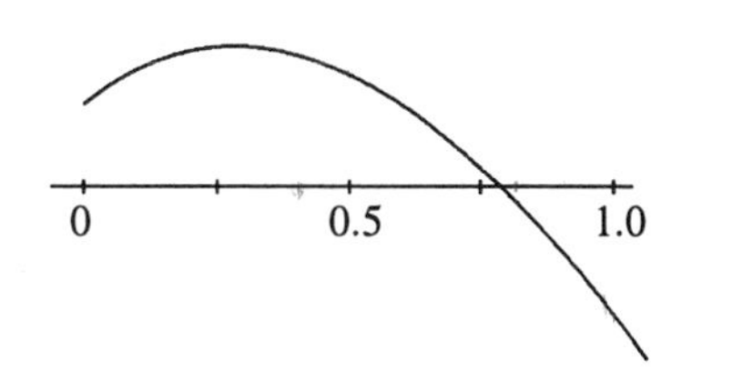

(a) For an initial bracket of [0.0, 1.0] what are the next three brackets using the bisection method on this function?

(b) If $x_1 = 0.0$ and $x_2 = 1.0$, mark on the graph the approximate location of x_3 using one step of the secant method.

(c) If $x_1 = 0.5$, mark on the graph the approximate location of x_2 and x_3 using two steps of Newton's method.

4.4 The polynomial $f(x) = x^3 - 2x - 5$ has a root α in $[2,3]$.

(a) Show that $[2,3]$ is a bracket for $f(x)$.

(b) Apply four steps of the bisection method to reduce the width of the bracket to $1/16$.

(c) Calculate x_3 and x_4 by the secant method starting with $x_1 = 3$ and $x_2 = 2$.

(d) Calculate x_2, x_3, and x_4 using Newton's method with $x_1 = 2$.

4.5 To find where $\sin x = x/2$ for $x > 0$,

(a) find a bracket for an appropriate function $f(x)$.

(b) Apply four steps of the bisection method to reduce the width of the bracket by $1/16$.

(c) Calculate x_3 and x_4 by the secant method starting with x_1 and x_2 equal to the bracket values.

(d) Calculate x_2, x_3, and x_4 using Newton's method with x_1 the midpoint of your bracket.

4.6 Given x_0 and x_1, show that the inverse secant rule, discussed at the end of Section 4.1, and the direct secant rule (4.6) produce the same iterates. In particular, with $G(y)$ the inverse function of $f(x)$, and $P(y)$ the linear polynomial that interpolates $G(y)$ at $f(x_i)$ and $f(x_{i-1})$, show that $P(0) = x_{i+1}$ is given by (4.6).

4.7 In the convergence proof for the secant rule, it was stated that if $\varepsilon = \max(\varepsilon_0, \varepsilon_1) < 1$, then the inequality

$$\varepsilon_{i+1} \leq \varepsilon_i \cdot \varepsilon_{i-1}$$

implied

$$\varepsilon_i \leq \varepsilon^{\delta_i},$$

and

$$\delta_i = \frac{1}{\sqrt{5}}\left[\left(\frac{1+\sqrt{5}}{2}\right)^{i+1} - \left(\frac{1-\sqrt{5}}{2}\right)^{i+1}\right].$$

Establish this.

4.8 The special function

$$\operatorname{erf}(x) = \frac{2}{\sqrt{\pi}}\int_0^x e^{-t^2}\,dt$$

is important in statistics and in many areas of science and engineering. Because the integrand is positive for all t, the function is strictly increasing and so has an inverse $x = \operatorname{erf}^{-1}(y)$. The inverse error function is an important function in its own right and can be evaluated for given y by solving the equation $y = \operatorname{erf}(x)$. The algorithm of the MATLAB function `erfinv.m` first forms a rational approximation to y that is accurate to about six figures. Two Newton iterations are then done to get a result to full accuracy. What is Newton's method for solving this equation? Why would you expect two iterations to be enough? (Don't forget to consider the multiplicity of the root.)

4.2 AN ALGORITHM COMBINING BISECTION AND THE SECANT RULE

It is a challenging task to fuse several methods into an efficient computational scheme. This section is devoted to a code, Zero, based on one written by Dekker [6] that does this. Roughly speaking, the code uses the secant rule unless bisection appears advantageous. A very similar code is found in the NAG library. Brent [3] added the use of quadratic inverse interpolation to Dekker's ideas. Brent's code is the basis for codes in MATLAB and the IMSL library.

Normal input to Zero is a subprogram for the evaluation of a continuous function $f(x)$ and arguments B and C for which $f(B)f(C) < 0$. Throughout the computation B and C are end points of an interval with $f(B)f(C) < 0$ that is decreasing in length. In favorable circumstances B is computed with the secant rule and is a much better approximate root than either C or the midpoint $M = (C+B)/2$ of the interval. To deal with unfavorable circumstances, the code interchanges the values of B and C as necessary so that $|f(B)| \leq |f(C)|$ holds. If at any time the computed $f(B)$ is zero, the computation is terminated and B is reported as a root.

The convergence test is a mixed relative–absolute error test. Two parameters ABSERR and RELERR are input and it is asked of each iterate whether

$$\left|\frac{C-B}{2}\right| \leq \max[\text{ABSERR}, |B| \times \text{RELERR}]. \tag{4.13}$$

For reasons discussed in Chapter 1, the code will not permit RELERR to be smaller than 10 units of roundoff, nor ABSERR to be zero. However, to understand what the test means, first suppose that RELERR is zero. The test is then asking if an interval believed to contain a root has a length no more than $2 \times$ ABSERR. If so, the midpoint M is no farther than ABSERR from a root and this is a pure absolute error test on M as an approximate root. However, it is believed that the quantity B reported as the answer is closer to a root than M is. Even if it is not, the test implies that B is within $2 \times$ ABSERR of a root. Similarly, if the parameter ABSERR is zero and if the test were

$$\left|\frac{C-B}{2}\right| \leq |M| \times \text{RELERR},$$

the test would be a pure relative error test for the approximate root M. Because it is believed that B is a better approximate root, it is used in the test rather than M. The `fzero.m` function in MATLAB has a similar, but somewhat simpler, test. The codes in the NAG and IMSL libraries have convergence tests that are broadly similar, but they also test the size of the residual and convergence can occur either because the root has been located to a specified accuracy or because the magnitude of the residual is smaller than a specified value.

Unless there is a reason to do otherwise, Zero uses the secant rule. A variable A is initialized to C. The two variables A, B are the two iterates used by the secant rule to calculate

$$D = B - f(B)\frac{B-A}{f(B)-f(A)}.$$

A danger with the secant rule (and Newton's method) is an interpolant that is horizontal or nearly so. The extreme case is a division by zero in this formula. This danger is avoided by requiring D to lie in the interval $[B, C]$ known to contain a root and checking this without performing the division. Pursuing the tactic further, the code requires that D lie in $[B, M]$ on the grounds that B ought to be a better approximation to the root than C, so if the secant rule is working properly, D ought to be closer to B than to C. If D does not lie in $[B, M]$, the midpoint M is used as the next iterate.

The performance of the code can be improved in some circumstances by selecting an iterate in a different way. If D is *too* close to B, a better tactic is to move a minimum distance away from B. The quantity $\max[\text{ABSERR}, |B| \times \text{RELERR}]$ is called TOL in the code. If $|D - B| < \text{TOL}$, then the value $B + \text{TOL} \times \text{sign}(C - B)$ is used instead of D. This choice cannot result in an iterate outside the interval $[B, C]$ since $|B - C| > 2 \times \text{TOL}$ (or else the error test would have been passed). If the root α is further from

B than TOL, the iterate chosen in this way is closer to the root than D. If it is closer, this iterate and B will bracket the root and the code will converge at the next test on the error because the length of the bracket is TOL.

There are circumstances in which the current iterate B is converging to a root, but the end point C is fixed. Because convergence is judged by the length of the interval $[B,C]$ and because the rate of convergence of the secant rule depends on using values from the neighborhood of the root, the code monitors the length of the interval. If four iterations have not resulted in a reduction by a factor of 1/8, the code bisects three times. This guarantees that the code will reduce the length of an interval containing a root by a factor of 1/8 in a maximum of seven function evaluations.

In summary, if the value D of the secant rule lies outside $[B,M]$ or if the overall reduction in interval length has been unsatisfactory, the code bisects the interval. If D is *too* close to B, a minimum change of TOL is used. Otherwise D is used. After deciding how to compute the next iterate, it is formed explicitly and replaces B. If $f(B) = 0$, the code exits. Otherwise, quantities are updated for the next iteration: the old B replaces A. The old C is kept or is replaced by the old B, whichever results in $f(B)f(C) < 0$.

If the code is given normal input [$f(x)$ continuous, $f(B)f(C) \leq 0$], then on normal return, either the computed $f(B) = 0$, or the computed $f(B)$ and $f(C)$ satisfy $f(B)f(C) < 0$, $|f(B)| \leq |f(C)|$, and the output values of B and C satisfy (4.13). In the latter case there is either a root of $f(x)$ in the interval $[B,C]$ or else one of the end points is so close to a root that the sign has been computed incorrectly in the working precision.

EXERCISES

4.9 The algorithm described combining the bisection method with the secant method is very efficient. Suppose that the initial B and C satisfy $|B - C| = 10^{10}$, and a root is sought with an absolute error of at most 10^{-5}.

(a) How many function evaluations does the bisection method use?

(b) What is the maximum number of function evaluations needed by the combined algorithm?

4.3 ROUTINES FOR ZERO FINDING

The algorithm of the preceding section has been implemented in a routine called Zero designed to compute a root of the nonlinear equation $F(x) = 0$. A typical invocation of Zero in C++ is

```
flag = Zero(f, b, c, abserr, relerr, residual);
```

in FORTRAN it is

```
CALL ZERO(F,B,C,ABSERR,RELERR,RESIDL,FLAG)
```

and

flag = Zero(f, &b, &c, abserr, relerr, &residual);

in C. In FORTRAN F, or f in C and C++, is the name of the function subprogram for evaluating $F(x)$. In FORTRAN it must be declared in an EXTERNAL statement in the program that calls ZERO. Normal input consists of a continuous function $F(x)$ and values B and C such that $F(B)F(C) \leq 0$. Both B and C are also output quantities, so they must be variables in the calling program. On output it is always the case that $|F(B)| \leq |F(C)|$.

The code attempts to bracket a root between B and C, with B being the better approximation, so that the convergence test

$$\left|\frac{B-C}{2}\right| \leq \max[\text{ABSERR}, |B| \times \text{RELERR}]$$

is satisfied. It makes no sense to allow RELERR $< u$, the unit roundoff of the computer used, because this is asking for a more accurate result than the correctly rounded true result. To provide a little protection near limiting precision, it is required that RELERR $\geq 10u$. If the desired root should be zero, or very close to zero, a pure relative error test is not appropriate. For this reason it is required that ABSERR > 0.

Normal output has either $F(B)F(C) < 0$ and the convergence test met, or $F(B) = 0$. This is signaled by FLAG = 0. At most 500 evaluations of F are allowed. If more appear to be necessary, FLAG is set to 1 and the code terminates before the convergence test is satisfied. The value FLAG $= -1$ indicates invalid input, that is, ABSERR ≤ 0 or RELERR $< 10u$, and FLAG $= -2$ means $F(B)F(C) > 0$. The value RESIDL (or residual in C and C++) is the final residual $F(B)$. Convergence is judged by the length of an interval known to contain a root. The algorithm is so robust that it can locate roots of functions that are only piecewise continuous. If it is applied to a function that has a pole of odd multiplicity, it might locate a pole rather than a root. This is recognized by a "large" residual and signaled by FLAG = 2.

Example 4.4. The function $F(x) = e^{-x} - 2x$ has $F(0) > 0$ and $F(1) < 0$, hence the equation $F(x) = 0$ has a root between $C = 0$ and $B = 1$. The sample program provided illustrates the use of the zero-finding routine. Note that a globally defined integer is used to count the number of F evaluations required. In FORTRAN this is accomplished via a COMMON statement (preferably labeled COMMON) and in C or C++ it is done by an appropriate placement of the variable declarations. The output is

```
Approximate root B =   3.517337121480649E-001
The residual F(B)  =  -2.430063176779660E-009
         7 evaluations of F were required.
```

■

Example 4.5. In Chapter 5 we discuss a problem that requires the solution of

$$\phi(\lambda) = \frac{x^2}{a^2+\lambda} + \frac{y^2}{b^2+\lambda} + \frac{z^2}{c^2+\lambda} - 1 = 0$$

for its smallest positive root. The particular values of the parameters used there are $x = y = z = 50$ and $a = 1, b = 2, c = 100$. Figure 4.5 is a rough sketch of ϕ for $a^2 <$

$b^2 < c^2$. It is drawn by looking at the behavior of ϕ as $\lambda \to -a^2, -b^2, -c^2, +\infty$, and $-\infty$. The portion of interest to us is between $-a^2$ and $+\infty$. Since as $\lambda \to -a^2$ from the right, $\phi(\lambda) \to +\infty$ and since as $\lambda \to +\infty$, $\phi(\lambda) \to -1$, the continuous function ϕ must have a zero larger than $-a^2$. Differentiating $\phi(\lambda)$ we get

$$\phi'(\lambda) = -\frac{x^2}{(a^2+\lambda)^2} - \frac{y^2}{(b^2+\lambda)^2} - \frac{z^2}{(c^2+\lambda)^2} < 0.$$

Because $\phi(\lambda)$ is strictly decreasing, there is only one zero λ_0 greater than $-a^2$. The root is simple since $\phi'(\lambda_0) \neq 0$.

Interestingly, $\phi(\lambda)$ can be scaled to a cubic polynomial

$$P(\lambda) = -(a^2+\lambda)(b^2+\lambda)(c^2+\lambda)\phi(\lambda).$$

This allows us to apply some bounds on the relative error of an approximate zero of a polynomial developed in Exercise 4.31.

The equation $\phi(\lambda) = 0$ was solved using the code Zero with relative and absolute error requests of 10^{-6} and 10^{-8}, respectively. Poor initial values of B and C were used to show the excellent rate of convergence. Table 4.1 displays successive values of B and C and tells which method was used by Zero in computing B.

Table 4.1. Solution of $\phi(\lambda) = 0$ by Zero

B	C	Method	$\phi(B)$
10000.	0	Input	−3.7512496(−1)
9998.7995	0	Secant	−3.7505744(−1)
4999.3997	9998.7995	Bisect	1.6629363(−1)
6535.1283	4999.3997	Secant	−8.4003349(−2)
6019.7152	4999.3997	Secant	−1.3682969(−2)
5919.4259	6019.7152	Secant	1.3608157(−3)
5928.4978	5919.4259	Secant	−2.0070584(−5)
5928.3659	5919.4259	Secant	−2.9023897(−8)

For the error bounds in Exercise 4.31, we computed $P(B) = 1.6261594 \times 10^4$ and $P'(B) = 8.5162684 \times 10^7$. The constant term in $P(\lambda)$ is $a_0 = -x^2b^2c^2 - y^2a^2c^2 - z^2a^2b^2 + a^2b^2c^2$, which in this case is -1.2497000×10^8. The bounds state that there is a root r_j of P such that

$$\left|\frac{B-r_j}{r_j}\right| \le \left|\frac{P(B)}{a_0}\right|^{1/3} = 5.1 \times 10^{-2}$$

and a root r_i such that

$$\left|\frac{B-r_i}{B}\right| \le 3\left|\frac{P(B)}{BP'(B)}\right| = 9.7 \times 10^{-8}.$$

The second error bound is quite good, but the first is pessimistic. Generally we do not know which of the two bounds will be better. An interesting point here is the size of the residual $P(B)$. The value B is supposed to be a root of $\phi(\lambda) = 0$ and $P(\lambda) = 0$. If its quality were judged by the size of the residual, B might be thought a poor

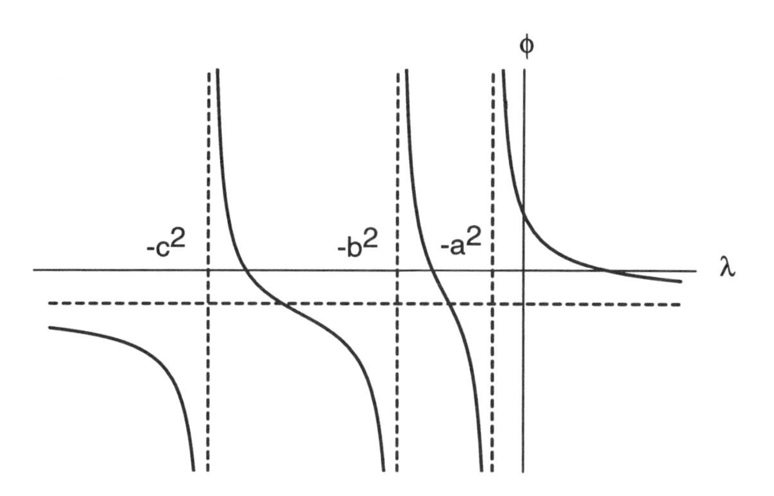

Figure 4.5 $\phi(\lambda)$ for Example 4.5.

approximation when solving the one equation and a good approximation when solving the other. This is despite the fact that it approximates the same root in both cases. To be more concrete, MATHCAD's default convergence test that the residual be smaller in magnitude than 10^{-3} would reject an approximate root of $P(\lambda) = 0$ that is actually quite accurate. This problem illustrates well the issue of scaling. The convergence test in Zero and in similar codes is reliable. The magnitude of the residual reported by the code is important only for detecting the convergence to a pole that is possible when the function input is not continuous. Otherwise its size is a statement about the scaling of the problem, not the quality of the approximate zero. ■

EXERCISES

Unless otherwise indicated, use ABSERR $= 10^{-8}$ and RELERR $= 10^{-6}$ for the computations with Zero.

4.10 Use the code Zero with an initial bracket of $[0, 1]$ to find the roots of the equation $F(x) = 0$, where $F(x)$ is given by each of the following.

(a) $\cos 2x$

(b) $(x-0.3)(x-0.6)$

(c) $(x-0.3)(x-0.6)(x-0.9)$

(d) $(x+1)(x-0.8)^7$

(e) $(x+1)[x^7 - 7(0.8)x^6 + 21(0.8)^2x^5 - 35(0.8)^3x^4 + 35(0.8)^4x^3 - 21(0.8)^5x^2 + 7(0.8)^6x - (0.8)^7]$

(f) $1/\cos 2x$

(g) $\begin{cases} +1, & 0 \le x \le 0.3 \\ -2, & 0.3 < x \le 1 \end{cases}$

(h) $(x-3)(x+1)$

Print out the approximate roots, FLAG, the number of function evaluations required, and the residual. Discuss the results. [Sketches of $F(x)$ will be helpful.]

4.11 A wire weighs 0.518 lb/ft and is suspended between two towers of equal height (at the same level) and 500 ft apart. If the sag in the wire is 50 ft, find the maximum tension T in the wire. The appropriate equations to be solved are

$$c + 50 = c \cosh \frac{500}{2c}$$

$$T = 0.518(c + 50).$$

4.12 For turbulent flow of fluid in a smooth pipe, the equation

$$1 = \sqrt{c_f}(-0.4 + 1.74 \ln(\text{Re}\sqrt{c_f}))$$

models the relationship between the friction factor c_f and the Reynold's number Re. Compute c_f for Re $= 10^4, 10^5, 10^6$. Solve for all values of the Reynold's number in the same run. Do this by communicating the parameter Re to the function subprogram using labeled COMMON in FORTRAN or a globally defined variable in C or C++.

4.13 In [12] the study of neutron transport in a rod leads to a transcendental equation that has roots related to the critical lengths. For a rod of length ℓ the equation is

$$\cot(\ell x) = \frac{x^2 - 1}{2x}.$$

Make a rough sketch of the two functions $\cot(\ell x)$ and $(x^2 - 1)/(2x)$ to get an idea of where they intersect to yield roots. For $\ell = 1$, determine the three smallest positive roots.

4.14 An equation determining the critical load for columns with batten plates is derived in [9, p. 151]. Suitable values of the physical parameters for experiments performed by Timoshenko lead to the problem

$$\frac{1}{180} = \left(\frac{1 - \cos(\pi/10)}{\cos(\pi/10) - \cos(z)}\right)\frac{\sin(z)}{z}$$

and the smallest positive root is desired. Make a rough sketch of the function to get an idea of where the root is. Scale to avoid difficulties with poles and the apparent singularity at 0, and then compute the root.

4.15 An equation for the temperature T at which o-toluidine has a vapor pressure of 500 mm Hg is found in [8, p. 424]. In degrees absolute T satisfies

$$21.1306 - \frac{3480.3}{T} - 5.081\log_{10} T = 0.$$

It is not obvious where the roots are, but a little analysis will help you locate them. Using brackets from your analysis, compute all the roots.

4.16 The geometrical concentration factor C in a certain solar energy collection model [10, p. 33] satisfies

$$C = \frac{\pi(h/\cos A)^2 f}{\frac{1}{2}\pi D^2\left(1 + \sin A - \frac{1}{2}\cos A\right)}.$$

Rescale the problem to avoid poles. Find the smallest positive root A if $h = 300, C = 1200, f = 0.8$, and $D = 14$.

4.17 In trying to solve the equations of radiative transfer in semi-infinite atmospheres, one encounters the nonlinear equation

$$\omega_0 = \frac{2k}{\ln[(1+k)/(1-k)]}$$

where the number $\omega_0, 0 < \omega_0 < 1$, is called an albedo. Show that for fixed ω_0, if k is a root, so is $-k$, and that there is a unique value of k with $0 < k < 1$ satisfying the equation. For $\omega_0 = 0.25, 0.50$, and 0.75, find the corresponding positive k values. Make some sketches to help you locate the roots.

4.18 Exercise 5.27 concerns a temperature distribution problem where it is necessary to find positive roots of

$$2xJ_1(x) - J_0(x) = 0,$$

where $J_0(x)$ and $J_1(x)$ are zeroth and first order Bessel functions of the first kind. Compute the three smallest positive roots.

4.19 An analysis of the Schrödinger equation for a particle of mass m in a rectangular potential well leads to discrete sets of values of the total energy E that are solutions of a pair of transcendental equations. One of these equations is

$$\cot\left(\frac{a}{\hbar}\sqrt{2mV_0}\sqrt{E/V_0}\right) = \sqrt{\frac{E/V_0}{1 - E/V_0}},$$

where

$$\hbar = \frac{h}{2\pi}, \; h = 6.625 \times 10^{-27} \text{ erg-sec},$$

is Planck's constant. Find the value of E that satisfies this equation. Use the following data, which correspond to a simplified model of the hydrogen atom:

$$m = 9.109 \times 10^{-28} \text{ g}$$
$$V_0 = 2.179 \times 10^{-11} \text{ erg}$$
$$a = 5.292 \times 10^{-9} \text{ cm}.$$

On some machines it may be necessary to scale some of the variables to avoid underflow. Also, be careful with your choice of ABSERR if you want an accurate answer.

4.20 The following problem concerns the cooling of a sphere. Suppose the sphere is of radius a and is initially at a temperature V. It cools by Newton's law of cooling with thermal conductivity k, thalpance ε, and diffusivity h^2 after being suddenly placed in air at 0°C. It can be shown that the temperature $\theta(r,t)$ at time $t > 0$ and radius r is

$$\theta(r,t) = \sum_{n=1}^{\infty} \frac{A_n}{r} e^{-\gamma_n^2 h^2 t} \sin \gamma_n r.$$

Here the γ_n are the (positive) roots of

$$\gamma_n \cos \gamma_n a - \left(\frac{1}{a} - \frac{\varepsilon}{k} \right) \sin \gamma_n a = 0$$

and

$$A_n = \frac{2\gamma_n V}{[\gamma_n a - \cos \gamma_n a \sin \gamma_n a]} \int_0^a r \sin \gamma_n r \, dr.$$

For a steel sphere cooling in air at 0°C, suppose the initial temperature is $V = 100°C$ and the radius is $a = 0.30$ meters. Appropriate physical constants are $h^2 = 1.73 \times 10^{-5}$, $\varepsilon = 20$, and $k = 60$. Find the three smallest values of $\gamma_n a$ and use them to compute A_1, A_2, and A_3. Approximate the temperature at $r = 0.25$ for $t = 10^k$ seconds, $k = 2, 3, 4, 5$.

4.21 When solving $f(x) = 0$, the subroutine Zero requires you to input B and C such that $f(B)f(C) \leq 0$. Often it is not obvious what to use for B and C, so many routines that are similar to Zero begin with a search for B and C that provide a change of sign. Write a routine Root that first finds a change of sign and then calls Zero to compute a root. The parameter list of Root should be the same as that of Zero except that B and C are replaced by arguments Z and SCALE. Here the Z input is a guess for a root. If all goes well, on output Z is to be the answer B obtained by Zero. The search algorithm you are to implement is essentially that of the `fzero.m` program of MATLAB. However, `fzero.m` begins searching with an increment $Z/20$ if $Z \neq 0$ and $1/20$ otherwise. In Root the initial search increment is to be an input variable SCALE. Initialize $DZ =$ SCALE, $B = Z - DZ$, $C = Z + DZ$. If $f(B)f(C) \leq 0$ (be sure to code this properly), call Zero to compute a root. Otherwise, double the increment, $DZ = 2 \times DZ$, expand the search to the left by $B = B - DZ$, and test again. If this does not result in a change of sign, expand the search to the right by $C = C + DZ$ and test again. If this does not result in a change of sign, double the increment and repeat. Limit the number of tries. Test your code using one of the examples in the text. After you are sure it works, you might want to use it in other exercises.

4.4 CONDITION, LIMITING PRECISION, AND MULTIPLE ROOTS

It is important to ask what limitations on accuracy are imposed by finite precision arithmetic. Since we seek a machine representable number $\bar{\alpha}$ that makes $f(\bar{\alpha})$ as nearly zero as possible, the details of the computation of f, the machine word length, and the roundoff characteristics play important roles. We have remarked that the computed function values may vary erratically in an interval about the zero and we have seen an example in Figure 1.2. Let us look at another example in more detail.

Example 4.6. Consider the polynomial $(x-1)^3$ that we evaluate in the form $((x-3)x+3)x-1$ in three-digit decimal chopped floating point arithmetic. For $x = 1.00, 1.01, \ldots, 1.17$ the computed function values are exactly zero with the exception of the value 0.0100 at $x = 1.01$, 1.11, 1.15. For $x = 1.18, 1.19, \ldots, 1.24$ all function values are 0.200 except for a value of exactly zero at $x = 1.20$ and a value of 0.0200 at $x = 1.23$. The reader might enjoy evaluating the function for x values less than 1 to explore this phenomenon. It is clear that these erratic values might cause the secant rule to be unstable. Evaluating the derivative shows that Newton's method can also be unstable at a multiple root like this one. ■

What effect on the accuracy of a root does inaccuracy in the function values have? To get some feeling for this, suppose that the routine for $f(x)$ actually returns a value

$\bar{f}(x)$, and for x a machine number near a root α, the best we can say is that

$$|f(x)-\bar{f}(x)| \leq \varepsilon$$

for a suitable ε. Suppose that z is a machine number and $\bar{f}(z)=0$. How much in error can z be? If α is of multiplicity m, then

$$f(z) \approx (z-\alpha)^m \frac{f^{(m)}(\alpha)}{m!}.$$

Since it is possible for $f(z)$ to be as large as ε, we could have

$$\mp\varepsilon \approx (z-\alpha)^m \frac{f^{(m)}(\alpha)}{m!},$$

so it is possible that

$$|z-\alpha| \approx \varepsilon^{1/m} \left| \frac{m!}{f^{(m)}(\alpha)} \right|^{1/m}. \tag{4.14}$$

For small ε and $m>1$ the term $\varepsilon^{1/m}$ is much larger than ε and there is a serious loss of accuracy. The other factor plays a role, but generally we must consider multiple roots to be ill-conditioned (sensitive to errors in the evaluation of f). The ill conditioning of the root of multiplicity 3 in $(x-1)^3=0$ is evident in Example 4.6. We saw there that $x=1.20$ led to a function value of exactly zero, and this is certainly a poor approximation to the root at 1.00. The essence of the matter is that at a multiple root, $f(x)$ is almost tangent to the horizontal axis so that shifting the curve vertically by a small amount shifts its intersection with the axis by a considerable amount. Exercise 4.24 is an example of this effect.

Even when $m=1$, the root can be poorly determined. As we have already seen, clusters of roots "look" like multiple roots from a distance, but we are now considering what happens close to a root. Even when well separated and simple, a root is poorly determined if $f(x)$ passes through zero with a small slope. More formally, the quantity

$$|z-\alpha| \approx \frac{\varepsilon}{|f'(\alpha)|}$$

can be large when the slope $|f'(\alpha)|$ is small. A famous example from Wilkinson [11, p. 43] illustrates this dramatically.

Example 4.7. Consider the polynomial equation

$$(x-1)(x-2)\cdots(x-19)(x-20)=0,$$

which has the roots $1,2,\ldots,19,20$. These roots are obviously simple and well separated. The coefficient of x^{19} is -210. If this coefficient is changed by 2^{-23} to become -210.000000119, the roots become those shown in Table 4.2.

Table 4.2.

Roots of $(x-1)(x-2)\cdots(x-20)-2^{-23}x^{19}=0$	
1.000000000	6.000006944
2.000000000	6.999697234
3.000000000	8.007267603
4.000000000	8.917250249
4.999999928	
10.095266145 ±0.643500904i	
11.793633881 ±1.652329728i	
13.992358137 ±2.518830070i	
16.730737466 ±2.812624894i	
19.502439400 ±1.940330347i	
20.846908101	

Notice that five pairs of roots have become complex with imaginary parts of substantial magnitude! There is really no remedy for this ill conditioning except to use more digits in the computations. ■

Multiple roots are awkward not only because of their ill conditioning but for other reasons, too. Bisection cannot compute roots of even multiplicity because there is no sign change. Its rate of convergence to roots of odd multiplicity is not affected by the multiplicity, but the other methods we have presented slow down drastically when computing multiple roots. If the derivative $f'(x)$ is available, something can be done about both these difficulties. Near α,

$$f(x) = (x-\alpha)^m g(x)$$

and

$$f'(x) = (x-\alpha)^{m-1} G(x),$$

where

$$G(x) = mg(x) + (x-\alpha)g'(x)$$

and

$$G(\alpha) = mg(\alpha) \neq 0.$$

This says that zeros of $f(x)$ of even multiplicity are zeros of $f'(x)$ of odd multiplicity, so they could be computed with a bisection code or Zero. Also, notice that

$$u(x) = \frac{f(x)}{f'(x)} = (x-\alpha)H(x) \text{ near } \alpha,$$

where

$$H(x) = g(x)/G(x), H(\alpha) = 1/m \neq 0,$$

so that $u(x)$ has only simple zeros. Because of this, solving $u(x) = 0$ with Zero is faster than solving $f(x) = 0$ and allows the code to compute zeros of $f(x)$ of even multiplicity. However, it must be appreciated that $u(x)$ has a pole wherever $f'(x) = 0$ and $f(x) \neq 0$.

EXERCISES

4.22 What is the value of the right-hand side of (4.14) for the root in Exercise 4.10a?

4.23 What is the value of the right-hand side of (4.14) for the root of $f(x) = (x-10)(3x-1)^2$ in $[0,1]$? Assume that $\varepsilon = 10^{-12}$.

4.24 The problem $f(x) = (x+1)(x-0.8)^7 = 0$ has 0.8 as a root of multiplicity 7. Evaluate the expression (4.14) for the condition of this root. Perturb $f(x)$ by 10^{-7} using the form in Exercise 4.10e; then solve

$$f(x) + 10^{-7} = 0$$

accurately with Zero (use ABSERR = RELERR = 10^{-10}). How much was the root 0.8 perturbed? Compare this to the result of (4.14) with $\varepsilon = 10^{-7}$. Repeat using the form in Exercise 4.10d.

4.5 NONLINEAR SYSTEMS OF EQUATIONS

A problem occurring quite frequently in computational mathematics is to find some or all of the solutions of a system of n simultaneous nonlinear equations in n unknowns. Such problems are generally much more difficult than a single equation. An obvious starting point is to generalize the methods we have studied for the case $n = 1$. Unfortunately, the bracketing property of the method of bisection does not hold for $n > 1$. There is a generalization of the secant method, but it is not at all obvious because of the more complicated geometry in higher dimensions. Newton's method, however, does extend nicely. Only the case $n = 2$ is examined because the notation is simpler and the basic ideas are the same for the general case.

Consider the system of equations

$$\begin{aligned} f(x,y) &= 0 \\ g(x,y) &= 0, \end{aligned} \tag{4.15}$$

which we occasionally write in vector form as

$$\mathbf{h}(\mathbf{w}) = 0,\ \mathbf{w} = \begin{pmatrix} x \\ y \end{pmatrix},\ \mathbf{h} = \begin{pmatrix} f \\ g \end{pmatrix}.$$

To solve (4.15), Newton's method for finding the root of a single nonlinear equation is generalized to two dimensions. The functions $f(x,y)$ and $g(x,y)$ are expanded about a point (x_0, y_0) using Taylor's theorem for functions of two variables (see the appendix for a statement of the theorem). Carrying only terms of degrees 0 and 1 gives the approximating system

$$\begin{aligned} f(x_0,y_0) + \frac{\partial f}{\partial x}(x_0,y_0)(x-x_0) + \frac{\partial f}{\partial y}(x_0,y_0)(y-y_0) &= 0 \\ g(x_0,y_0) + \frac{\partial g}{\partial x}(x_0,y_0)(x-x_0) + \frac{\partial g}{\partial y}(x_0,y_0)(y-y_0) &= 0, \end{aligned} \tag{4.16}$$

which is linear in the variables $x - x_0$ and $y - y_0$. The next approximation (x_1, y_1) to the solution of (4.16) is found by solving the linear equations

$$\frac{\partial f}{\partial x}(x_0,y_0)\Delta x_0 + \frac{\partial f}{\partial y}(x_0,y_0)\Delta y_0 = -f(x_0,y_0)$$
$$\frac{\partial g}{\partial x}(x_0,y_0)\Delta x_0 + \frac{\partial g}{\partial y}(x_0,y_0)\Delta y_0 = -g(x_0,y_0)$$

for $\Delta x_0 = x_1 - x_0$ and $\Delta y_0 = y_1 - y_0$ and forming

$$x_1 = \Delta x_0 + x_0 \text{ and } y_1 = \Delta y_0 + y_0.$$

In general (x_{k+1}, y_{k+1}) is obtained from (x_k, y_k) by adding a correction

$$(\Delta x_k, \Delta y_k) = (x_{k+1} - x_k, y_{k+1} - y_k)$$

obtained by solving a linear system. To summarize, we have derived

---

Newton's method for two equations and two unknowns:

$$\begin{pmatrix} \frac{\partial f}{\partial x}(x_k,y_k) & \frac{\partial f}{\partial y}(x_k,y_k) \\ \frac{\partial g}{\partial x}(x_k,y_k) & \frac{\partial g}{\partial y}(x_k,y_k) \end{pmatrix} \begin{pmatrix} \Delta x_k \\ \Delta y_k \end{pmatrix} = -\begin{pmatrix} f(x_k,y_k) \\ g(x_k,y_k) \end{pmatrix},$$

---

or in matrix form,

$$J(\mathbf{w_k})\Delta\mathbf{w_k} = -h(\mathbf{w_k}).$$

The matrix J is called the *Jacobian matrix* of the system of equations composed of f and g.

Example 4.8. Set up Newton's method for obtaining solutions to the equations

$$f(x,y) = x^2 + xy^3 - 9 = 0$$
$$g(x,y) = 3x^2y - y^3 - 4 = 0.$$

Since

$$\frac{\partial f}{\partial x} = 2x + y^3, \ \frac{\partial f}{\partial y} = 3xy^2$$
$$\frac{\partial g}{\partial x} = 6xy, \ \frac{\partial g}{\partial y} = 3x^2 - 3y^2,$$

the system to be solved at each iteration is

$$\begin{pmatrix} 2x_k + y_k^3 & 3x_k y_k^2 \\ 6x_k y_k & 3x_k^2 - 3y_k^2 \end{pmatrix} \begin{pmatrix} \Delta x_k \\ \Delta y_k \end{pmatrix} = -\begin{pmatrix} x_k^2 + x_k y_k^3 - 9 \\ 3x_k^2 y_k - y_k^3 - 4 \end{pmatrix}.$$

The following table gives some numerical results for different starting points (x_0, y_0). In all cases the iterations were terminated when the quantity $\max(\|\mathbf{h}\|, \|\Delta\mathbf{w}\|/\|\mathbf{w}\|)$ was less than 10^{-6}.

(x_0, y_0)	Solution (x, y)	Number of Iterations
$(1.2, 2.5)$	$(1.3364, 1.7542)$	4
$(-2.0, 2.5)$	$(-0.9013, -2.0866)$	9
$(-1.2, -2.5)$	$(-0.9013, -2.0866)$	4
$(2.0, -2.5)$	$(-3.0016, 0.1481)$	19

These computations show that this system of equations has at least three solutions. Which solution is found depends on the starting guess (x_0, y_0). ■

As with Newton's method for a function of one variable, it can be shown that if $\mathbf{h}$ is twice continuously differentiable near a root $\mathbf{a}$ of $\mathbf{h}(\mathbf{w}) = 0$, if the Jacobian matrix at $\mathbf{a}$, $J(\mathbf{a})$, is not singular, and if $\mathbf{w_0}$ is sufficiently close to $\mathbf{a}$, then Newton's method will converge to $\mathbf{a}$ and the convergence will be quadratic.

Advanced references like [7] develop Newton's method much further. A serious practical difficulty is to find an initial approximation sufficiently close to the desired root $\mathbf{a}$ that the method will converge. A knowledge of the problem and continuation can be very helpful in this. A general approach is to connect finding a root $\mathbf{w}$ of $\mathbf{h}(\mathbf{w}) = 0$ with minimizing the residual $R(\mathbf{w}) = f^2(\mathbf{w}) + g^2(\mathbf{w})$. Clearly this function has a minimum of zero at any root of $\mathbf{h}(\mathbf{w}) = 0$. The idea is to regard the change $\Delta\mathbf{w_k}$ computed from Newton's method as giving a direction in which we search for a value λ such that the iterate

$$\mathbf{w_{k+1}} = \mathbf{w_k} + \lambda \Delta \mathbf{w_k}$$

results in a smaller value of the residual:

$$R(\mathbf{w_{k+1}}) < R(\mathbf{w_k}).$$

This is always possible because until we reach a root,

$$\left[\frac{\partial}{\partial \lambda} R(\mathbf{w_k} + \lambda \Delta \mathbf{w_k})\right]_{\lambda=0} = -2R(\mathbf{w_k}) < 0.$$

There are many practical details to be worked out. For example, it is not necessary, or even desirable, to find the value of λ that minimizes the residual. Methods of this kind are called damped Newton methods. A careful implementation will often converge when Newton's method on its own will not and will have nearly the same efficiency when both converge.

EXERCISES

4.25 Use Newton iteration to find a solution (good to at least an absolute error of 10^{-4} in magnitude) near $(0.5, 1.0, 0.0)$ of the nonlinear system

$$2x^2 - x + y^2 - z = 0$$
$$32x^2 - y^2 - 20z = 0$$
$$y^2 - 14xz = 0.$$

4.26 We seek the three parameters α, β, and γ in the model

$$f(x) = \alpha e^{\beta x} + \gamma x$$

by interpolating the three data points $(1, 10)$, $(2, 12)$, and $(3,18)$. Use Newton iteration to solve for the parameters to three significant figures.

4.6 CASE STUDY 4

The Lotka–Volterra equations

$$\frac{dx}{dt} = a(x - xy)$$
$$\frac{dy}{dt} = -c(y - xy)$$

describing the populations of predator and prey are studied at length in most modern books on ordinary differential equations (see, e.g., Boyce and DiPrima [2]). Those emphasizing a modeling approach, like Borrelli and Coleman [1], give considerable attention to the formulation of the model and conclusions that can be drawn. Although the population equations do not have an analytical solution, the equation

$$\frac{dy}{dx} = -\frac{c(y - xy)}{a(x - xy)}$$

for trajectories in the phase plane can be solved because it is separable. An easy calculation detailed in the books cited shows that the solutions satisfy the conservation law

$$e^{cx+ay} = Kx^c y^a,$$

where K is an arbitrary constant. The trajectories reveal the qualitative properties of the solutions, so there is great interest in their computation. Davis [5] exploits the conservation law to compute trajectories by solving nonlinear algebraic equations, the subject of this case study.

Following Volterra, Davis considered evaluation of the trajectory for parameters $a = 2, c = 1$ and initial condition $(x, y) = (1, 3)$. In this case there is a periodic solution of the differential equations, that is, the populations of predator and prey are sustainable for all time. In the phase plane this periodic solution corresponds to a trajectory that is a closed curve, as seen in Figure 4.6. The initial condition determines the constant K for the solution of interest. A little manipulation of the conservation law leads to

$$f(x) = xe^{-x} = y^{-2}e^{2y}/K = (3/y)^2 e^{2y-7} = \gamma.$$

Points (x, y) on the trajectory are computed for a sequence of y by forming the corresponding $\gamma > 0$ and solving this algebraic equation for x. Davis gives a series for the smallest positive root,

$$x = \gamma + \gamma^2 + \frac{3}{2}\gamma^3 + \cdots + \frac{n^{n-1}}{n!}\gamma^n + \cdots = \sum_{n=1}^{\infty} t_n.$$

For positive γ we are summing positive terms, so as we learned in Chapter 1 we can expect to evaluate the series accurately in a relative sense provided that it converges at

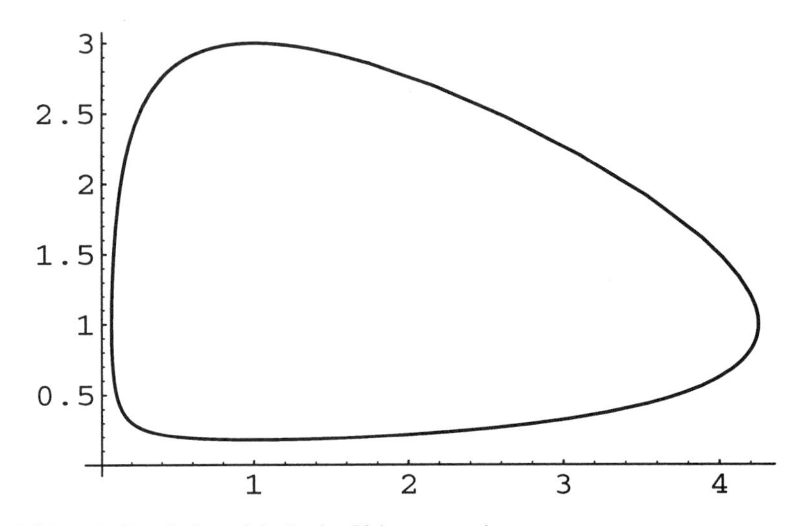

Figure 4.6 A periodic solution of the Lotka–Volterra equation.

a reasonable rate. The ratio test for convergence is illuminating. A well-known limit for e shows that the ratio

$$\frac{t_{n+1}}{t_n} = \gamma\left(1+\frac{1}{n}\right)^{n-1}$$

has the limit

$$\lim_{n\to\infty}\frac{t_{n+1}}{t_n} = \lim_{n\to\infty}\gamma\left(1+\frac{1}{n}\right)^{-1}\left(1+\frac{1}{n}\right)^{n} = \gamma\cdot 1\cdot e.$$

The ratio test amounts to comparing the rate of convergence of the series to the rate of convergence of a geometric series, in this case $\sum_{n=1}^{\infty}(\gamma e)^n$. For values of γ rather less than $1/e$, the series converges quickly. It was pointed out in Chapter 1 that $n!$ grows rapidly as n increases, making integer overflow a dangerous possibility when evaluating the series, and the factor n^n grows even more rapidly. It is better to compute the coefficients by

$$t_{n+1} = t_n\gamma\left(1+\frac{1}{n}\right)^{n-1},$$

which is nicely scaled.

The function $f(x) = xe^{-x}$ and its derivative, $f'(x) = (1-x)e^{-x}$, are so simple that properties of the equation $f(x) = \gamma$ are easily determined. The function vanishes at $x = 0$ and tends to 0 as $x \to \infty$. It strictly increases to its maximum of e^{-1} at $x = 1$ and strictly decreases thereafter. These facts tell us that for $0 < \gamma < e^{-1}$, the equation $f(x) = \gamma$ has exactly two roots. One lies in $(0,1)$ and the other in $(1,\infty)$. Both are simple. The roots merge to form the double root $x = 1$ when $\gamma = e^{-1}$ and there is no root at all for $\gamma > e^{-1}$. This is easily understood geometrically. The trajectory is a closed curve and the two roots represent the left and right portions of the curve for given y. We have to expect numerical difficulties solving for $x(y)$ when $\gamma = e^{-1}$ because the curve has a horizontal tangent then.

If we were solving this equation for a given γ as an isolated problem, a good way to proceed would be to use Zero. Because the code converges very quickly from poor guesses, we might try a "large" interval so as to increase the chance of locating the root that is larger than 1. We might, for example, choose $[B,C] = [1,1000]$. This presents no difficulty for the code, but makes a point raised in Chapter 1. The function $\exp(-x)$ underflows for x as large as 1000, causing problems with some operating systems. Davis solves for x when $y = 1$. It is a matter of a few minute's work to alter the example code provided with Zero to solve this problem using a system that deals with underflows by setting them to 0. With a relative error tolerance of 10^{-6} and an absolute error tolerance of 10^{-8}, only 15 evaluations of the function were needed to compute the root $x \approx 4.24960$. With this value of y, the constant $\gamma \approx 0.060641$. The code reports the residual of the approximate root to be $r \approx 8 \times 10^{-9}$. This is a situation illuminated by backward error analysis: the computed root is the exact root of $f(x) = \gamma + r$, a problem very close to the one posed. Here we see that the x coordinate computed by Zero is the exact value for a slightly different y coordinate.

When computing a trajectory, we are solving a sequence of problems, indeed, more than a hundred in generating the figure. The first derivative is readily available, the roots are simple, and a good initial guess is available. The circumstances suggest that writing a code based on Newton's method would be worth the trouble. Before discussing the computation of the trajectory, let's do a couple of experiments with Newton's method. First we compute the smaller root when $y = 1$. Recall that when solving $F(x) = f(x) - \gamma = 0$, the method is

$$x_{i+1} = x_i - \frac{F(x_i)}{F'(x_i)}.$$

Because the method is quadratically convergent for these simple roots, the difference between successive iterates provides a convenient and reliable way to test for convergence. Of course, for this test to work it is necessary that the iterate be close enough to the root that the method really is converging quadratically fast and not so close that finite precision corrupts the estimate. Because γ is relatively small in this case, we can compute an accurate value for x using the series to see how well the procedure works. With $x_0 = \gamma$, the table shows that convergence appears to be quadratic right from the start and the estimated error is quite close to the true error, except at limiting precision. (It is quite possible that the root computed in this manner is more accurate than the reference value computed with the series.) Because there is a natural measure of scale provided by the constant γ, the small residual tells us that after only a few iterations, we have a very accurate solution in the sense of backward error analysis.

Iterate	x	Residual	Est. error	Error
0	$0.606415e-1$	$-0.36e-02$	$0.40e-02$	$0.41e-02$
1	$0.646775e-1$	$-0.15e-04$	$0.17e-04$	$0.17e-04$
2	$0.646944e-1$	$-0.26e-09$	$0.30e-09$	$0.30e-09$
3	$0.646944e-1$	$0.45e-17$	$-0.51e-17$	$-0.28e-16$

This is what we expect of Newton's method, but things do not always go so well. Suppose now that we try to compute the larger root starting with a guess of 8. In order to measure the error, we computed the root accurately in a preliminary computation. The first difficulty we encountered is that the method does not converge to this root. Remember that we can expect convergence to the root nearest the guess only when the guess is "sufficiently close." The error reported in the table is the difference between the iterate and an accurate value for the smaller root.

Iterate	x	Residual	Est. error	Error
0	$0.800000e+1$	$-0.58e-01$	$-0.25e+02$	$-0.79e+01$
1	$-0.166814e+2$	$-0.29e+09$	$0.94e+00$	$0.17e+02$
2	$-0.157380e+2$	$-0.11e+09$	$0.94e+00$	$0.16e+02$
3	$-0.147977e+2$	$-0.40e+08$	$0.94e+00$	$0.15e+02$
⋮	⋮	⋮	⋮	⋮
23	$0.620850e-1$	$-0.23e-02$	$0.26e-02$	$0.26e-02$
24	$0.646873e-1$	$-0.62e-05$	$0.70e-05$	$0.70e-05$
25	$0.646944e-1$	$-0.45e-10$	$0.51e-10$	$0.51e-10$
26	$0.646944e-1$	$0.45e-17$	$-0.51e-17$	$0.00e+00$

As the residual makes clear, we went from a reasonable initial guess to approximations that are terrible in the sense of backward error analysis. This kind of thing did not happen with Zero, even though it was given much worse guesses, because of the way it combines the secant rule and bisection. All goes well near the root, but convergence is very slow when an iterate is large and negative. Indeed, the estimated error is the change in the iterate and the table shows it to be nearly constant then. This is easy to see analytically from

$$x_{i+1} - x_i = \frac{x_i e^{-x_i} - \gamma}{(1 - x_i)e^{-x_i}}.$$

Examination of the sizes of the terms when $x_i \ll -1$ shows that the change is approximately $x_i/(1-x_i)$. This can be further approximated as 1, agreeing as well as might be expected with the values 0.94 seen in the table. These approximations make it clear that if we should turn up an approximation $x_i \ll -1$, the iterates are going to increase slowly to the positive roots.

Davis used this algebraic approach only for computing a couple of points on the trajectory. If we want to compute the closed trajectory of the figure, we need to do continuation efficiently and deal with the fact that the differential equation for the phase is singular when $x = 1$. First let's look for a moment at the solution of $x \exp(-x) = \gamma$ for a sequence of values γ. Suppose we have found a root x corresponding to a given γ and want to compute a root corresponding to $\gamma + \delta$ for some "small" δ. One possibility for an initial guess is simply x. Often this works well enough, but for this equation a short calculation shows that

$$\frac{dx}{d\gamma} = \frac{e^x}{1-x}.$$

A rather better initial guess is then $x + \delta dx/d\gamma$. Notice the change in character of the problem that shows up here when $x = 1$. This is the kind of difficulty that we overcome when tracing a trajectory by exploiting additional information at our disposal.

The program that computed the figure accepts the constants a, c and the initial point (x, y). Using these data, it computes the constant K. The conservation law allows us to solve for x when y is given, or to solve for y when x is given, and the code selects the more appropriate at each step. The differential equations for the populations tell us that for a small increment δ in t, the change in x is about $\delta dx/dt$ and the change in y is about $\delta dy/dt$. If $|dy/dt| \leq |dx/dt|$, the code uses the value $x + \delta dx/dt$ and solves for $y(x)$. Otherwise, it uses the value $y + \delta dy/dt$ and solves for $x(y)$. This amounts to changing the coordinate system in which the curve is viewed so as to avoid difficulties with vertical tangents. After choosing which equation to solve, Newton's method is easily applied and converges very rapidly. Superlinear convergence is used to test for convergence to a specified relative error tolerance.

REFERENCES

1. R. Borrelli and C. Coleman, *Differential Equations: A Modeling Approach*, Prentice Hall, Englewood Cliffs, N.J., 1987.
2. W. Boyce and R. DiPrima, *Elementary Differential Equations and Boundary Value Problems*, Wiley, New York, 1992.
3. R. Brent, *Algorithms for Minimization without Derivatives*, Prentice Hall, Englewood Cliffs, N.J., 1973.
4. P. David and J. Voge, *Propagation of Waves*, Pergamon Press, New York, 1969.
5. H. Davis, *Introduction to Nonlinear Differential and Integral Equations*, Dover, New York, 1962.
6. T. Dekker, "Finding a zero by means of successive linear interpolation," in *Constructive Aspects of the Fundamental Theorem of Algebra*, B. Dejon and P. Henrici, eds., Wiley, London, 1969.
7. J. Dennis, Jr., and R. Schnabel, *Numerical Methods for Unconstrained Optimization and Nonlinear Equations*, Prentice Hall, Englewood Cliffs, N.J., 1983.
8. J. Eberhardt and T. Sweet, "The numerical solution of equations in chemistry," *J. Chem. Ed.*, 37 (1960), pp. 422–430.
9. S. Timoshenko, *Theory of Elastic Stability*, McGraw Hill, New York, 1961.
10. L. Vant-Hull and A. Hildebrandt, "Solar thermal power systems based on optical transmission," *Solar Energy*, 18 (1976), pp. 31–40.
11. J. Wilkinson, *Rounding Errors in Algebraic Processes*, Dover, Mineola, N.Y., 1994.
12. G.M. Wing, *An Introduction to Transport Theory*, Wiley, New York, 1962.

MISCELLANEOUS EXERCISES FOR CHAPTER 4

4.27 A semi-infinite medium is at a uniform initial temperature $T_0 = 70°$F. For time $t > 0$, a constant heat flux density q = 300 Btu/hr sq ft is maintained on the surface $x = 0$. Knowing the thermal conductivity k = 1.0 Btu/hr/ft/°F and the thermal diffusivity α = 0.04

sq ft/hr, the resulting temperature $T(x,t)$ is given by

$$T(x,t) = T_0 + \frac{q}{k}\left[2\sqrt{\frac{\alpha t}{\pi}}e^{-x^2/4\alpha t} - x\left(1 - \operatorname{erf}\left(\frac{x}{2\sqrt{\alpha t}}\right)\right)\right],$$

where

$$\operatorname{erf}(y) = \frac{2}{\sqrt{\pi}}\int_0^y e^{-z^2}\,dz$$

is the error function. Find the times t required for the temperature at distances $x = 0.1, 0.2, \ldots, 0.5$ to reach a preassigned value $T = 100°$F. Use ABSERR $= 10^{-8}$ and RELERR $= 10^{-6}$. The function erf(y) is available in many FORTRAN and some C and C++ libraries.

4.28 Write a code like Zero based upon bisection and Newton's method. Are there advantages to using Newton's method instead of the secant rule?

4.29 Modify Zero so as to input $f'(x)$ along with $f(x)$. The code is to compute roots via the function $u(x) = f(x)/f'(x)$ as described in the text. This makes the modified code faster for multiple roots and permits the computation of roots of even multiplicity.

4.30 Given

$$\begin{aligned}
a_{11}x_1 \sin\theta + a_{12}x_2 + a_{13}x_3\cos\theta &= b_1\\
a_{21}x_1 + a_{22}x_2\cos\theta + a_{23}x_3 &= b_2\\
a_{31}x_1\cos\theta + a_{32}x_2 + a_{33}x_3\sin\theta &= b_3\\
a_{41}x_1 + a_{42}x_2\sin\theta + a_{43}x_3 &= b_4,
\end{aligned}$$

devise an algorithm using the codes Zero and Factor/Solve to solve for x_1, x_2, x_3 and θ. Sketch a program in FORTRAN or C or C++ to implement your scheme. Do not worry about input/output nor an initial bracket, but *do* define $F(x)$ carefully.

4.31 In parts (a) and (b) below, error bounds are derived for an approximate root σ (real or complex) of the polynomial equation

$$P(x) = x^n + a_{n-1}x^{n-1} + \cdots + a_1x + a_0 = 0.$$

In each case we require an accurate value of $P(\sigma)$. Since root solvers may make this residual about as small as possible in the working precision, it is necessary to compute $P(\sigma)$ in higher precision. Let $r_1, r_2, \ldots, r_n$ be the roots of $P(x) = 0$.

(a) The theory of equations tells us that $P(x)$ can be factored in the form

$$P(x) = (x - r_1)(x - r_2)\cdots(x - r_n).$$

Show that

$$a_0 = (-1)^n r_1 r_2 \cdots r_n$$

and then that

$$\left|\frac{P(\sigma)}{a_0}\right| \geq \min_j \left|\frac{\sigma - r_j}{r_j}\right|^n.$$

This implies that

$$\min_j \left|\frac{\sigma - r_j}{r_j}\right| \leq \left|\frac{P(\sigma)}{a_0}\right|^{1/n},$$

which says that there is some zero that is approximated with a relative error of no more than

$$\left|\frac{P(\sigma)}{a_0}\right|^{1/n}.$$

This bound is very pessimistic when σ approximates well a zero that is much larger than some other zero. To understand this assertion, work out a numerical example for a quadratic with $\sigma \approx r_1$ and $|r_1| \gg |r_2|$. Then argue that the assertion is true in general.

(b) Show that

$$\frac{P'(\sigma)}{P(\sigma)} = \sum_{j=1}^{n} \frac{1}{\sigma - r_j}$$

by differentiating $\ln P(x)$. This then implies that

$$\left|\frac{P'(\sigma)}{P(\sigma)}\right| \leq n \cdot \frac{1}{\min_j |\sigma - r_j|}$$

and

$$\min_j |\sigma - r_j| \leq n \left|\frac{P(\sigma)}{P'(\sigma)}\right|.$$

This is an absolute error bound, but we get the following relative error bound easily:

$$\min_j \left|\frac{\sigma - r_j}{\sigma}\right| \leq n \left|\frac{P(\sigma)}{\sigma P'(\sigma)}\right|.$$

How is this error bound related to the error estimate derived for Newton's method?

4.32 The book [4, p. 65] contains a cubic equation for a parameter s in the context of corrections for the earth's curvature in the interference zone. The equation

$$s^3 - \frac{3}{2}s^2 - \frac{s}{2}\left(\frac{1+u}{v^2} - 1\right) + \frac{1}{2v^2} = 0$$

depends on two parameters, u and v, which are obtained from the heights of the towers, the distance between stations, and the radius of the earth. Representative values are $v = 1/291, u = 30$. The smallest positive root is the one of interest, but calculate them all. The residuals of the larger roots are quite large. Are they inaccurate? Compare with Exercise 4.1. Use the computable error bounds of Exercise 4.31 to bound the errors of the roots.

CHAPTER 5

NUMERICAL INTEGRATION

Approximating $\int_a^b f(x)\,dx$ numerically is called numerical integration or quadrature. Most of this chapter is concerned with finite intervals $[a,b]$, but there is some discussion of integrals with a and/or b infinite. Sometimes it is useful to introduce a weight function $w(x) \geq 0$ and so approximate integrals of the form $\int_a^b f(x)w(x)\,dx$. There are a number of reasons for studying numerical integration. The antiderivative of f may not be known or may not be elementary. The integral may not be available because the function f is defined by values in a table or by a subprogram. Or, definite integrals must be approximated as part of a more complicated numerical scheme, such as one for the solution of differential equations by finite elements by means of variational or Galerkin methods.

A basic principle in numerical analysis is that if we cannot do what we want with a given function $f(x)$, we approximate it with a function for which we can. Often the approximating function is an interpolating polynomial. Using this principle we shall derive some basic quadrature rules and study their errors. When approximating functions we found that piecewise polynomial interpolants had advantages over polynomial interpolants, and the same is true in this context. In a way piecewise polynomial interpolants are more natural for quadrature because using such a function amounts to breaking up the interval of integration into pieces and approximating by a polynomial on each piece. A key idea in quadrature is to take account of the behavior of $f(x)$ when splitting up the interval. This "adaptive" quadrature is described in Section 5.2 and a code is discussed in the following section. Adaptive quadrature is the main topic of the chapter, but some attention is given to the integration of tabular data and to the integration of functions of two independent variables. Particular attention is paid to preparing problems for their effective solution by codes of the kind developed here.

5.1 BASIC QUADRATURE RULES

To approximate

$$\int_a^b f(x)w(x)\,dx \tag{5.1}$$

suppose that values of f are available at N distinct points $\{x_1,\ldots,x_N\}$. Let $P_N(x)$ be the polynomial that interpolates f at these points. The Lagrangian form of $P_N(x)$ leads easily to the approximation

$$\int_a^b f(x)w(x)\,dx \approx \int_a^b P_N(x)w(x)\,dx = \int_a^b \sum_{i=1}^{N} f(x_i)L_i(x)w(x)\,dx$$
$$= \sum_{i=1}^{N} f(x_i)\int_a^b L_i(x)w(x)\,dx = \sum_{i=1}^{N} A_i f(x_i). \tag{5.2}$$

It is assumed here that the *weights* A_i exist. This is equivalent to the existence of the integrals

$$\int_a^b x^j w(x)\,dx \quad \text{for} \quad j = 0,1,\ldots,N-1.$$

In the case of most interest in this chapter, namely $w(x) \equiv 1$, a and b finite, there is no doubt about this. However, if the interval is infinite (e.g., $\int_0^\infty f(x)\,dx$), the approach fails because none of the x^j has an integral over this interval.

The fundamental difficulty with the approach in the case of $\int_0^\infty f(x)\,dx$ is that it is based on approximating $f(x)$ by a polynomial, and polynomials do not have finite integrals over infinite intervals. For the integral of $f(x)$ to exist, it must tend to zero rapidly as $x \to \infty$. A useful device is to isolate the behavior that is different from a polynomial in a weight function that is handled analytically in the formula. For example, if we introduce the weight function $w(x) = e^{-x}$ and define $F(x) = f(x)e^x$, the integral can be rewritten as $\int_0^\infty F(x)e^{-x}\,dx$. It is straightforward to obtain formulas for integrals of the form $\int_0^\infty F(x)e^{-x}\,dx$ because the integrals $\int_a^b x^j e^{-x}\,dx$ exist for all j. Whether this device results in a good approximation to $\int_0^\infty f(x)\,dx$ is a question about whether $F(x)$ behaves more like a polynomial than $f(x)$.

Infinite intervals are one kind of problem that presents difficulties. An integrand that is singular also presents difficulties because it does not behave like a polynomial. Often a weight function is a good way to deal with such problems. For example, in the solution of plane potential problems by boundary element methods, it is necessary to approximate a great many integrals of the form

$$\int_0^1 F(x)\ln x\,dx$$

(and subsequently to solve a system of linear equations to produce a numerical solution to an integral equation of potential theory). The function $\ln x$ can be used as a weight function because it is nonpositive over the interval and the integrals

$$\int_a^b x^j \ln(x)\,dx$$

exist for all j (so the weight function $w(x)$ in (5.1) can be taken to be $-\ln x$). Similarly to what was done with the example of an integral over an infinite interval, if we wish to compute $\int_0^1 f(x)\,dx$ and $f(x)$ behaves like $\ln(x)$ as $x \to 0$, we could introduce $\ln(x)$ as a weight function and write $F(x) = f(x)/\ln(x)$. By "behaves like" as $x \to 0$ is meant

$$\lim_{x \to 0} \frac{f(x)}{\ln(x)} \to c.$$

From here on this will be written as $f(x) \sim c\ln(x)$. Because $F(x)$ has a finite limit at $x = 0$, it is better approximated by a polynomial than $f(x)$, which is infinite there.

A formula of the form

$$\sum_{i=1}^{N} A_i f(x_i) \tag{5.3}$$

for approximating (5.1) is called a *quadrature formula* or *quadrature rule.* The scheme for generating rules just described leads to *interpolatory* quadrature rules. Such a rule will integrate exactly any polynomial of degree less than N. This is because if $f(x)$ is a polynomial of degree less than N, then by the uniqueness of interpolation, $P_N(x) \equiv f(x)$ and the rule is constructed so as to integrate $P_N(x)$ exactly. In general, we say that a quadrature formula (5.3) has the *degree of precision* $d \geq 0$ if it integrates exactly any polynomial of degree at most d, but not one of degree $d+1$. We shall find that a judicious selection of the interpolation points $\{x_i\}$ in the construction of (5.2) leads to formulas with a degree of precision greater than $N-1$. Generally we have in mind $\{x_i\}$ that lie in $[a,b]$, but in some important applications this is not the case. For example, the very important Adams formulas for the solution of ordinary differential equations are based on quadrature rules that use the end points a and b as nodes, but all other x_i lie outside the interval. The same is true of a method for integrating tabular data that is considered later in this chapter.

The following theorem develops some bounds on the error of a formula with degree of precision d. It is stated using the notation $||f||$ for the maximum over $[a,b]$ of $|f(x)|$. Also, as in Chapter 3, M_q is used for $||f^{(q)}||$. Finally, the absolute error of the quadrature formula is denoted by $E(f)$, that is,

$$E(f) = \int_a^b f(x)w(x)\,dx - \sum_{i=1}^{N} A_i f(x_i).$$

Theorem 5.1. *If the quadrature formula (5.2) has degree of precision d, then for any polynomial $p(x)$ of degree $q \leq d$,*

$$|E(f)| \leq ||f-p|| \left(\int_a^b w(x)\,dx + \sum_{i=1}^{N} |A_i| \right). \tag{5.4}$$

If each $A_i > 0$, then

$$|E(f)| \leq 2||f-p|| \int_a^b w(x)\,dx. \tag{5.5}$$

Proof. For $p(x)$ any polynomial of degree $q \leq d$,

$$|E(f)| \leq \left| \int_a^b p(x)w(x)\,dx + \int_a^b (f(x)-p(x))w(x)\,dx \right.$$

$$-\sum_{i=1}^{N} A_i p(x_i) - \sum_{i=1}^{N} A_i (f(x_i) - p(x_i))\Big|$$

$$\le E(p) + \int_a^b |f-p| w(x)\,dx + \sum_{i=1}^{N} |A_i||f-p|$$

$$\le \|f-p\| \left(\int_a^b w(x)\,dx + \sum_{i=1}^{N} |A_i| \right),$$

where we have used the fact that $E(p) = 0$. (Why?) This is (5.4). For (5.5), when each $A_i > 0$ the absolute values in (5.4) can be dropped. Because the quadrature formula integrates constants exactly, applying it to $f(x) = 1$ shows that

$$\sum_{i=1}^{N} A_i \cdot 1 = \int_a^b w(x) \cdot 1\,dx,$$

which results in (5.5). ■

Corollary 5.1. *If $f(x)$ has $d+1$ continuous derivatives on $[a,b]$, then*

$$|E(f)| \le \left(\frac{b-a}{2}\right)^{d+1} \frac{M_{d+1}}{(d+1)!} \left(\int_a^b w(x)\,dx + \sum_{i=1}^{N} |A_i| \right). \tag{5.6}$$

If each $A_i > 0$, then

$$|E(f)| \le \left(\frac{b-a}{2}\right)^{d+1} \frac{M_{d+1}}{(d+1)!}\, 2 \int_a^b w(x)\,dx. \tag{5.7}$$

Proof. Since the bounds of Theorem 5.1 hold for any $p(x)$, we can use the $p(x)$ coming from Taylor's theorem (see the appendix) with $x_0 = (a+b)/2$ and $n = q$:

$$f(x) = p(x) + R_{q+1}(x),$$

where

$$p(x) = f(x_0) + \frac{(x-x_0)}{1!} f'(x_0) + \cdots + \frac{(x-x_0)^q}{q!} f^{(q)}(x_0)$$

and

$$R_{q+1}(x) = \frac{(x-x_0)^{q+1}}{(q+1)!} f^{(q+1)}(z)$$

for some z between x_0 and x. This implies that

$$\|f-p\| = \max_{a \le x \le b} \left| \frac{(x-x_0)^{q+1}}{(q+1)!} f^{(q+1)}(z) \right| \le \left(\frac{b-a}{2}\right)^{q+1} \frac{M_{q+1}}{(q+1)!}. \tag{5.8}$$

Substituting this with $q = d$ into (5.4) or (5.5) yields (5.6) or (5.7). ■

When we studied polynomial interpolation, we learned that interpolants of high degree are likely to oscillate and provide unsatisfactory fits. The situation now is different because it is the area under the curve that is being approximated, and it seems at least possible that the oscillations will average out. This is the importance of the special case of formulas with all $A_i > 0$. At least as far as the bound of the Theorem 5.1 goes, increasing the degree of precision with such formulas can only help. Unfortunately, the interpolatory quadrature formulas for $\int_a^b f(x)\,dx$ based on $\{x_i\}$ equally spaced in $[a,b]$, which are called *Newton–Cotes* formulas, have some A_i that are negative for even modest degrees. The results of these formulas may not converge to the value of the integral as the degree is increased. However, we shall take up another family of formulas of arbitrarily high degree of precision for which all the A_i are positive.

In the bounds (5.4), (5.5) we can choose *any* polynomial $p(x)$ of any degree $q \leq d$. For finite a, b there is a polynomial $p^*(x)$ of degree at most d that is as close as possible to f in the sense that

$$||f - p^*|| = \min_{\substack{p \text{ a poly.} \\ \text{of deg. } \leq q}} ||f - p||.$$

The code that accompanies this chapter, called Adapt, is based on two formulas with $A_i > 0$ for all i. One has $d = 5$ and the other, $d = 11$. In the bound (5.5), in the one case we have $||f - p^*||$ for the best possible approximation by a polynomial p_5^* of degree 5 and in the other, p_{11}^* of degree 11. According to the bound, the formula of degree 11 cannot be worse because the polynomial of degree 5 can be considered a polynomial of degree 11 with some zero coefficients. It would be remarkable if it were not the case that $||f - p_{11}^*||$ is quite a bit smaller than $||f - p_5^*||$, hence that the formula of degree 11 is quite a bit more accurate than the formula of degree 5.

A more detailed analysis of the error shows that for many formulas with $w(x) \equiv 1$, including *all* those considered in this book, the error $E(f)$ can be expressed as

$$E(f) = c\left(\frac{b-a}{2}\right)^{d+2} f^{(d+1)}(\xi) \tag{5.9}$$

for some ξ in (a,b). Note that this is an equality rather than a bound.

The result (5.9) is a traditional one, but when it involves a derivative of high order, it causes people to doubt whether the formula is practical. For instance, the formula of degree 11 mentioned earlier satisfies (5.9) with $f^{(12)}$. It is hard to come to any understanding of such a high order derivative, and a natural question is, What happens if you use the formula and the derivative does not exist? We appreciate now that the form (5.9) is just a consequence of the method of analysis. The inequality (5.8) provides bounds when f has only $q+1$ continuous derivatives, and the bound based on best approximation does not directly assume *any* continuous derivatives. There is no reason to fear a formula of high degree of precision because of an expression like (5.9); other expressions for its error are applicable when the function is less smooth.

If a quadrature formula has the degree of precision d, then

$$E(x^j) = 0, \; j = 0, 1, \ldots, d, \tag{5.10}$$

and

$$E(x^{d+1}) \neq 0. \tag{5.11}$$

If we *assume* that the error has the form (5.9), it is easy to find c from

$$E(x^{d+1}) = c\left(\frac{b-a}{2}\right)^{d+2}(d+1)!.$$

The equations (5.10), (5.11) furnish another way to generate quadrature rules. The approach is known as the *method of undetermined coefficients*. In this approach the coefficients $\{A_i\}$ are regarded as unknowns that are found by satisfying the system of linear equations (5.10) for d as large as possible. Before giving examples, we note that It is often convenient to apply the method of undetermined coefficients to a standard interval $[-1,1]$ and then transform to a general interval $[a,b]$ by a simple change of variable. If we have

$$\int_{-1}^{1} f(x)\,dx = \sum_{i=1}^{N} A_i f(x_i) + c f^{(d+1)}(\xi),$$

let

$$t = \frac{b-a}{2}(x+1) + a.$$

Then $dt = (b-a)dx/2$ and

$$\int_a^b f(t)\,dt = \frac{b-a}{2}\int_{-1}^{1} f\left(\frac{b-a}{2}x_i + a\right)dx$$

$$= \frac{b-a}{2}\sum_{i=1}^{N} A_i f\left(\frac{b-a}{2}x_i + \frac{a+b}{2}\right) + \frac{b-a}{2}E(f).$$

Since

$$\frac{d}{dx} = \frac{dt}{dx}\frac{d}{dt} = \frac{b-a}{2}\frac{d}{dt},$$

it follows that

$$\frac{d^{d+1}}{dx^{d+1}} = \left(\frac{b-a}{2}\right)^{d+1}\frac{d^{d+1}}{dt^{d+1}},$$

so that the change of variable yields

$$\int_a^b f(t)\,dt = \sum_{i=1}^{N}\left(\frac{b-a}{2}A_i\right) f\left(\frac{b-a}{2}x_i + \frac{a+b}{2}\right) + \left(\frac{b-a}{2}\right)^{d+2} f^{(d+1)}(\xi). \quad (5.12)$$

Example 5.1. We seek a quadrature formula of the form

$$\int_{-1}^{1} f(x)\,dx = A_1 f(-1) + A_2 f(1) + E(f).$$

In the method of undetermined coefficients

$$\begin{aligned} f(x) &= 1 \text{ implies } 2 = A_1 + A_2 \\ f(x) &= x \text{ implies } 0 = -A_1 + A_2. \end{aligned}$$

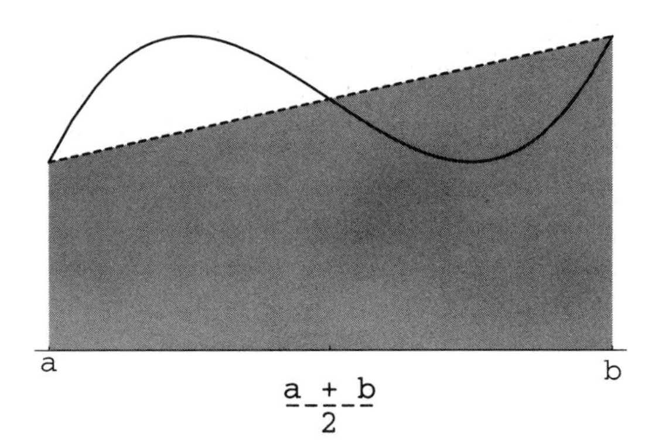

Figure 5.1 trapezoid rule.

Hence, $A_1 = A_2 = 1$. We also observe that, by construction, $d \geq 1$. Then $f(x) = x^2$ yields

$$\frac{2}{3} = A_1 + A_2 + E(x^2).$$

Since $E(x^2) \neq 0$ this tells us that $d = 1$ and $c = E(x^2)/2! = -2/3$, which is to say that

$$\int_{-1}^{1} f(x)\,dx = f(-1) + f(1) - \frac{2}{3} f''(\xi)$$

for some ξ in $(-1, 1)$.

For a general interval $[a, b]$ we apply the result of the change of variable formula (5.12). We have (in braces) the

---

trapezoid rule:

$$\int_{a}^{b} f(x)\,dx = \left\{ \frac{b-a}{2} (f(a) + f(b)) \right\} - \frac{1}{12}(b-a)^3 f''(\xi_t)$$

---

for some ξ_t in (a, b). The name of the quadrature rule comes from the fact that it amounts to approximating the area under the curve $f(x)$ by the area of a trapezoid. See Figure 5.1 for an illustration. ■

Example 5.2. Let us find the most accurate formula of the form

$$\int_{-1}^{1} f(x)\,dx = A_1 f(-1) + A_2 f(0) + A_3 f(1) + E(f).$$

In the method of undetermined coefficients, we try to integrate powers of x exactly to as high a degree as possible.

$$\begin{array}{rclcrcrcrcr} f(x) &=& 1 & \text{implies} & 2 &=& A_1 &+& A_2 &+& A_3 \\ f(x) &=& x & \text{implies} & 0 &=& -A_1 & & &+& A_3 \\ f(x) &=& x^2 & \text{implies} & \frac{2}{3} &=& A_1 & & &+& A_3. \end{array}$$

This set of equations determines the coefficients to be $A_1 = A_3 = \frac{1}{3}, A_2 = \frac{4}{3}$. To find the degree of precision d, we check the next higher power,

$$f(x) = x^3 \text{ implies } 0 = -A_1 + A_3 + E(x^3) = E(x^3)$$

and find that d is at least 3, higher than we might have expected. For the next power,

$$f(x) = x^4 \text{ implies } \frac{2}{5} = A_1 + A_3 + E(x^4) = \frac{2}{3} + E(x^4),$$

so $d = 3$ and $E(x^4) = (2/5) - (2/3) = -(4/15)$. Then $c = -(4/15)/4! = -1/90$ and

$$\int_{-1}^{1} f(x)\,dx = \frac{1}{3}f(-1) + \frac{4}{3}f(0) + \frac{1}{3}f(1) - \frac{1}{90}f^{(4)}(\xi)$$

for some ξ. As in (5.12), for a general $[a,b]$ this gives us

---

Simpson's rule:

$$\int_a^b f(x)\,dx = \left\{\frac{b-a}{6}\left[f(a) + 4f\left(\frac{a+b}{2}\right) + f(b)\right]\right\} - \frac{1}{2880}(b-a)^5 f^{(4)}(\xi). \quad (5.13)$$

See Figure 5.2 for an illustration. ∎

Both these formulas are Newton–Cotes formulas because the nodes are equally spaced in $[a,b]$. The procedure was to select the nodes $\{x_i\}$ and then solve a system of linear equations for the $\{A_i\}$. This is typical when the nodes are specified in advance. But what if the $\{x_i\}$ are allowed to be unknown as well? With twice as many unknowns, $2N$, at our disposal, we might hope to find formulas with a much higher degree of precision, perhaps even $2N-1$; that is, we might hope to get a $(2N-1)$st degree formula that uses only N evaluations of f. Unfortunately, the system of equations for $\{A_i\}$ and $\{x_i\}$ is then not linear. It is not obvious that there are real solutions, and if there are, how to obtain them. Gauss elegantly solved the problem for general N, even with rather general weight functions and infinite intervals. The resulting formulas are known as *Gaussian quadrature formulas*. Special cases can be worked out in an elementary way.

Example 5.3. For $N = 1$ the Gaussian formula has the form

$$\int_{-1}^{1} f(x)\,dx = A_1 f(x_1) + E(f).$$

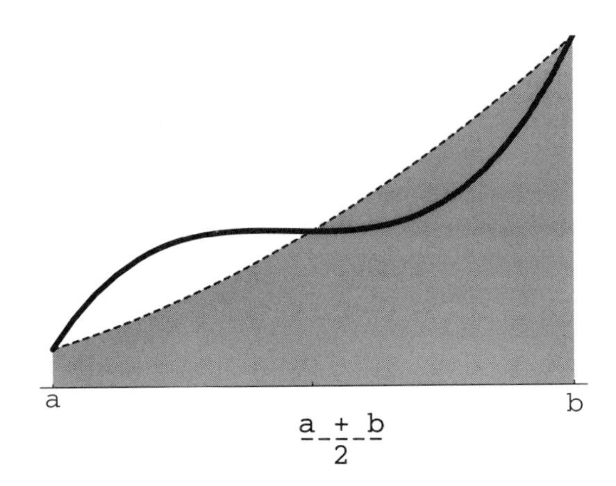

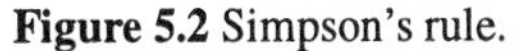

Figure 5.2 Simpson's rule.

In the method of undetermined coefficients,

$$\begin{aligned} f(x) &= 1 \text{ implies } 2 = A_1 \\ f(x) &= x \text{ implies } 0 = A_1 x_1, \end{aligned}$$

hence, $A_1 = 2$ and $x_1 = 0$. To determine the error, we try

$$f(x) = x^2 \text{ implies } \tfrac{2}{3} = 2 \times 0 + E(x^2),$$

and find $d = 1$, $c = (\frac{2}{3})/2 = \frac{1}{3}$, and

$$\int_{-1}^{1} f(x)\,dx = 2f(0) + \frac{1}{3} f''(\xi).$$

On $[a, b]$ this becomes

$$\int_{a}^{b} f(x)\,dx = \left\{(b-a) f\left(\frac{a+b}{2}\right)\right\} + \frac{1}{24}(b-a)^3 f''(\xi). \tag{5.14}$$

This formula is known as the *midpoint rule.* ■

Example 5.4. For $N = 3$, the form is

$$\int_{-1}^{1} f(x)\,dx = A_1 f(x_1) + A_2 f(x_2) + A_3 f(x_3) + E(f).$$

Because of the symmetry of the interval $[-1, 1]$, it is plausible that $A_1 = A_3$, $x_2 = 0$, and $x_1 = -x_3$, so let us try

$$\int_{-1}^{1} f(x)\,dx = A_1 f(x_1) + A_2 f(0) + A_1 f(-x_1) + E(f).$$

Now,

$$
\begin{array}{rclrrclcll}
f(x) & = & 1 & \text{implies} & 2 & = & 2A_1 & + & A_2 & \\
f(x) & = & x & \text{implies} & 0 & = & A_1x_1 & - & A_1x_1 & = 0 \text{ (automatic)} \\
f(x) & = & x^2 & \text{implies} & \frac{2}{3} & = & 2A_1x_1^2 & & & \\
f(x) & = & x^3 & \text{implies} & 0 & = & A_1x_1^3 & - & A_1x_1^3 & = 0 \text{ (automatic)} \\
f(x) & = & x^4 & \text{implies} & \frac{2}{5} & = & 2A_1x_1^4. & & &
\end{array}
$$

At this point we have three equations in the three unknowns A_1, A_2, and x_1. The last two equations require that $x_1^2 = 3/5, A_1 = 5/9$ and the first that $A_2 = 8/9$. To find the error, we try

$$f(x) = x^5 \text{ implies } 0 = A_1x_1^5 - A_1x_1^5 + E(x^5) = E(x^5).$$

This implies that $d \geq 5$. Finally

$$
\begin{aligned}
f(x) = x^6 \text{ implies } \frac{2}{7} = 2A_1x_1^6 + E(x^6) &= \left(\frac{10}{9}\right)\left(\frac{3}{5}\right)^3 + E(x^6) \\
&= \frac{6}{25} + E(x^6).
\end{aligned}
$$

This says that $d = 5$ and $c = \left(\frac{2}{7} - \frac{6}{25}\right)/6!$. Collecting the results,

$$\int_{-1}^{1} f(x)\,dx = \frac{1}{9}\left[5f\left(-\sqrt{\frac{3}{5}}\right) + 8f(0) + 5f\left(\sqrt{\frac{3}{5}}\right)\right] + \frac{1}{15{,}750}f^{(6)}(\xi).$$

On $[a,b]$ the resulting quadrature rule is called the

---

three-point Gaussian quadrature formula,

$$\frac{b-a}{18}\left[5f\left(\frac{a+b}{2} - \frac{b-a}{2}\sqrt{\frac{3}{5}}\right) + 8f\left(\frac{a+b}{2}\right) + 5f\left(\frac{a+b}{2} + \frac{b-a}{2}\sqrt{\frac{3}{5}}\right)\right], \tag{5.15}$$

---

and its error is

$$\frac{(b-a)^7}{2{,}016{,}000}f^{(6)}(\xi). \tag{5.16}$$

See Figure 5.3 for an illustration. ■

For larger N the method of undetermined coefficients is impractical for deriving Gaussian quadrature rules. Besides, the questions of the existence of formulas and the best possible degree of precision are left open in this approach. Gauss used the theory of orthogonal polynomials to answer these questions. We cannot develop the theory here (see [8, pp. 327–331]), but it is possible to see how high degrees of precision can

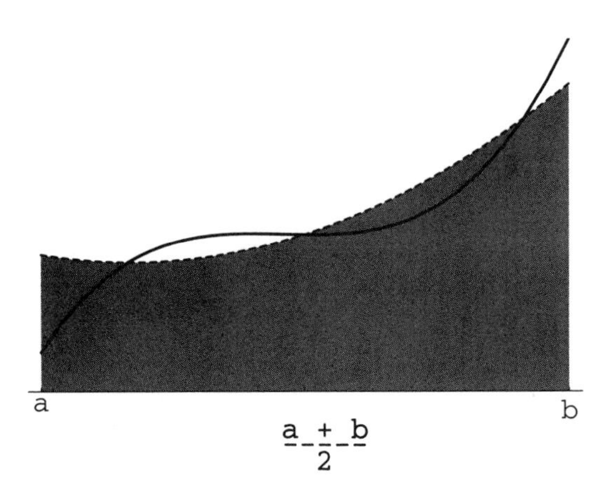

Figure 5.3 Three-point Gaussian quadrature.

be achieved. With reasonable conditions on $w(x)$ and $[a,b]$, it is known that there is a sequence of polynomials $\theta_{N+1}(x)$, $N = 0, 1, \ldots$, such that $\theta_{N+1}(x)$ is of degree N and

$$\int_a^b x^j \theta_{N+1}(x) w(x)\, dx = 0 \quad \text{for} \quad j < N. \tag{5.17}$$

When $w(x) \equiv 1$, $a = -1$, and $b = 1$, these polynomials are the Legendre polynomials (see [8, p. 202]). It is also known that the N distinct roots of $\theta_{N+1}(x)$ are real and lie in (a,b). Suppose that an interpolatory quadrature formula (5.2) is based on interpolation at the roots of $\theta_{N+1}(x)$. If $f(x)$ is any polynomial of degree $2N-1$, it can be written as

$$f(x) = q(x)\theta_{N+1}(x) + r(x),$$

where the quotient $q(x)$ and remainder $r(x)$ polynomials are of degree at most $N-1$. Then

$$\begin{aligned}\int_a^b f(x)w(x)\, dx &= \int_a^b q(x)\theta_{N+1}(x)w(x)\, dx + \int_a^b r(x)w(x)\, dx \\ &= \int_a^b r(x)w(x)\, dx,\end{aligned}$$

where the first integral vanishes because of (5.17). For any choice of the nodes $\{x_i\}$, the formula of (5.2) integrates a polynomial of degree N exactly, so

$$\int_a^b r(x)w(x)\, dx = \sum_{i=1}^{N} A_i r(x_i).$$

The formula applied to $f(x)$ has

$$\sum_{i=1}^{N} A_i f(x_i) = \sum_{i=1}^{N} A_i q(x_i)\theta_{N+1}(x_i) + \sum_{i=1}^{N} A_i r(x_i).$$

Now we use the fact that the x_i are roots of $\theta_N(x)$ to see that

$$\sum_{i=1}^{N} A_i f(x_i) = \sum_{i=1}^{N} A_i r(x_i) = \int_a^b r(x)w(x)\,dx = \int_a^b f(x)w(x)\,dx.$$

Since any polynomial $f(x)$ of degree $2N-1$ is integrated exactly, this formula has a degree of precision that is at least $2N-1$.

There are computationally convenient ways to derive Gaussian quadrature rules, and formulas may be found in specialized books. Gaussian formulas are valuable because they provide the highest degree of precision for the number of values of $f(x)$. An important fact about Gaussian formulas is that the A_i are all positive. As discussed in connection with the error bounds, this means that we can use formulas of a very high degree of precision, even for integrands that are not smooth. Gaussian formulas incorporating weight functions are especially valuable tools for dealing with integrands that are singular and intervals that are infinite. Whether or not there is a weight function, the nodes of a Gaussian formula all lie in the interior of $[a,b]$. This means that the formula does not use $f(a)$ or $f(b)$. We shall see that this is quite helpful in dealing with singular integrands.

So far we have been considering procedures based on approximating $f(x)$ over the whole interval $[a,b]$. Just as with polynomial interpolation, the error depends strongly on the length of the interval. This suggests that we break up the interval and so approximate the function by a piecewise polynomial function rather than a single polynomial. The simplest approach is to split the interval into pieces specified in advance. If we partition $[a,b]$ into $a = x_1 < x_2 < \cdots < x_{n+1} = b$, then $\int_a^b f(x)\,dx = \sum_{i=1}^{n} \int_{x_i}^{x_{i+1}} f(x)\,dx$, and we can apply standard quadrature rules to each of the n integrals. The resulting formula is known as a *composite* or *compound* rule. Traditionally the $\{x_i\}$ have been chosen to be equally spaced in $[a,b]$ and the same formula used on each piece, but this is not necessary.

Example 5.5. Composite Trapezoid Rule. The composite trapezoid rule approximates $I = \int_a^b f(x)\,dx$ by splitting $[a,b]$ into n pieces of length $h = (b-a)/n$ and applying the trapezoid rule to each piece. With the definition $x_i = a + ih$, this is

$$I \approx T_n = \frac{h}{2}\left[f(x_0) + f(x_1)\right] + \cdots + \frac{h}{2}\left[f(x_{n-1}) + f(x_n)\right],$$

which simplifies to

$$T_n = h\left[\frac{1}{2}f(a) + f(x_1) + \cdots + f(x_{n-1}) + \frac{1}{2}f(b)\right].$$

Figure 5.4 illustrates the composite trapezoid rule. ■

An ingenious use of repeated integration by parts establishes the Euler–Maclaurin sum formula. It states that if $f^{(2\nu)}(x)$ is continuous on $[a,b]$, then

$$I = T_n - \sum_{k=1}^{\nu-1} \frac{h^{2k}}{(2k)!} B_{2k}\left[f^{(2k-1)}(b) - f^{(2k-1)}(a)\right] - \frac{nh^{2\nu+1}}{(2\nu)!} B_{2\nu} f^{(2\nu)}(\xi)$$

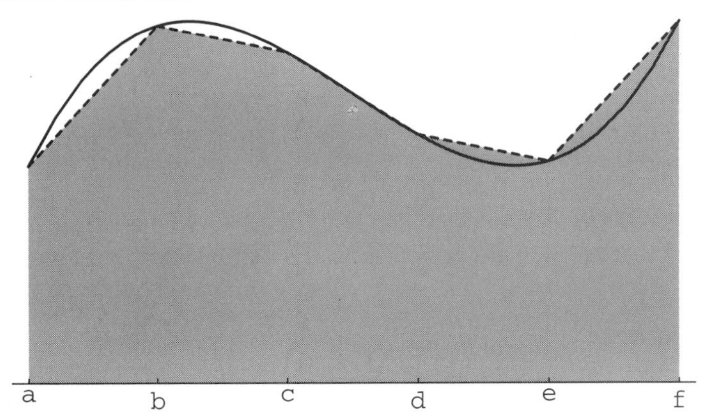

Figure 5.4 Composite trapezoid rule.

for some ξ in $[a,b]$. The coefficients B_{2k} appearing here are known as the Bernoulli numbers. The first few terms of the error expansion are

$$I = T_n - \frac{h^2}{12}\left[f^{(1)}(b) - f^{(1)}(a)\right] + \frac{h^4}{720}\left[f^{(3)}(b) - f^{(3)}(a)\right] - \cdots.$$

The basic trapezoid rule applied to an interval of length h has an error that goes to zero like h^3. When the $n = 1/h$ terms are combined, the error of the approximation to the integral goes to zero like h^2. However, notice that if it should happen that $f^{(1)}(b) = f^{(1)}(a)$, the formula is more accurate than usual. If in addition other derivatives have the same values at the ends of the interval, the formula is still more accurate. When integrating a periodic function over a multiple of a period, all the derivatives at the ends of the interval are equal and this formula is extraordinarily accurate. In fact, if the periodic function is analytic, so that it has derivatives of all orders, $T_n \to I$ faster than any power of h! Although rather special, this is extremely important in the context of Fourier analysis.

The error of T_n can be estimated by comparing it to the more accurate result T_{2n} obtained by halving each subinterval. A convenient way to evaluate the formula is

$$\begin{aligned} T_{2n} &= \frac{h}{2}\left[\frac{1}{2}f(a) + f(x_{1/2}) + f(x_1) + \cdots + f(x_{n-1}) + f(x_{n-1/2}) + \frac{1}{2}f(b)\right] \\ &= \frac{1}{2}(T_n + M_n), \end{aligned}$$

where

$$M_n = h \sum_{k=1}^{n} f(a + (k - 1/2)h).$$

It is important to note that all the evaluations of f made in forming T_n are reused in T_{2n}.

There is a way of exploiting the error expansion of the composite trapezoid rule due to Romberg that is popular for general integrands. The idea is to combine T_n and

T_{2n} in such a way as to obtain a higher order result. According to the error expansion,

$$I = T_n - \frac{h^2}{12}\left[f^{(1)}(b) - f^{(1)}(a)\right] + \frac{h^4}{720}\left[f^{(3)}(b) - f^{(3)}(a)\right] - \cdots$$
$$= T_{2n} - \frac{(h/2)^2}{12}\left[f^{(1)}(b) - f^{(1)}(a)\right] + \frac{(h/2)^4}{720}\left[f^{(3)}(b) - f^{(3)}(a)\right]\cdots.$$

A little manipulation then shows that

$$I = \frac{2^2 T_{2n} - T_n}{2^2 - 1} + \left(\frac{2^2 - 2^4}{2^2 - 1}\right)\frac{(h/2)^4}{720}\left[f^{(3)}(b) - f^{(3)}(a)\right] - \cdots.$$

The formula

$$\frac{2^2 T_{2n} - T_n}{2^2 - 1}$$

is of higher order than each of the individual formulas. As it turns out, this formula is the composite Simpson's rule. Romberg developed a computationally convenient way of successively combining results so as to increase the order by two with each computation of a composite trapezoid rule. The process is called *extrapolation.*

Romberg integration can be very effective. It adapts the order of the method to the problem. It does, however, depend on the integrand being smooth throughout the interval. Also, it evaluates at the ends of the interval, which is sometimes inconvenient. MATHCAD uses Romberg integration for quadrature. If there is a singularity at an end of an interval or if the process does not converge, the code switches to a variant based on the midpoint rule that does not evaluate at the ends of the intervals and divides intervals by 3 rather than 2. ■

EXERCISES

5.1 Use the method of undetermined coefficients to derive Newton's $\frac{3}{8}$-rule:

$$\int_{-1}^{1} f(x)\,dx = A_1 f(-1) + A_2 f\left(-\frac{1}{3}\right) + A_3 f\left(\frac{1}{3}\right) + A_4 f(1) + c f^{(d+1)}(\xi).$$

Calculate A_1, A_2, A_3, A_4, d, and c in the usual manner.

5.2 Use the method of undetermined coefficients to find the two-point Gaussian quadrature formula with its associated error. Begin with

$$\int_{-1}^{1} f(x)\,dx = A_1 f(-x_1) + A_1 f(x_1) + E(f)$$

and calculate A_1 and x_1 in the usual manner. Assuming $E(f) = c f^{(d+1)}(\xi)$, find d and c. What is the corresponding formula and associated error on the general interval $[a,b]$?

5.3 Implement the composite trapezoid rule and apply it to

$$\int_0^{\pi} \frac{dx}{4 + \sin(20x)}.$$

Of course, you must choose h small enough that samples are taken in each period. Approximate the integral for a number of values of h that tend to 0. According to the theory of Example 5.5, the approximations T_n ought to converge extremely fast. Is that what you find?

5.2 ADAPTIVE QUADRATURE

In this section a code is developed that approximates $I = \int_a^b f(x)\,dx$ to an accuracy specified by the user. This is done by splitting the interval $[a,b]$ into pieces and applying a basic formula to each piece. The interval is split in a way adapted to the behavior of $f(x)$. A fundamental tool is an error estimate. With the capability of estimating the error of integrating over a subinterval, we ask if the error is acceptable, and if it is not, the subinterval is split again. As we have seen, for a formula of even modest degree of precision, reducing the length of the interval increases substantially the accuracy of the approximation. Proceeding in this manner, the formula is applied over long subintervals where $f(x)$ is easy to approximate and over short ones where it is difficult. Codes like the one developed here are in very wide use, being found, for example, in the collection of state-of-the-art codes QUADPACK [12], libraries like NAG and IMSL, and computing environments like MATLAB.

When the code is supplied absolute and relative error tolerances ABSERR and RELERR, it attempts to calculate a value ANSWER such that

$$|I - \text{ANSWER}| \leq \max(\text{ABSERR}, \text{RELERR} \times |I|).$$

The computational form of this uses an error estimate

$$\text{ERREST} \approx \int_a^b f(x)\,dx - \text{ANSWER}$$

and replaces I by ANSWER:

$$|\text{ERREST}| \leq \max(\text{ABSERR}, \text{RELERR} \times |\text{ANSWER}|). \tag{5.18}$$

We cannot hope to get a more accurate approximate integral than the correctly rounded true value, so it makes no sense to take RELERR $< u$, the unit roundoff of the computer used. Indeed, we require RELERR $\geq 10u$ so that we do not work with error estimates that are nothing but roundoff. We also require ABSERR > 0 so as to deal with the rare situation that $I = 0$.

The method employed in the code breaks the interval $[a,b]$ into pieces $[\alpha,\beta]$ on which the basic quadrature rule is sufficiently accurate. To decide this we must be able to estimate the error of the rule. This is done with a basic principle of numerical analysis, namely to estimate the error of a result by comparing it to a more accurate result. Besides the approximation

$$Q \approx \int_\alpha^\beta f(x)\,dx,$$

another approximation $\widehat{Q}$ is formed that is believed to be more accurate. Then

$$\int_\alpha^\beta f(x)\,dx - Q = \widehat{Q} - Q + \left[\int_\alpha^\beta f(x)\,dx - \widehat{Q}\right]$$

says that when $\widehat{Q}$ is more accurate than Q, the error in Q is approximately equal to $\widehat{Q} - Q$.

To keep down the cost of estimating the error, we use evaluations of $f(x)$ in both formulas. As a simple example, we might take Q to be the trapezoid rule and $\widehat{Q}$ to be Simpson's rule. The trapezoid rule is based on the values $f(\alpha)$ and $f(\beta)$. The error estimate is computed with Simpson's rule, which needs only the one additional value $f((\alpha+\beta)/2)$. Simpson's rule is considerably more accurate than the trapezoid rule, giving us reason to hope that the estimate will be a good one.

The code Adapt uses for Q the three-point Gaussian quadrature formula of Example 5.4 that has degree of precision $d_1 = 5$. A formula of much higher degree of precision is used for $\widehat{Q}$. The error analysis based on best possible polynomial approximation gives us confidence that $\widehat{Q}$ *will* be more accurate than Q. It would be possible to use another Gaussian rule for $\widehat{Q}$, but the N-point Gaussian rule for $N \neq 3$ uses N completely different $\{x_i\}$ (except possibly $x = 0$). To keep the number of f evaluations to a minimum, another approach is taken. In 1964 Kronrod derived for each N-point Gaussian rule a corresponding rule of degree $3N+1$ or $3N+2$ (depending on whether N is even or odd). The idea was to start with the N Gaussian nodes and add $N+1$ others chosen to obtain the highest possible degree of precision. Formulas for $N \leq 40$ are tabulated in [10]. The $N = 3$ case is

$$\int_{-1}^{1} f(x)\,dx = \{A_1 f(x_1) + A_2 f(x_2) + A_3 f(x_3) + A_4 f(0)$$
$$+ A_3 f(-x_3) + A_2 f(-x_2) + A_1 f(-x_1)\} + 5.84 * 10^{-13} f^{(12)}(\xi),$$

where

$$x_1 = \sqrt{\frac{3}{5}} \quad \text{(as for three-point Gaussian quadrature)}$$
$$x_2 = 0.9604912687080202$$
$$x_3 = 0.4342437493468026$$
$$A_1 = 0.2684880898683334$$
$$A_2 = 0.1046562260264672$$
$$A_3 = 0.4013974147759622$$
$$A_4 = 0.4509165386584744.$$

The basic idea of adaptive quadrature is to approximate the integral over a subinterval (α, β). If the approximation is not accurate enough, the interval is split into (usually) two parts and the integral approximated over each subinterval. Eventually, accurate approximations are computed on all the subintervals of $[a, b]$ that are added up to approximate the integral over $[a, b]$ or the cost is deemed too high and the computation terminated. The error terms of the formulas $Q, \widehat{Q}$ quantify the benefit of splitting an interval; recall from (5.16) that

$$\int_{\alpha}^{\beta} f(x)\,dx - Q = (4.96 \times 10^{-7})(\beta - \alpha)^7 f^{(6)}(\xi_1).$$

The corresponding result for the Kronrod formula of seven points is

$$\int_{\alpha}^{\beta} f(x)\,dx - \widehat{Q} = (7.14 \times 10^{-17})(\beta - \alpha)^{13} f^{(12)}(\xi).$$

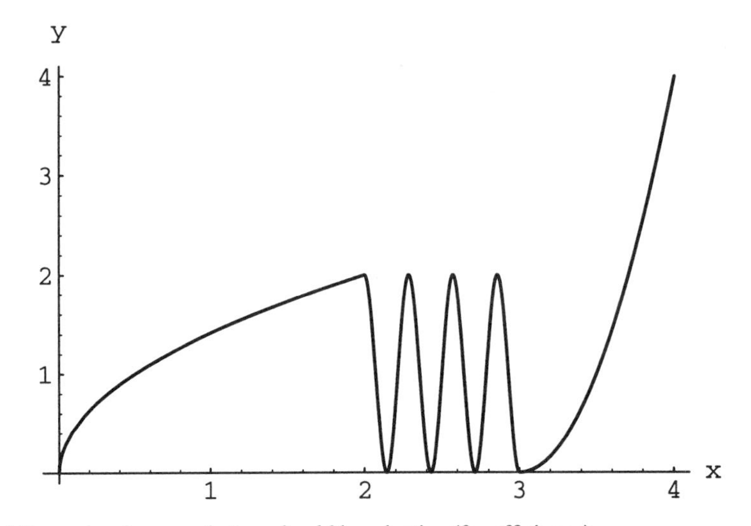

Figure 5.5 Example where quadrature should be adaptive (for efficiency).

Clearly, halving $\beta - \alpha$ will generally result in *much* more accurate approximations. This process will resort to short intervals only where $f(x)$ changes rapidly and long ones elsewhere. For example, the function $f(x)$ graphed in Figure 5.5 is likely to require short intervals near 0 and [2, 3], but not elsewhere. A process like the one described is called *adaptive* because where $f(x)$ is evaluated in the approximation of its integral depends on its behavior.

It is important to understand that quadrature formulas of the kind taken up in this chapter sample f at only a finite number of points, and if these points are not representative of the behavior of the curve $f(x)$, the result will not be accurate. What is even worse is that the error estimate comes from comparing the result to that of another formula of the same kind, so it is possible that both are inaccurate. Because of this *any* quadrature formula and error estimate of the kind taken up here is doomed to be completely wrong on some problems. As a concrete example, for the Gauss–Kronrod ($N = 3$) case, let

$$f(x) = x^2(x^2 - x_1^2)^2(x^2 - x_2^2)^2(x^2 - x_3^2)^2.$$

Clearly, $f(x) \geq 0$ on $[-1,1]$, so $\int_{-1}^{1} f(x)\,dx$ is a positive number. Yet both formulas see only $f(x_i) = 0$, hence calculate $Q = \widehat{Q} = 0$. The result is terribly wrong and the error estimate does not detect this. Applying a quadrature code blindly can get you into trouble!

The core of an adaptive quadrature code is a function that approximates $\int_{\alpha_1}^{\beta_1} f(x)\,dx$ by Q and estimates the error of the approximation by $E_1 := \widehat{Q} - Q$. Suppose we want $I = \int_a^b f(x)\,dx$ to an accuracy

$$\text{TOL} := \max(\text{ABSERR}, \text{RELERR} \times |\text{ANSWER}|).$$

A state-of-the-art code like those of QUADPACK [12] proceeds as follows. At a given stage of the process the interval $[a,b]$ is partitioned into subintervals $a = \alpha_1 < \beta_1 =$

$\alpha_2 < \beta_2 = \alpha_3 < \cdots < \beta_N = b$; there is an estimate Q_j available for the integral of $f(x)$ over each subinterval $[\alpha_j, \beta_j]$ and an estimate E_j available of the error of Q_j. By adding up the Q_j and E_j, an approximation $Q =$ ANSWER to I is available along with an estimate $E =$ ERREST of its error. If the current approximation to I is not good enough, the subinterval $[\alpha_j, \beta_j]$ with the *largest* error $|E_j|$ is selected for improvement. It is split into two pieces $[\alpha_j, (\alpha_j+\beta_j)/2]$, $[(\alpha_j+\beta_j)/2, \beta_j]$, and approximate integrals over these pieces and estimates of their error are computed. The two subintervals and the associated quantities replace the subinterval $[\alpha_j, \beta_j]$ and its associated quantities. This global adaptive procedure is extremely efficient, but at a cost of keeping track of all the subintervals and the information associated with them. In the code Adapt the adaptation is more local and the implementation a little simpler. The difference in Adapt is that when a subinterval is integrated with enough accuracy, it is no longer a candidate for improvement. Thus a queue is kept of subintervals on which the estimated error of the integral over the subinterval is considered to be too large. At a typical stage of the process, the code tests whether the current approximation to I is sufficiently accurate to satisfy the user's error request. If it is not, the *next* subinterval $[\alpha_j, \beta_j]$ is removed from the queue, split in half, and approximations to the integral over each half, along with error estimates, are computed. They are placed at the end of the queue. If an approximation to an integral $\int_\alpha^\beta f(x)\,dx$ is estimated to have an error no more than $|(\beta-\alpha)/(b-a)| \times$ TOL, it is accepted and $[\alpha, \beta]$ is not examined again. This is more stringent than necessary because adding up approximations

$$\int_a^b f(x)\,dx = \sum \int_{\alpha_j}^{\beta_j} f(x)dx \approx \sum Q_j$$

that satisfy this condition will result in an error such that

$$|error| = |\sum E_i| \le \sum \left|\frac{\beta-\alpha}{b-a}\right| \times \text{TOL} = \text{TOL}.$$

The global adaptation used in the collection QUADPACK [12] subdivides until $|\sum E_i| \le$ TOL. The more local adaptation of Adapt subdivides until $\sum |E_j| \le$ TOL. Because $\sum |E_j|$ is always at least as big as $|\sum E_j|$, and perhaps much larger, the local process works harder but should provide answers that are more accurate than the specified tolerance TOL.

Let us now formalize the algorithm. To start the process, form the approximation Q to the integral over $[a,b]$ and its estimated error ERREST. If |ERREST| $\le$ TOL, we take ANSWER $= Q$ as the result and return. If ERREST does not pass the error test, a, b, Q, and $E =$ ERREST are placed in the queue and the following loop is entered:

remove α, β, Q, E from top of queue
compute $QL \approx \int_\alpha^{(\alpha+\beta)/2} f(x)\,dx$ with estimated error EL
compute $QR \approx \int_{(\alpha+\beta)/2}^{\beta} f(x)\,dx$ with estimated error ER
ANSWER := ANSWER + $((QL+QR)-Q)$
ERREST := ERREST + $((EL+ER)-E)$
if EL is too big, add α, $(\alpha+\beta)/2$, QL, EL to end of queue
if ER is too big, add $(\alpha+\beta)/2$, β, QR, ER to end of queue.

This procedure is repeated until one of the following events occurs:

1. The queue becomes empty.
2. $|\text{ERREST}| \leq \text{TOL}$.
3. The queue becomes larger than the space allocated.
4. Too many f evaluations are made.

The first two cases represent a success. The last two represent a failure in the sense that the code has failed to achieve the desired accuracy in the work allotted. It may be that the estimated error, although larger than specified by means of the tolerances, is acceptable and the answer reported will suffice. Even when it is not, an inaccurate value of the integral may be useful as a indication of the size of the integral when selecting appropriate tolerances.

Notice the way quantities have been grouped in the computation of ANSWER and ERREST. The quantity Q and the sum $QL+QR$ both approximate the integral of f over $[\alpha, \beta]$. They normally agree in a few leading digits, so their difference involves cancellation and is computed without arithmetic error. Because the difference is normally a small correction to ANSWER, it is possible to make a great many corrections without accumulation of arithmetic error. If the quantities are not grouped, the correction of a "small" ANSWER may be inaccurate.

5.3 CODES FOR ADAPTIVE QUADRATURE

The code presented here uses the three-point Gaussian rule to estimate integrals along with the seven-point Kronrod rule to estimate errors. In FORTRAN the routine Adapt has a typical call

```
CALL ADAPT(F,A,B,ABSERR,RELERR,ANSWER,ERREST,FLAG)
```

and in C++,

```
flag = Adapt(f, a, b, abserr, relerr, answer, errest);
```

and it is

```
flag = Adapt(f, a, b, abserr, relerr, &answer, &errest);
```

in the C version.

The first argument, F or f, is the name of the function that provides values of the integral. In FORTRAN F must be declared in an EXTERNAL statement. The next four variables are input parameters: A and B are the (finite) end points of the integration interval, ABSERR and RELERR are the required absolute and relative error tolerances. The remaining variables are output quantities: ANSWER contains the approximation to the integral and ERREST an estimate for the error of the approximation. The value of FLAG is 0 for a normal return with (5.18) satisfied. FLAG > 0

signals that there was not enough storage for the queue, or that too many function evaluations were needed. Illegal input (ABSERR ≤ 0 or RELERR $< 10u$) is indicated by FLAG $= -1$. In C and C++ the value of flag is the return value of the function Adapt.

Example 5.6. Let us try the code on a problem with an analytical integral, for example, $\int_0^1 e^x \, dx$ for which $I = e - 1 = 1.71828182845904$. Its output for a requested accuracy of ABSERR $= 10^{-5}$ and RELERR $= 10^{-8}$ follows.

```
FLAG =              0
Approximate value of the integral =        1.718281004372522
Error in ANSWER is approximately     8.240865232136876E-007
         7 evaluations of F were required.
```

Routine Adapt evaluates f at least seven times, so the code found this task to be very easy. It is seen that ERREST approximates the true error of $0.82408652 \times 10^{-6}$ extremely well. ■

Example 5.7. A more interesting problem is to estimate $\int_0^1 x^{1/7}/(x^2+1) \, dx$, again with ABSERR $= 10^{-5}$, RELERR $= 10^{-8}$. Although continuous at $x = 0$, the integrand has infinite derivatives there, so it does not behave like a polynomial near this point. An adaptive code deals with this by using short intervals near the point. Application of the Adapt code results in

```
FLAG =              0
Approximate value of the integral =    6.718072986891337E-001
Error in ANSWER is approximately     -6.448552584540975E-005
       119 evaluations of F were required.
```

The techniques of the next section can be used to produce an accurate value of $I = 0.671800032402$ for the integral, which tells us that the actual error is about -0.73×10^{-5}. In comparison to Example 5.6, this problem required many more evaluations of f, and ERREST is not nearly as accurate. It should be appreciated that all we need is a rough approximation to the error, so this one is perfectly acceptable. It is instructive to see how Adapt samples the integrand in this example. In Figure 5.6 ordered pairs (x, y) are plotted (as are pairs $(x, 0)$ along the x-axis); x is a sampling point used by Adapt and y the corresponding value of the integrand. Notice how the points congregate around the singularity at zero as the theory predicts they should. ■

EXERCISES

Unless otherwise indicated, use as tolerances ABSERR $= 10^{-8}$ and RELERR $= 10^{-6}$ for the computing exercises.

5.4 To test out Adapt try the following integrands.

(a) $4/(1+x^2)$

(b) $x^{1/10}$

(c) $1/[(3x-2)^2]$

(d) $1 + \sin^2 38\pi x$

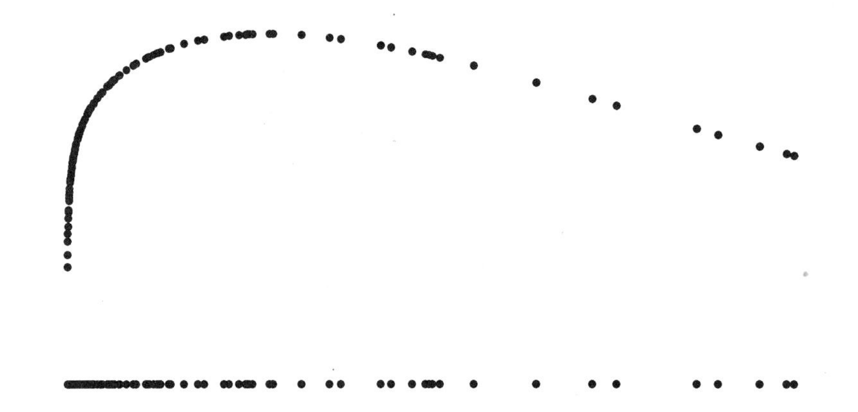

Figure 5.6 Sampling of the integrand of Example 5.8 by Adapt.

(e) $f(x) = \begin{cases} 0 & 0 \le x \le 0.1 \\ 2 & 0.1 < x < 0.6 \\ -1 & 0.6 \le x \le 1 \end{cases}$

(f) $\left|x - \frac{1}{4}\right|^{-1/2}$

Use ABSERR $= 10^{-12}$, RELERR $= 10^{-6}$, and $A = 0$, $B = 1$ for all parts. Which problems require the most function evaluations? Why? What are the exact answers? What are the actual errors? How good an estimate is ERREST? On which problems is it better than others? Why?

5.5 The integrand in Exercise 5.4d has period $1/38$. Rewrite this as $38\int_0^{1/38}(1 + \sin^2 38\pi x)\,dx$ and use Adapt on this with the same tolerances as in Exercise 5.4. Is this approach faster? More accurate?

5.6 If a package is not available for evaluating the zeroth order Bessel function $J_0(x)$, then an alternative is based on the formula

$$J_0(x) = \frac{1}{\pi}\int_0^{\pi}\cos(x\sin\theta)\,d\theta.$$

Use the formula to evaluate $J_0(x)$ for $x = 1.0, 2.0, 3.0$. Compare with tabulated results (e.g., [7]).

5.7 The function $y(x) = e^{-x^2}\int_0^x e^{t^2}\,dt$ is called Dawson's integral. Tabulate this function for $x = 0.0$, 0.1, 0.2, 0.3, 0.4, 0.5. To avoid unnecessary function evaluations, split the integral into pieces.

5.8 Derive a step function $f(x)$ (a function that is constant on subintervals of $[a, b]$) for which Adapt returns FLAG = 0, ERREST = 0, yet ANSWER is totally wrong. Explain.

5.9 A sphere of radius R floats half submerged in a liquid. If it is pushed down until the diametral plane is a distance p $(0 < p \le R)$ below the surface of the liquid and is then released, the period of the resulting vibration is

$$T = 8R\sqrt{\frac{R}{g(6R^2 - p^2)}}\int_0^{2\pi}\frac{dt}{\sqrt{1 - k^2\sin^2 t}},$$

where $k^2 = p^2/(6R^2 - p^2)$ and $g = 32.174$ ft/sec^2. For $R = 1$ find T when $p = 0.50, 0.75, 1.00$.

5.10 A population is governed by a seasonally varying ability of the environment to support it. A simple model of this is the differential equation

$$P'(t) = kP(t)\left[M\left(1 - r\cos\frac{\pi}{6}t\right) - P(t)\right],$$

where t measures time in months, $P(t)$ is the population at time t, and the remaining parameters are known constants. This equation will be solved numerically in the next chapter (Exercise 6.19); here we note that the problem can be transformed into

$$P(t) = \frac{P(0)F(t)}{1 + kP(0)\int_0^t F(s)\,ds},$$

where $F(t) = \exp[kM(t - (6r)/\pi\sin(\pi t/6))]$. Assume that $k = 0.001$, $M = 1000$, $r = 0.3$, $P(0) =$

250 and use Adapt efficiently to table $P(t)$ for $t = 0, 3, 6, 9, \ldots, 36$.

5.11 Examine the effect of noise on Adapt as follows. For the input function $F(x)$ use

$$T1 := f(x) \times 10^n$$
$$T2 := T1 + f(x)$$
$$F(x) := T2 - T1$$

where $f(x)$ is the original integrand (use some of the examples in Exercise 5.4). With ABSERR = RELERR $\approx \sqrt{u}$ (u the unit roundoff) try $n \approx -\frac{1}{4}\log_{10} u$ and then $n \approx -\frac{3}{4}\log_{10} u$. What is the behavior? Does the algorithm appear to be stable?

5.12 To solve the nonlinear two-point boundary value problem

$$y'' = e^y - 1, y(0) = 0, y(1) = 3$$

using standard initial value problem codes (e.g., Rke), it is necessary to find the missing initial condition $y'(0)$. Observing that $y'' = \exp(y) - 1$ can be written in the form

$$\frac{d}{dx}\left[\frac{(y')^2}{2} - e^y + y\right] = 0,$$

we can integrate to obtain

$$\frac{(y')^2}{2} - e^y + y = c, \text{ a constant.}$$

Since $y(0) = 0$, this says $y'(0) = \sqrt{2c+2}$. Solving for $y'(x)$ (by separation of variables) yields

$$\sqrt{2}x = \int_0^y \frac{dy}{\sqrt{c + e^y - y}},$$

which, when evaluated at $x = 1$, becomes

$$\sqrt{2} = \int_0^3 \frac{dy}{\sqrt{c + e^y - y}}.$$

Use Adapt and Zero to find c and then $y'(0)$.

5.4 SPECIAL DEVICES FOR INTEGRATION

Quadrature formulas approximate integrals using a finite number of samples. If the samples are not representative, the result may be inaccurate despite an estimated error that is acceptable. Put differently, the approximate integral and its error estimate are based on an assumption that the integrand changes smoothly between samples. Adaptive codes generally do very well at recognizing the behavior of integrands, but $f(x)$ with sharp peaks or many oscillations in the interval present special difficulties. Sometimes it is necessary to assist a code by breaking up the interval of integration yourself so as to ensure that the code will take samples in critical areas. A contrived example will help make the point.

The family of integrals

$$I_n = \int_0^\pi \sin^{2n}(x)\,dx$$

can be evaluated readily with the recurrence

$$I_n = \frac{2n-1}{2n} I_{n-1}, \; I_0 = \pi.$$

When n is large, the integrand has a sharp peak at the midpoint of the interval. With tolerances of ABSERR $= 10^{-5}$ and RELERR $= 10^{-11}$, for $n = 200$ Adapt returned ANSWER $= 0.1252567600019366$ and an estimated error of about -3.654×10^{-6} at a cost of 203 evaluations of the integrand. The true error is about -3.654×10^{-6}. The code had no difficulty producing an accurate answer and an excellent error estimate because it samples at the midpoint and "sees" the peak. However, if the interval is split into $[0, 2.6]$ and $[2.6, \pi]$, the results are quite different. Integrating over the two intervals and adding the results provides an approximate integral of $4.111202459491848 \times 10^{-7}$ and an estimated error of about -1.896×10^{-7} at a cost of 14 evaluations of the integrand. The code has been completely fooled! This happened because the initial samples did not reveal the presence of the peak. The code took the minimum number of samples from each interval, namely seven, showing that it believed the problem to be very easy. When this happens it is prudent to consider whether you agree that the problem is very easy and if not, to break up the interval into pieces that will cause the code to "see" the behavior of the integrand. Of course, one must be careful how to do this breaking into pieces, as $[0, 2.6]$ and $[2.6, \pi]$ won't do for this problem.

Adapt is based on approximation by polynomials over finite intervals. As a consequence it may have to resort to a great many subintervals to integrate functions that do not behave like a polynomial near a critical point or to integrate functions over an infinite interval. Gaussian quadrature formulas with a suitable weight function are a good way to handle such problems. Specific formulas can be found in sources like [13]. Substantial collections of quadrature codes such as those of QUADPACK, NAG, and IMSL contain specialized routines for a variety of difficulties. In this section we discuss some techniques for preparing problems to make them more suitable for a general-purpose code like Adapt. With a little mathematical effort, a problem that the code cannot solve might be put into a form that it can solve, and problems that it can solve are solved more efficiently.

OSCILLATORY INTEGRANDS

If the integral $f(x)$ is periodic with period p, that is, $f(x+p) = f(x)$ for all x, and $b-a$ is some integer multiple of p, the integration should be performed over just one period (or sometimes less) and symmetry used for the remainder. For example, for the integral in Exercise 5.3,

$$\int_0^{\pi} \frac{1}{4+\sin(20x)}\,dx = 10\int_0^{\pi/10} \frac{1}{4+\sin(20x)}\,dx.$$

If you have worked Exercise 5.3 you should have found that the composite trapezoid rule is very efficient; consequently, if you have many functions to be integrated over a period, it is advisable to use a special-purpose code based on the composite trapezoidal rule.

Oscillatory nonperiodic functions are more difficult. Generally the integral should be split into subintervals so that there are few oscillations on each subinterval. Adapt may do this automatically, but the computation can be made more reliable and possibly less expensive by a little analysis. As an example, let us estimate

$$\int_0^{\pi} \frac{\sin(20x)}{1+x^2}\,dx.$$

This integral could be rewritten many ways, one of which is

$$\sum_{j=1}^{20} \int_{(j-1)\pi/20}^{j\pi/20} \frac{\sin(20x)}{1+x^2}\,dx.$$

Proceeding in this manner, Adapt is called 20 times, but on each interval the integrand has constant sign and varies smoothly between samples. This is a reasonable way to deal with a single integral of this kind, but in contexts like Fourier analysis, where many functions having the form of a periodic function times a nonperiodic function need to be integrated, it is better to use special-purpose formulas such as product quadrature. This particular example is treated in Case Study 3 where the use of the general-purpose code Adapt is contrasted with Filon's method for finite Fourier integrals.

INFINITE INTEGRATION INTERVAL

Adapt cannot be applied directly to $\int_a^\infty f(x)\,dx$ because it deals only with finite intervals. One way to apply the code is to use the definition

$$\int_a^\infty f(x)\,dx = \lim_{b\to\infty} \int_a^b f(x)\,dx.$$

The idea is to determine an analytical bound for the tail $|\int_b^\infty f(x)\,dx|$. With it an end point b can be selected large enough that $\int_a^\infty f(x)\,dx$ equals $\int_a^b f(x)\,dx$ to the accuracy required. It does not matter if b is rather larger than necessary, so a crude bound for the tail will suffice.

Another way to get to a finite interval is to change the variable of integration. For example, to estimate $\int_1^\infty (e^{-x}\sin x)/x\,dx$, the new variable $s = 1/x$ yields the equivalent integral $\int_0^1 (e^{-1/s}\sin 1/s)/s\,ds$ on a finite interval. Generally this trades one difficulty (an infinite interval) for another (a singularity at an end point). For this particular example, $\lim_{s\to\infty}(e^{-1/s}\sin 1/s)/s = 0$, so the integrand is continuous at $s = 0$ and Adapt can be applied directly. If the original problem were $\int_0^\infty (e^{-x}\sin x)/x\,dx$, then $s = x^{-1}$ would not help because it produces the same infinite interval. The choice $s = e^{-x}$ leads to $-\int_0^1 (\sin \ln s)/\ln s\,ds$ for which the integrand approaches 0 as $s \to 0$. On the integration interval $(-\infty,\infty)$ the transformation $s = \arctan x$ is often useful.

SINGULAR INTEGRANDS

By a singularity we mean a point where the function or a low order derivative is discontinuous. There are two cases: (1) finite jumps and (2) infinite values. In either case, if there is a known singularity at a point c in (a,b), a basic tactic is to split the interval into (a,c) and (c,b). In this way we can arrange that we approximate integrals with integrands that are finite except possibly at one end.

If $f(x)$ is finite at the singularity (as for step functions), Adapt can be used on each piece and the results added to produce an approximation to $\int_a^b f(x)\,dx$. This is clear enough, but it must be kept in mind that a function that is smooth to the eye may not be smooth to the code. Surprisingly often someone integrates a spline with a quadrature routine. Of course, such an integration should be done analytically, but it may be convenient to resort to an adaptive quadrature routine. Just remember that splines have discontinuities in low order derivatives. If a routine like Adapt is applied to each piece of a piecewise polynomial function, it will have no difficulty. Indeed, it will be exact if the degree of precision of the basic formula is sufficiently high. However, if no attention is paid to the lack of smoothness at the knots, the code will have to deal with it automatically. This can represent quite a substantial, and completely unnecessary, expense as the code locates the knots and resorts to short intervals to deal with them.

Infinite discontinuities require some study. First we reduce the problem to computing an integral $\int_a^b f(x)\,dx$ with a singularity at one end, let us say a for definiteness. Then we have to sort out the behavior of the integrand at a to convince ourselves that the integral even exists. We have already seen problems for which the singularity is logarithmic, $f(x) \sim c\ln(x)$ as $x \to a$ and (in Example 5.7) algebraic, $f(x) \sim x^{1/7}$ as $x \to 0$. In the case of an algebraic singularity, $f(x) \sim c(x-a)^\gamma$ as $x \to a$, it is necessary that $\gamma > -1$ for the integral to exist. The behavior at the singular point tells us what kind of weight function to introduce if we wish to deal with the difficulty by means of a special formula.

One way to deal with singularities using a general-purpose code like Adapt is to introduce a new variable with the aim of removing, or at least weakening, the singularity. This can be done quite generally for singularities of the form $f(x) \sim c(x-a)^\gamma$. Let us try a new variable t with $x = a + t^\beta$. Then

$$\int_a^b f(x)\,dx = \int_0^{(b-a)^{1/\beta}} f(a+t^\beta)\beta t^{\beta-1}\,dt = \int_0^{(b-a)^{1/\beta}} G(t)\,dt.$$

The new integrand $G(t) \sim ct^{\beta\gamma}\beta t^{\beta-1}$. It will be nonsingular if $\beta(\gamma+1)-1 \geq 0$. By choosing $\beta = 1/(\gamma+1)$ (recall that $\gamma > -1$), the function $G(t) \sim c\beta$ as $t \to 0$ and we can apply our favorite code to the new problem. If the code evaluates at the ends of the interval of integration, we have to code the subprogram for $G(t)$ carefully so that the proper limit value is returned when the argument is $t = 0$. For such codes we might prefer to take, say, $\beta = 2/(\gamma+1)$ so that $G(t) \sim c\beta t$ and the limit value is simply $G(0) = 0$.

To illustrate the technique, suppose we want to compute

$$\int_0^1 \frac{e^x}{\sqrt[3]{x}}\,dx.$$

Often series expansions are an easy way to understand the behavior of an integrand. Here the difficulty is at $x = 0$. Using the series for $\exp(x)$,

$$\frac{e^x}{\sqrt[3]{x}} = \frac{1+x+x^2/2+\cdots}{\sqrt[3]{x}} = \frac{1}{\sqrt[3]{x}} + x^{2/3} + \cdots.$$

Thus $f(x) \sim x^{-1/3}$ and $\gamma = -1/3$. The integral exists because $\gamma > -1$. If we take $\beta = 2/(\gamma+1) = 3$, the integral becomes

$$\int_0^1 \frac{e^x}{\sqrt[3]{x}}\,dx = \int_0^1 \frac{e^{t^3}}{\sqrt[3]{t^3}} 3t^2\,dt = \int_0^1 3te^{t^3}\,dt,$$

which presents no difficulty at all.

Example 5.8. As a more substantial example, suppose we wish to approximate

$$\int_0^1 \frac{x^{7/4} e^x}{\sinh^2(x)}\, dx.$$

Using the series expansion

$$\sinh(x) = x + \frac{x^3}{3!} + \cdots$$

we have

$$\begin{aligned} f(x) &= \frac{x^{7/4}(1 + x + \cdots)}{\left(x + \frac{x^3}{3!} + \cdots\right)^2} \\ &= x^{7/4} \frac{1 + x + \cdots}{x^2(1 + x^2/6 + \cdots)^2} \\ &= x^{-1/4} \frac{1 + x + \cdots}{1 + x^2/3 + \cdots} \\ &= x^{-1/4} + \cdots. \end{aligned}$$

The integrand $f(x) \sim x^{-1/4}$ as $x \to 0$, so the integral exists and a suitable change of variables would be $x = t^{4/3}$. Since Adapt does not evaluate at endpoints, it can be applied in a straightforward way to the transformed integral

$$\int_0^1 \frac{4}{3} \frac{\exp(t^{4/3})}{\sinh^2(t^{4/3})} t^{8/3}\, dt.$$

Applying Adapt directly to the original problem with ABSERR $= 10^{-10}$ and RELERR $= 10^{-8}$, results in ANSWER $= 1.913146656196971$ and an estimated error of about 1.485×10^{-8} at a cost of 1841 evaluations of f. On the other hand, applying the code to the problem after the change of variables results in ANSWER $=$ 1.913146663755082 and an estimated error of about 1.003×10^{-8} at a cost of 147 evaluations of f. ■

A technique called *subtracting out the singularity* is a valuable alternative to a change of variables. The integrand is split into two terms, one containing the singularity that is integrated analytically and another that is smoother and is approximated numerically. The technique is an alternative to using special formulas based on a weight function suggested earlier for the integrals arising in plane potential computations:

$$\int_0^1 f(x) \ln(x)\, dx.$$

Before it was assumed implicitly that $f(x)$ is well behaved at $x = 0$. Then

$$f(x) \ln(x) \sim f(0) \ln(x),$$

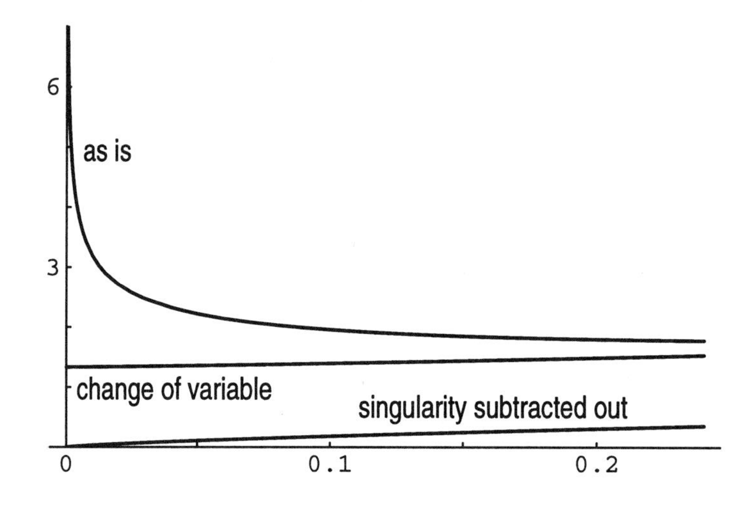

Figure 5.7 Integrands for Examples 5.8 and 5.9 near $x = 0$.

and we can write

$$\int_0^1 f(x)\ln(x)\,dx = \int_0^1 f(0)\ln(x)\,dx + \int_0^1 [f(x) - f(0)]\ln(x)\,dx.$$

The first integral is done analytically and the second numerically. The integral to be computed numerically is easier than the original one. This is seen by expanding $f(x) = f(0) + f'(0)x + \cdots$ and observing that the integrand behaves like $f'(0)x\ln(x)$ as $x \to 0$.

It is easier to apply this technique than to change variables, but a good change of variables will be more effective because it deals with all the singular behavior rather than just the leading term(s) that are subtracted out.

Example 5.9. Let us compare the techniques for the problem in Example 5.8. In subtracting the singularity, we write

$$\int_0^1 \frac{x^{7/4}e^x}{\sinh^2(x)}\,dx = \int_0^1 x^{-1/4}\,dx + \int_0^1 \left[\frac{x^{7/4}e^x}{\sinh^2(x)} - x^{-1/4}\right]dx.$$

With the same tolerances as before, the output from Adapt is

$$\text{ANSWER} = 1.913146679435873, \ \text{ERREST} = -5.476 \times 10^{-9}$$

at a cost of 287 calls to f. See Figure 5.7 for a plot near $x = 0$ of the original integrand and those from Examples 5.8 and 5.9. ■

We end this section with a more realistic problem.

Example 5.10. A conducting ellipsoidal column projecting above a flat conducting plane has been used as a model of a lightning rod [2]. When the ellipsoid is given by the equation

$$\frac{x^2}{a^2} + \frac{y^2}{b^2} + \frac{z^2}{c^2} = 1,$$

the potential function is known to be

$$V(x,y,z) = -z + Az \int_\lambda^\infty \frac{du}{\sqrt{(a^2+u)(b^2+u)(c^2+u)^3}}.$$

The constant A depends only on the shape of the rod and is given by

$$A = \frac{1}{\int_0^\infty \left(1/\sqrt{(a^2+u)(b^2+u)(c^2+u)^3}\right) du}.$$

The quantity λ is a function of the place where we want to evaluate the potential. It is the unique nonnegative root of the equation

$$\phi(\lambda) = \frac{x^2}{a^2+\lambda} + \frac{y^2}{b^2+\lambda} + \frac{z^2}{c^2+\lambda} - 1 = 0.$$

Suppose that for a tall, slender rod described by $a = 1$, $b = 2$, and $c = 100$, we seek the potential V at $x = y = z = 50$. As we saw in the preceding chapter, the root λ of ϕ is approximately 5928.359. To compute

$$\int_\lambda^\infty \frac{du}{\sqrt{(a^2+u)(b^2+u)(c^2+u)^3}} \tag{5.19}$$

we note that the integrand tends to zero like $u^{-5/2}$ as $u \to \infty$, so the integral is well defined and the change of variables $u = 1/t$ is satisfactory. This substitution produces

$$\int_0^{1/\lambda} \frac{(1/t^2)\,dt}{\sqrt{(a^2+1/t)(b^2+1/t)(c^2+1/t)^3}}.$$

This is acceptable mathematically but is in poor computational form. Here it is easy to rearrange to obtain

$$\int_0^{1/\lambda} \sqrt{\frac{t}{(a^2t+1)(b^2t+1)(c^2t+1)^3}}\,dt.$$

Using Adapt with RELERR $= 10^{-5}$ and ABSERR $= 10^{-14}$ produces

$$\text{ANSWER} = 0.5705983 \times 10^{-6}, \text{ and ERREST} = -0.52 \times 10^{-11},$$

which requires 105 f evaluations. The integrand is approximately $\sqrt{t}$ near $t = 0$, which suggests that a better change of variables could have been made (e.g., $u = 1/w^2$). Then the integral is

$$\int_0^{1/\sqrt{\lambda}} \frac{2w^2\,dw}{\sqrt{(a^2w^2+1)(b^2w^2+1)(c^2w^2+1)^3}}.$$

This is a somewhat easier integral to evaluate. Routine Adapt produces

$$\text{ANSWER} = 0.5705883 \times 10^{-6}, \text{ and ERREST} = 0.46 \times 10^{-11},$$

which requires 35 f evaluations. Note that the two results agree to the requested five figures.

These transformations cannot be applied directly to computing A because the interval remains infinite. However, since (5.19) has to be computed anyway, we could split

$$\int_0^\infty \frac{du}{\sqrt{(a^2+u)(b^2+u)(c^2+u)^3}}$$

into $\int_0^\lambda + \int_\lambda^\infty$. The first integral is computed to be 0.7218051×10^{-5} with an error estimate of 0.31×10^{-10} and requires 357 f evaluations, while the sum using the second value from above is 0.7788639×10^{-5}. This yields $A = 0.1283921 \times 10^6$ and finally $V = -46.3370$. ■

EXERCISES

Unless otherwise indicated, use as tolerances ABSERR $= 10^{-8}$ and RELERR $= 10^{-6}$ for the computing exercises.

5.13 Evaluate $\int_0^1 (\sin x/\sqrt{x})\,dx$ using Adapt

(a) on the problem as is,

(b) using one of the series techniques, and

(c) using a change of variable.

Compare the results.

5.14 Repeat Exercise 5.13 with $\int_0^1 (e^x/\sqrt{x})\,dx$.

5.15 The integral

$$\int_0^1 x^2 \sin\frac{1}{x}\,dx$$

can be difficult to estimate because of the large number of oscillations as $x \to 0$ (although the amplitudes are approaching zero, too).

(a) Use Adapt for this problem with RELERR $= 10^{-12}$ and the three absolute tolerances ABSERR $= 10^{-3}$, 10^{-6}, and 10^{-9}.

(b) The change of variable $t = 1/x$ produces the equivalent integral $\int_1^\infty (\sin t/t^4)\,dt$. Approximate this by $\int_1^b (\sin t/t^4)\,dt$ where b is chosen sufficiently large. How large should b be in order to guarantee that

$$\left| \int_1^\infty \frac{\sin t\,dt}{t^4} - \int_1^b \frac{\sin t\,dt}{t^4} \right| \leq \text{ABSERR}/2?$$

Try Adapt with RELERR $= 10^{-12}$ and ABSERR $= \frac{1}{2}\times 10^{-3}$, $\frac{1}{2}\times 10^{-6}$, and $\frac{1}{2}\times 10^{-9}$.

(c) Compare the answers in (a) and (b). Discuss efficiency and accuracy.

5.16 The integral

$$\int_0^1 \frac{x^{1/4}(e^x-1)^2}{\sin^3(x)}\,dx$$

has a rather nasty singularity at $x = 0$. Analyze the nature of the singularity and argue informally that the integral exists.

(a) Use Adapt for this problem as it is. How is this even possible?

(b) Treat the singularity using a technique from Section 5.4, then use Adapt.

(c) Compare the answers in (a) and (b). Discuss efficiency and accuracy.

5.17 The exponential integral

$$E_1(t) = \int_1^\infty e^{-tx}\frac{dx}{x},$$

$t > 0$, arises in the study of radiative transfer and transport theory [9]. Some manipulation shows that

$$E_1(t) = \int_1^\infty e^{-x}\frac{dx}{x} + \int_t^1 e^{-x}\frac{dx}{x}$$

$$= -\left\{ \int_1^\infty e^{-x} \frac{dx}{x} - \int_0^1 (1 - e^{-x}) \frac{dx}{x} \right\}$$
$$+ \int_t^1 \frac{dx}{x} + \int_0^t (1 - e^{-x}) \frac{dx}{x}.$$

The expression in braces is known to have the value $\gamma \approx 0.5772157$, the Euler–Mascheroni constant. The second term integrates analytically to $-\ln t$. Hence,

$$E_1(t) = -\gamma - \ln t + \int_0^t (1 - e^{-x}) \frac{dx}{x}.$$

Evaluate $E_1(t)$ for $t = 1.0, 2.0, 3.0$. Does there appear to be any difficulty caused by the behavior of the integrand at $x = 0$?

5.18 In studying the conduction properties of a V-grooved heat pipe, a heat flux constant $C(\theta)$ satisfies

$$C(\theta) = \frac{\sin^2 \theta}{16} \int_\theta^{\pi/2} \sin t \left[\cot \theta - \frac{\pi - 2t - \sin 2t}{2 \cos^2 t} \right]^3 dt,$$

(θ in radians). Compute values of C for $\theta = \pi/6, \pi/4, \pi/3, 5\pi/12, 17\pi/36$. It can be shown mathematically that $C(\theta)$ is strictly decreasing and $0 < C(\theta) < 1$. The denominator in the integrand vanishes at $\pi/2$. Use series expansions to sort out the behavior of the integrand at this point, and if it causes difficulty, fix it.

5.19 Debye's specific heat law gives the molar specific heat of a solid, C_v, as a function of its temperature, T:

$$C_v(T) = \frac{9R}{U_D^3} \int_0^{U_D} \frac{x^4 e^x}{(e^x - 1)^2} dx,$$

where $R = 1.986$ calories/mole is the molar gas constant, $U_D = \theta_D / T$, and θ_D is the Debye temperature that is characteristic of each substance. For diamond $\theta_D = 1900$K, evaluate $C_v(T)$ at the temperatures indicated in the accompanying table. Compare with the experimental values given. Does the behavior of the integrand at $x = 0$ appear to cause problems?

T(K)	C_v (calories/mole)
100	0.06
200	0.56
300	1.46
400	2.42
500	3.17
1000	5.09

5.20 The Gibb's effect describes the behavior of a Fourier series near a discontinuity of the function approximated. The magnitude of the jump can be related to the integral

$$I = -\int_\pi^\infty \frac{\sin s}{s} ds,$$

which is about 0.281. Routine Adapt cannot be applied directly because it requires a finite interval. One way of dealing with this is to drop the tail of the integral, that is, approximate the integral by

$$-\int_\pi^b \frac{\sin s}{s} ds.$$

This does not work well in this instance because the integrand decays slowly as $s \to \infty$. To see this, work out an analytical bound for the error made in dropping the tail, that is, bound

$$-\int_b^\infty \frac{\sin s}{s} ds.$$

Integration by parts leads to

$$I = \left.\frac{\cos s}{s}\right|_\pi^\infty + \int_\pi^\infty \frac{\cos s}{s^2} dx$$
$$= -\frac{\cos \pi}{\pi} + \int_\pi^\infty \frac{\cos s}{s^2} dx.$$

The integral arising here is easier to approximate because the integrand decays faster. Integration by parts can be repeated to get integrals that are even easier to approximate. Approximate the original integral accurately enough to verify that its magnitude is about 0.281. Do this by applying Adapt to the integral that arises after integrating by parts a few times. To apply the code you will have to drop the tail. There are two sources of error in the computation. One is the error made by Adapt, which you can control by the tolerances that you specify. The other error is due to dropping the tail. You can control this error by working out an analytical bound for the error and choosing a value b large enough to guarantee the accuracy that you need.

5.5 INTEGRATION OF TABULAR DATA

The problem addressed here is the approximation of $\int_a^b f(x)\,dx$ given only (x_n, y_n) for $1 \le n \le N$, where $y_n = f(x_n)$. Adaptive quadrature routines cannot be used because they automatically choose the points where f is to be evaluated and it is unlikely that such points will correspond to the data points $\{x_n\}$. The basic approach is the same as in Section 5.1: approximate $f(x)$ by a piecewise polynomial function $F(x)$, which is then integrated exactly.

Since the complete cubic interpolating spline worked so well as an approximating function in Chapter 3, it is a natural choice for $F(x)$. For the sake of simplicity, let us assume that $a = x_1$ and $b = x_N$. Then, using the notation of Chapter 3 for the spline,

$$\int_a^b S(x)\,dx = \sum_{n=1}^{N-1} \int_{x_n}^{x_{n+1}} S(x)\,dx$$
$$= \sum_{n=1}^{N-1} \left(a_n h_n + b_n \frac{h_n^2}{2} + c_n \frac{h_n^3}{3} + d_n \frac{h_n^4}{4} \right).$$

Substituting the expressions for a_n, b_n, and d_n in terms of the data and c_n leads to

$$\int_a^b S(x)\,dx = \sum_{n=1}^{N-1} \left\{ f_n h_n + \left[\frac{f_{n+1} - f_n}{h_n} - \frac{2}{3} c_n h_n - \frac{1}{3} c_{n+1} h_n \right] \frac{h_n^2}{2} + c_n \frac{h_n^3}{3} + \frac{c_{n+1} - c_n}{3h_n} \frac{h_n^4}{4} \right\}$$
$$= \sum_{n=1}^{N-1} \left\{ \frac{h_n}{2}(f_n + f_{n+1}) - \frac{h_n^3}{12}(c_n + c_{n+1}) \right\}. \tag{5.20}$$

An algorithm based on this technique first uses SPCOEF/Spline_coeff from Chapter 3 to get $\{c_n\}_{n=1}^N$ and then evaluates (5.20). In terms of $h = \max_n(x_n - x_{n-1})$, the complete cubic spline $S(x)$ provides an approximation to a smooth function f that is accurate to $O(h^4)$. Accordingly, if the sample points x_i are sufficiently dense in $[x_1, x_N]$ and the function is sufficiently smooth, we can expect $\int_{x_1}^{x_N} S(x)\,dx$ to be an $O(h^4)$ estimate of $\int_{x_1}^{x_N} f(x)\,dx$.

The cubic spline is familiar and it is easy enough to manipulate once the linear system for $\{c_n\}_{n=1}^N$ has been solved, but there are other interesting possibilities. The linear system arises because $S(x)$ is required to have two continuous derivatives on $[x_1, x_N]$. This smoothness is unnecessary to the approximation of $\int_{x_1}^{x_N} f(x)\,dx$. The shape-preserving spline of Section 3.5.2 is less smooth but is attractive here for several reasons. The reasons and the following formula are left as an exercise:

$$\int_a^b S(x)\,dx = \sum_{n=1}^{N-1} \left\{ \frac{h_n}{2}(f_n + f_{n+1}) + \frac{h_n^2}{12}(b_n - b_{n+1}) \right\}. \tag{5.21}$$

A widely used scheme is based on local quadratic interpolation. To approximate $f(x)$ by a quadratic over $[x_n, x_{n+1}]$, we must interpolate it at three points. One possibility is to interpolate at $\{x_{n-1}, x_n, x_{n+1}\}$. Another is to interpolate at $\{x_n, x_{n+1}, x_{n+2}\}$. There is no obvious reason to prefer one to the other and the computation is inexpensive, so a reasonable way to proceed is to compute both and average the two results. This provides a symmetrical formula that smooths out mild measurement errors in the tabulated values of f. Of course, at the ends $n = 1$ and $n = N-1$, only one quadratic is used. The formula for a typical (interior) interval $[x_n, x_{n+1}]$ is

$$\int_{x_n}^{x_{n+1}} f(x)\,dx \approx \frac{h_n}{12}\left\{ -\frac{h_n^2}{h_{n-1}(h_{n-1}+h_n)} f_{n-1} + \left(\frac{3h_{n-1}+h_n}{h_{n-1}} + \frac{3h_{n+1}+2h_n}{h_{n+1}+h_n}\right) f_n + \left(\frac{3h_{n-1}+2h_n}{h_{n-1}+h_n} + \frac{3h_{n+1}+h_n}{h_{n+1}}\right) f_{n+1} - \frac{h_n^2}{h_{n+1}(h_n+h_{n+1})} f_{n+2} \right\}. \tag{5.22}$$

Reference [4] contains further discussion and a FORTRAN code AVINT.

There is no obvious way to obtain a good estimate of the error of these quadrature rules, much less to control it. The difficulty is inherent to functions defined solely by tables.

EXERCISES

5.21 In performing an arginine tolerance test, a doctor measures glucose, insulin, glucagon, and growth hormone levels in the blood over a 1-hour time period at 10-minute intervals to obtain the following data:

time	glucose	insulin
0	102	11
10	114	26
20	122	36
30	132	47
40	115	39
50	107	27
60	100	15

time	glucagon	growth hormone
0	188	1.70
10	1300	1.70
20	2300	1.20
30	2600	2.50
40	1800	7.25
50	840	8.10
60	460	8.00

The doctor is interested in the integrated effect of each response. For example, if the glucose curve is represented by $g(t)$, then $\int_0^{60} g(t)\,dt$ is desired. Compute one of the integrals by

(a) the method (5.20) based on splines,

(b) the method (5.21) based on splines, and

(c) the method based on averaging quadratics (5.22).

Compare the results.

5.22 Supply the details of the derivation of (5.22). Work out the special forms for the end points.

5.23 Derive the formula (5.21) for the integration of tabular data using the shape-preserving spline. When might you prefer this to formula (5.20) based on the complete cubic spline? Consider the cost of computing the spline and how well the spline might fit the data.

5.6 INTEGRATION IN TWO VARIABLES

Definite integrals in two or more variables are generally much more difficult to approximate than those in one variable because the geometry of the region causes trouble. In this section we make a few observations about the common case of two variables, especially as it relates to finite elements.

Integration over a rectangle,

$$I(f) = \int_{a_1}^{b_1} \int_{a_2}^{b_2} f(x,y)\, dy\, dx,$$

can be handled easily with the formulas for one variable by treating I as an iterated integral. Thus we first approximate

$$I(f) \approx \sum_{i=1}^{N_1} A_i \int_{a_2}^{b_2} f(x_i,y)\, dy$$

with one quadrature rule using N_1 points $\{x_i\}$ and then

$$I(f) \approx \sum_{i=1}^{N_1} A_i \left(\sum_{j=1}^{N_2} B_j f(x_i,y_j) \right)$$

using another rule of N_2 points $\{y_j\}$. This approach generalizes to

$$I(f) = \int_{a}^{b} \int_{x_1(y)}^{x_2(y)} f(x,y) dy dx.$$

It is even possible to use an adaptive code for the integrals in one variable, but the matter is a little complicated. In Fritsch, Kahaner, and Lyness [6] it is explained how to go about this; for pitfalls to be avoided, see [11, Section 9].

Degree of precision now refers to polynomials in two variables, so a formula of, for example, degree 2 must integrate all polynomials of the form

$$a_{0,0} + a_{1,0}x + a_{0,1}y + a_{2,0}x^2 + a_{1,1}xy + a_{0,2}y^2$$

exactly on the region of interest. Just as in the case of one variable, we can derive quadrature formulas by interpolating $f(x,y)$ and integrating the interpolant. This is quite practical for integration over a square or a triangle and is often done. The scheme for rectangles based on iterated integrals can be quite efficient when the formulas for integration in one variable are Gaussian. They are not necessarily the best that can be done. As in the one-dimensional case, formulas can be derived that use the smallest number of evaluations of $f(x,y)$ to obtain a given degree of precision. Nevertheless, the most efficient formulas may not be the most attractive in practice. The approach based on interpolation can be very convenient when the interpolation is done at points interesting for other reasons, as in finite elements. The iterated integral approach can be very convenient because of its simplicity and generality.

In one dimension the transformation of any finite interval $[a,b]$ to a standard one such as $[-1,1]$ is trivial. In two dimensions the matter is far more important and difficult. Now an integral over a general region R,

$$\iint_R f(x,y)\,dy\,dx,$$

must be broken up into pieces that can be transformed into a standard square or triangle. Discretizing a region R in this way is an important part of any finite element code. If the region is decomposed into triangles (with straight sides), the easy transformation was stated in Chapter 3. An integral over a general triangle T becomes an integral over the standard, reference triangle T_*,

$$\iint_T f(x,y)\,dy\,dx = \iint_{T_*} f(x^*,y^*)|D^*|dy^*dx^*.$$

Here $|D^*|$ is the determinant of the Jacobian of the transformation. It relates the infinitesimal area $dy\,dx$ in the one set of variables to that in the other set. For the affine transformation from one triangle (with straight sides) to another, this matter is easy. In the general case it is necessary to investigate whether the transformation is a proper one, meaning that the image covers all the triangle T_* and has no overlaps.

The main point of this section is that the basic ideas of the one variable case carry over to several variables. There is the additional and very serious complication of splitting up the region of integration and transforming properly each piece to a standard region. The whole area of integration of functions of more than one variable is still the subject of research.

EXERCISES

5.24 Given the triangular domain T with vertices $(0,0)$, $(1,0)$, and $(0,1)$, we would like to approximate $\int\int f(x,y)\,dx\,dy$ over T.

(a) Derive a quadrature approximation of the form

$$A_1 f(0,0) + A_2 f(0,1) + A_3 f(1,0),$$

where the coefficients are chosen to make the approximation exact for $f = 1$, x, and y.

(b) Derive a corresponding composite quadrature formula for the subdivision obtained by cutting T into four subtriangles by connecting the midpoints of the edges of T.

5.7 CASE STUDY 5

This case study has two parts, one devoted to a problem involving singular integrands and the other to problems with integrands that have sharp peaks. As usual, our aim is to understand the difficulties and what might be done to make it possible to solve the problems effectively with standard codes. The first part is an example of a problem requiring more than one code for its solution. The second part develops a technique of classical applied mathematics and uses it for insight.

Reference [5] discusses the motion of a particle of mass m at position q in a potential field $V(q)$. In suitable circumstances the motion is a libration (or oscillation or vibration). For an energy level E, it is found that if the equation $E - V(q) = 0$ has simple roots $q_1 < q_2$, the period T of the motion is given by

$$T = \sqrt{2m} \int_{q_1}^{q_2} \frac{dq}{\sqrt{E - V(q)}}.$$

The integrand is singular at both end points, so the first thing we must do is sort out its behavior there. Near a simple root q_i, a Taylor series expansion tells us that as $q \to q_i$,

$$\frac{1}{\sqrt{E - V(q)}} \sim \frac{1}{\sqrt{-V'(q_i)(q - q_i)}},$$

hence the singularity is integrable. The argument shows why the theory requires that the roots be simple. If q_i were a multiple root, say double, the integrand would behave like

$$\frac{1}{\sqrt{-0.5V''(q_i)(q - q_i)^2}} = \frac{1}{|q - q_i| \sqrt{-0.5V''(q_i)}}$$

and the integral would not exist. Percival and Richards [5] write, "Gaussian numerical integration is a very effective method of obtaining the periods of libration in practice." Gaussian quadrature is generally quite effective, but a crucial point in the present situation is that it can be applied directly because it does not evaluate the integrand at the ends of the interval. The formulas of Adapt have the same property.

As a simple example let us take $V(q) = (q + 1)(q - 0.8)^7$ and $E = -4$. As can be deduced easily from the derivative $V'(q) = (8q + 6.2)(q - 0.8)^6$ or by inspection of Figure 5.8, $V(q)$ has a minimum at $q = -0.775$, and for $V(-0.775) < E < 0$, the equation $E - V(q) = 0$ has two simple roots, one in $[-1, -0.775]$ and the other in $[-0.775, 0.8]$. To evaluate the integral, we must first compute these roots numerically. The roots should be located very accurately because they define the integral that we wish to approximate. Moreover, if they are not determined accurately, a quadrature code might evaluate the integrand at an argument q for which the integrand is not defined because of taking the square root of a negative number or dividing by zero. A problem like this is very easy for Zero, so we take ABSERR $= 10^{-14}$ and RELERR $= 10^{-14}$. Even with these stringent tolerances the code requires only 21 function evaluations to find that $q_1 \approx -0.9041816$ with a residual of about 1.3×10^{-15} and $q_2 \approx -0.5797068$ with a residual of about 1.9×10^{-15}. Because E provides a natural measure of scale for this equation, the residuals and a backward error analysis make it clear that the roots are very accurate, although of course we already know that because Zero reported that it had obtained the specified accuracy.

Using the computed values for the end points, it is possible to approximate the integral by a simple call to Adapt. With ABSERR $= 10^{-6}$ and RELERR $= 10^{-6}$ this results in an answer of about 0.444687 with an estimated error of about 9.0×10^{-7} at a cost of 3171 function evaluations. This is relatively expensive for Adapt with its (arbitrary) limit of 3577 evaluations. If we wanted to do a number of integrals of this kind, some preparation of the problem would be worthwhile. We have already worked out the dominant behavior of the integrand at the end points and the value of the derivative appearing in the expression is readily available, so the method of subtracting out the singularity would be easy to apply. However, because the singularity is algebraic, it

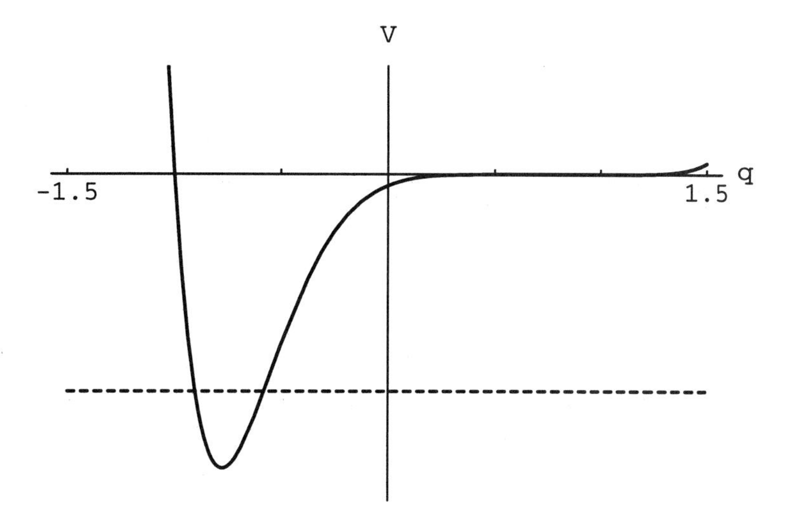

Figure 5.8 The potential $V(q) = (q+1)(q-0.8)^7$.

is easy to change variables and this technique is generally more effective. We need to break the interval of integration so that there is at most one singularity in each piece and when present, it is an end point. A natural choice is $[q_1, -0.775]$ and $[-0.775, q_2]$. In the notation used earlier in the chapter, $\gamma = -1/2$, so we might choose $\beta = 2$, that is, introduce the variable $t^2 = q - q_1$ in the portion to the left and $t^2 = q_2 - q$ in the portion to the right. Adapt is used twice with the same tolerances as before and the integrals are added to obtain an answer of about 0.444688. The estimated errors can also be added since they are estimates of the error rather than the magnitude of the error, and this gives an estimated error of about 8.2×10^{-7}. The value for the integral is consistent with that of the direct approach, but it costs a total of only 56 function evaluations.

Let us turn now to a different kind of difficulty, integrands with sharp peaks. Many important problems of applied mathematics are solved by transform methods that lead to the task of evaluating integrals. A method developed by Laplace illustrates the use of asymptotic methods for this purpose. A family of integrals depending on a parameter $s \gg 1$ of the form

$$L(s) = \int_a^b e^{sf(x)}\,dx$$

is considered. If the function $f(x)$ has a unique maximum at x_0 with $a < x_0 < b$, the integrand has a sharp peak at x_0, and the greater s is, the sharper the peak. The idea is first to approximate $f(x)$ near x_0 by

$$f(x) \approx f(x_0) + 0.5f''(x_0)(x - x_0)^2$$

and then to approximate the integral by integrating this function. Because the integrand decays so rapidly for large s, the approximation to the integral is scarcely affected by extending the interval of integration to $(-\infty, \infty)$. This is done in order to integrate the approximating function analytically. It amounts to approximating the integrand by the familiar bell-shaped curve of a normal distribution with mean x_0 and standard deviation $\sigma = \sqrt{1/(-sf''(x_0))}$. The result of these approximations is Laplace's formula,

$$L(s) \approx \int_{-\infty}^{\infty} e^{s[f(x_0)+0.5f''(x_0)(x-x_0)^2]}\,dx = e^{sf(x_0)}\sqrt{\frac{2\pi}{-sf''(x_0)}}.$$

A classic application of this formula is to the Stirling approximation of the gamma function seen in Example 1.10. Some manipulation of the integral definition of the gamma function puts it into the form required:

$$\Gamma(s) = \frac{\Gamma(s+1)}{s} = s^{-1}\int_0^{\infty} t^s e^{-t}\,dt = s^{-1}\int_0^{\infty} e^{-t+s\ln t}\,dt = s^s\int_0^{\infty} e^{s(-x+\ln x)}\,dx.$$

(Here $t = xs$.) Laplace's formula with $f(x) = \ln x - x$ and $x_0 = 1$ then gives

$$\Gamma(s) \approx \left(\frac{s}{e}\right)^s \sqrt{\frac{2\pi}{s}}.$$

Laplace's formula and the class of integrals it approximates illustrate how the approximations of classical applied mathematics differ from those of numerical analysis. A general-purpose code like Adapt that accepts any smooth integrand is likely to fail when presented an integrand with a sharp peak because it is not likely to "see" the peak. By this we mean that the code is not likely to take samples from the small subinterval where the peak occurs, so it, in effect, approximates the integrand by a function that does not have a peak. The approach of applied mathematics is much less general because it requires that the location of the peak be supplied along with information about its width. An advantage of the approach is that it provides important qualitative information about how the integral depends on a parameter. On the other hand, the accuracy of the formula depends on the parameter, and for a given value of s it might not be enough. As is often the case, when used to obtain a numerical approximation to an integral, the approaches are complementary with the one working better as the parameter increases and the other working worse, and vice versa. Let us take advantage of the insight provided by Laplace's method to compute an accurate value for an integral with a sharp peak using a general-purpose code. Clearly we ought first to locate the peak, then get some information about its width, and finally break up the interval appropriately. D. Amos [1] does exactly this to evaluate integral representations of statistical distributions. In each case an evaluation of the function amounts to the numerical integration of a bell-shaped integrand with a single maximum at x_0. First Newton's method is used to locate x_0. A scale σ is then determined using the Laplace method. Finally quadratures over intervals of length σ to the left and right of x_0 are summed until a limit of integration is reached or the truncation error is small enough. The basic idea is simple, but the generality of the distributions allowed and the refinements needed for fast evaluation of the functions make the details too complex to describe here. To illustrate the approach, we refine the Stirling approximation for the computation of a large factorial, a simple example that allows the easy computation of a reference value.

As pointed out in Chapter 1, if we are to evaluate the gamma function for a large argument, we must scale it to avoid overflow. The Stirling approximation derived above suggests that we approximate

$$\Gamma(s)\left(\frac{e}{s}\right)^s = \int_0^\infty e^{s(1-x+\ln x)}\,dx,$$

which is both simple and well scaled. If, for example, we take $s = 201$, we can obtain a reference value from

$$\Gamma(201)\left(\frac{e}{201}\right)^{201} = 200!\left(\frac{e}{201}\right)^{201} = \left(\frac{e}{201}\right)\prod_{k=1}^{200}\left(\frac{ke}{201}\right) \approx 0.17687724.$$

The Stirling approximation is $\sqrt{2\pi/201}$, which is about 0.17680393 and would be adequate for many purposes. An attempt to approximate the integral simply by truncating the semi-infinite interval to $[0, 100]$ and calling Adapt with ABSERR $= 10^{-8}$ and RELERR $= 10^{-8}$ results in a value of 0 at a cost of seven function evaluations. If you should find that Adapt requires only seven function evaluations, the smallest number it makes, you should give some thought to your problem. Either the integral is very easy or the scale of the problem has been missed entirely. For this integrand $x_0 = 1$ and $\sigma = \sqrt{1/201} \approx 0.0705$. The derivation of the Laplace formula suggests that the bulk of the integral is accounted for by the interval $[x_0 - 3\sigma, x_0 + 3\sigma]$, so we compute it first and then add on approximations to the integrals over subintervals of length σ on each side. The table shows the approximations computed in this way for the interval $[1 - k\sigma, 1 + k\sigma]$ along with the *cumulative* cost in function evaluations.

k	result	nfeval
3	0.17628738	105
4	0.17685170	133
5	0.17687653	147
6	0.17687723	161
7	0.17687724	175
true	0.17687724	

The last interval here was about $[0.5, 1.5]$, making the point that only a small portion of the interval of integration is important. Notice that the code is doing the minimal number of function evaluations (14 for the two pieces) for the larger k. The quadratures are very cheap then because the integrand is small and there is an absolute error tolerance.

REFERENCES

1. D. Amos, "Evaluation of some cumulative distribution functions by numerical quadrature," *SIAM Review*, 20 (1978), pp. 778–800.
2. H. Bateman, *Partial Differential Equations of Mathematical Physics*, Cambridge University Press, London, 1964.
3. B. Carnahan, H. Luther, and J. Wilkes, *Applied Numerical Methods*, Wiley, New York, 1969.

4. P. Davis and P. Rabinowitz, *Methods of Numerical Integration*, 2nd ed., Academic Press, New York, 1984.
5. I. Percival and D. Richards, *Introduction to Dynamics*, Cambridge University Press, Cambridge, Mass., 1982.
6. F. Fritsch, D. Kahaner, and J. Lyness, "Double integration using one-dimensional adaptive quadrature routines: A software interface problem," *ACM Trans. on Math. Software*, 7 (1981), pp. 46–75.
7. *Handbook of Mathematical Functions*, M. Abramowitz and I. Stegun, eds., Dover, Mineola, N.Y., 1964.
8. E. Issacson and H. Keller, *Analysis of Numerical Methods*, Dover, Mineola, N.Y., 1994.
9. V. Kourganoff, *Basic Methods in Transfer Problems, Relative Equilibrium and Newton Diffusion*, Dover, New York, 1963.
10. A. Kronrod, *Nodes and Weights of Quadrature Formulas*, (Trans.) Consultants Bureau, New York, 1965.
11. J. Lyness, "When not to use an automatic quadrature routine," *SIAM Review*, 25 (1983), pp. 63–87.
12. R. Piessens, E. de Doncker-Kapenga, C. Überhuber, and D. Kahaner, *QUADPACK: Subroutine Package for Automatic Integration*, Springer-Verlag, New York, 1983.
13. A. Stroud and D. Secrest, *Gaussian Quadrature Formulas*, Prentice Hall, Englewood Cliffs, N.J., 1966.

EXERCISES

Unless otherwise indicated, use ABSERR $= 10^{-8}$ and RELERR $= 10^{-6}$ for the computing exercises.

5.25 The potential inside the unit circle due to a specified potential $f(\theta)$ on the boundary is given by Poisson's integral:

$$\varphi(r,\theta) = \frac{1}{2\pi}\int_0^{2\pi} \frac{1-r^2}{1-2r\cos(\theta-\theta')+r^2} f(\theta')\,d\theta'.$$

There is difficulty evaluating this integral as $r \to 1$, since for $\theta' = \theta$,

$$\frac{1-r^2}{1-2r\cos(\theta-\theta')+r^2} = \frac{1+r}{1-r}.$$

This is not too severe because the term is large only if r is very close to 1, but in principle there should be no difficulty at all since as $r \to 1, \varphi(r,\theta) \to f(\theta)$ (see Bateman [2, pp. 239–241]). Realizing that

$$1 = \frac{1}{2\pi}\int_0^{2\pi} \frac{1-r^2}{1-2r\cos(\theta-\theta')+r^2}\,d\theta',$$

derive the form

$$\varphi(r,\theta) = f(\theta) + \frac{1}{2\pi}\int_0^{2\pi} \frac{1-r^2}{1-2r\cos(\theta-\theta')+r^2} [f(\theta')-f(\theta)]\,d\theta',$$

and argue that it should have somewhat better numerical properties. Explore this by evaluating $\phi(r,\theta)$ for r approaching 1 with $f(\theta) = \sin\theta$. The analytical solution is then just $\varphi(r,\theta) = r\sin\theta$.

5.26 The potential in a conducting strip of width b with potential zero on the bottom edge and a specified potential $F(x)$ on the upper edge is

$$\varphi(x,y) = \frac{\sin(\beta)}{b}\int_{-\infty}^{\infty} \frac{F(\xi)\,d\xi}{\cosh[(\xi-x)\pi/b]+\cos(\beta)},$$

where $\beta = \pi y/b$. Suppose that an experimenter applies the potential $F(x) = 1$ for $|x| \le 0.99$ and $F(x) = \exp[-100(|x|-0.99)]$ for $|x| \ge 0.99$. When $b = \pi$, compute and plot the potential along the middle of the strip, $\varphi(x,\pi/2)$.

Realizing that

$$\cosh[(\xi-x)\pi/b] \ge 1,$$

bound the effect on $\varphi(x,y)$ for $y \ne 0$ of replacing the infinite interval by a finite one:

$$\int_{-z}^{z} \frac{F(\xi)\,d\xi}{\cosh[(\xi-x)\pi/b]+\cos(\pi/b)}.$$

For a suitable choice of z, use this instead of the infinite interval. Show for the given $F(x)$ that $\varphi(x,y) = \varphi(-x,y)$, so only a plot for $x \geq 0$ is necessary.

5.27 This exercise is representative of a great many computations arising in the use of the classical separation of variables technique for solving field problems. Typically, one must compute many roots of nonlinear equations and integrals. The temperature distribution in a cylinder of radius a and height b with the bottom held at a temperature zero, the top at a temperature $f(r)$, and the side dissipating heat according to Newton's law of cooling, can be represented by a series. If the thermal conductivity of the cylinder is k and the thalpance is ε, then the temperature $\varphi(r,z)$ is

$$\varphi(r,z) = \sum_{n=1}^{\infty} A_n \frac{\sinh(q_n z)}{\sinh(q_n b)} J_0(q_n r).$$

The numbers q_n are the positive roots of the equation

$$\frac{k}{\varepsilon a} q_n a J_1(q_n a) - J_0(q_n a) = 0,$$

where the function $J_0(x)$ and $J_1(x)$ are Bessel functions of the first kind of orders zero and one, respectively. The coefficients A_n are given by

$$A_n = \frac{2}{a^2 \left[1 + (kq_n/\varepsilon)^2\right] J_1^2(q_n a)} \cdot \int_0^a r f(r) J_0(q_n r)\, dr.$$

The roots q_n depend only on the geometry and the material. Once they have been computed, one can consider virtually any temperature distribution $f(r)$ by computing the quantities A_n. For $k/\varepsilon a = 2$, we give the problem of solving for $q_n a$ for $n = 1,2,3$ in Exercise 4.18. If you have not done that problem, the roots are 0.940770563949734, 3.95937118501253, and 7.08638084796766. Then for $a = 1$, compute A_1, A_2, A_3 for $f(r) = \exp(-r) - \exp(-1)$.

5.28 If $6 = \int_0^1 \sqrt{C + 12e^{\pi t^2}}\, dt$, what is C? To answer this, apply Zero with modest tolerances to compute the root C of $f(C) = 6 - \int_0^1 \sqrt{C + 12e^{\pi t^2}}\, dt$. Evaluate the integral appearing in the function with Adapt and more stringent tolerances. A little analysis will provide an appropriate bracket to use for the root solver. Show analytically that the root is simple. This is an example of computing the roots of a function that is relatively expensive to evaluate. How many evaluations did Zero need?

5.29 Reference [3] contains the following problem. The length L of a carbon dioxide heat exchanger with input temperature T_I and output T_O satisfies

$$L = \frac{m}{\pi D} \int_{T_I}^{T_O} \frac{C_p(T)}{h(T)(T_S - T)}\, dT$$

where $m = 22.5, D = 0.495, T_S = 550$ (temperature of the CO_2),

$$C_p(T) = 0.251 + 0.346 * 10^{-4} T - \frac{14,400}{(T+460)^2}$$

$$h(T) = \frac{0.023k}{D}\left[\left(\frac{4m}{\pi D \mu(T)}\right)^2 \frac{\mu(T) C_p(T)}{k}\right]^{0.4}$$

$$\mu(T) = 0.0332\left(\frac{T+460}{460}\right)^{0.935}$$

$$k = 0.011.$$

For $T_I = 60$, and $L = 10$, use Adapt and Zero to compute T_O.

5.30 The code Adapt uses a queue to hold the subintervals whose quadrature errors are deemed too large to pass the tolerance test. Do you think there would be any difference in performance if a stack were used instead? Explain. Modify Adapt so that it uses a stack and test it on some difficult problems to see what actually happens.

CHAPTER 6

ORDINARY DIFFERENTIAL EQUATIONS

Historically, ordinary differential equations have originated in chemistry, physics, and engineering. They still do, but nowadays they also originate in medicine, biology, anthropology, and the like. Because differential equations are more complex theoretically than the other topics taken up in this book and because their numerical solution is less well understood, the art of numerical computing plays an important role in this chapter. There is a subdiscipline of numerical analysis called mathematical software that colors all the work of this book. It is particularly evident in this chapter with its more complex task.

6.1 SOME ELEMENTS OF THE THEORY

The simplest kind of ordinary differential equation problem is to find a function $y(x)$ with a derivative continuous on $[a,b]$ such that

$$y'(x) = f(x), \qquad a \leq x \leq b,$$

for a given continuous function $f(x)$. From elementary calculus it is known that such a function $y(x)$ exists—the indefinite integral of $f(x)$. However, if $y(x)$ satisfies the differential equation, so does $y(x) + c$ for any constant c. To specify a particular solution, some more information about $y(x)$ is required. The most common kind of additional information supplied is the value A of $y(x)$ at the initial point a. Then

$$y(x) = A + \int_a^x f(t)\,dt$$

is the unique solution to the initial value problem consisting of the differential equation satisfied by $y(x)$ and the initial value of $y(x)$.

The general first order ordinary differential equation has f depending on y as well as x. It is assumed that $f(x,y)$ is continuous for $a \leq x \leq b$ and all y. A solution $y(x)$ is a function of x with a continuous derivative for $a \leq x \leq b$ that satisfies the equation

$$y'(x) = f(x, y(x)) \tag{6.1}$$

for each x in the interval $[a,b]$. Typically the solution desired is specified by its value at the initial point of the interval:

$$y(a) = A. \tag{6.2}$$

Equation (6.2) is called an *initial condition,* and the combination of (6.1) and (6.2) is called an *initial value problem* for an ordinary differential equation.

In elementary treatments of differential equations, the initial value problem has a unique solution that exists throughout the interval of interest and that can be obtained by analytical means (more familiarly called a "trick"). However, for most problems that are not contrived, an analytical solution is impossible to obtain or is less satisfactory than a numerical solution. Matters are also complicated by the fact that solutions can fail to exist over the desired interval of interest. Problems with solutions that "blow up" place a special burden on a numerical procedure, although we might well expect a general-purpose code to compute such solutions until overflow occurs. Problems that have more than one solution are especially troublesome. Difficulties with existence and uniqueness will be excluded at the level of the theory to be developed here. A simple condition that guarantees that these difficulties will not occur can be formulated in terms of how $f(x,y)$ depends on y.

The function $f(x,y)$ satisfies a *Lipschitz condition* in y if for all x in the interval $[a,b]$ and for all u, v,

$$|f(x,u) - f(x,v)| \leq L|u - v| \tag{6.3}$$

with L a constant, hereafter called a *Lipschitz constant.* The inequality assumes a more familiar form if f has a continuous partial derivative in its second variable, for then

$$|f(x,u) - f(x,v)| = \left|\frac{\partial f}{\partial y}(x,w)\right||u - v|$$

for some w between u and v. If $\partial f/\partial y$ is bounded in magnitude for all arguments, then f satisfies a Lipschitz condition and any constant L such that

$$\left|\frac{\partial f}{\partial y}(x,w)\right| \leq L$$

for all x in $[a,b]$ and all w is a Lipschitz constant. If the partial derivative is not bounded, it is not hard to show that the inequality (6.3) cannot hold for all u, v and all x in $[a,b]$, so f does not satisfy a Lipschitz condition.

Example 6.1. The function $f(x,y) = x^2 \cos^2 y + y \sin^2 x$ defined for $|x| \leq 1$ and all y satisfies a Lipschitz condition with constant $L = 3$. To see this, differentiate with respect to y to get

$$\frac{\partial f}{\partial y} = -2x^2 \cos y \sin y + \sin^2 x,$$

and so for the range of x allowed

$$\left|\frac{\partial f}{\partial y}\right| \leq 2 \times 1 \times 1 \times 1 + 1 = 3.$$

■

Example 6.2. The function $f(x,y) = \sqrt{|y|}$ does not satisfy a Lipschitz condition because it has a continuous partial derivative for $y > 0$ that is not bounded as $y \to 0$:

$$\frac{\partial f}{\partial y} = \frac{1}{2\sqrt{y}}.$$

■

An important special case of (6.1) is that of a linear differential equation, an equation of the form $f(x,y) = g(x)y + h(x)$. The function $f(x,y)$ being continuous in (x,y) is then equivalent to $g(x)$ and $h(x)$ being continuous in x. Because

$$\frac{\partial f}{\partial y} = g(x),$$

and because a continuous function $g(x)$ is bounded in magnitude on any finite interval $[a,b]$, a linear equation is Lipschitzian in nearly all cases of practical interest.

Example 6.3. Dawson's integral is the function

$$y(x) = e^{-x^2} \int_0^x e^{t^2}\, dt.$$

You should verify that it is a solution of the initial value problem for the linear differential equation

$$\begin{aligned} y' &= 1 - 2xy \\ y(0) &= 0. \end{aligned}$$

On the interval $[0,b]$ for any $b \neq 0$, the function $f(x,y) = 1 - 2xy$ is continuous and Lipschitzian with Lipschitz constant $L = 2|b|$. ■

Sufficient conditions for existence and uniqueness can now be stated formally. For a proof, see [3].

Theorem 6.1. *Let $f(x,y)$ be continuous for all x in the finite interval $[a,b]$ and all y and satisfy (6.3). Then for any number A, the initial value problem $y' = f(x,y)$, $y(a) = A$ has a unique solution $y(x)$ that is defined for all x in $[a,b]$.*

So far we have spoken of a single equation in a single unknown $y(x)$. More commonly there are several unknowns. By a system of m first order differential equations in m unknowns is meant

$$\begin{aligned} Y_1' &= F_1(x, Y_1, Y_2, \ldots, Y_m) \\ Y_2' &= F_2(x, Y_1, Y_2, \ldots, Y_m) \\ &\vdots \\ Y_m' &= F_m(x, Y_1, Y_2, \ldots, Y_m). \end{aligned} \tag{6.4}$$

Along with the equations (6.4) there are initial conditions

$$\begin{aligned} Y_1(a) &= A_1 \\ Y_2(a) &= A_2 \\ &\vdots \\ Y_m(a) &= A_m. \end{aligned} \tag{6.5}$$

This can be written in tidy fashion using vector notation. If we let

$$\mathbf{Y}(x) = \begin{bmatrix} Y_1(x) \\ Y_2(x) \\ \vdots \\ Y_m(x) \end{bmatrix}, \quad \mathbf{A} = \begin{bmatrix} A_1 \\ A_2 \\ \vdots \\ A_m \end{bmatrix}, \quad \mathbf{F}(x,\mathbf{Y}) = \begin{bmatrix} F_1(x,\mathbf{Y}) \\ F_2(x,\mathbf{Y}) \\ \vdots \\ F_m(x,\mathbf{Y}) \end{bmatrix}, \tag{6.6}$$

then (6.4) and (6.5) become

$$\begin{aligned} \mathbf{Y}' &= \mathbf{F}(x,\mathbf{Y}) \\ \mathbf{Y}(a) &= \mathbf{A}. \end{aligned} \tag{6.7}$$

We again refer to the combination of (6.4) and (6.5) as an initial value problem. Using vector notation makes the case of m unknowns look like the case of one unknown. One of the fortunate aspects of the theory of the initial value problem is that the theory for a system of m first order equations is essentially the same as for a single one. Proofs for systems just introduce vectors and their norms where there are scalars and absolute values in the proofs for a single equation. For the vector function $\mathbf{F}(x,\mathbf{Y})$ to satisfy a Lipschitz condition, it is sufficient that each $F_i(x,Y_1,Y_2,\ldots,Y_m)$ satisfy a Lipschitz condition with respect to each variable Y_j; that is, there are constants L_{ij} such that

$$|F_i(x,Y_1,\ldots,Y_{j-1},u,Y_{j+1},\ldots,Y_m) - F_i(x,Y_1,\ldots,Y_{j-1},v,Y_{j+1},\ldots,Y_m)|$$

$$\leq L_{ij}|u-v|$$

for each i, j. With this, the natural analog of Theorem 6.1 holds. Since the theory of numerical methods for a system of equations is also essentially the same as for a single equation, we content ourselves with treating the case of a single equation in detail and just state the analog for systems.

Most computer codes require the problem to be provided in the standard form (6.4) and (6.5), but equations arise in a great variety of forms. For example, second order equations, that is, equations of the form

$$y'' = g(x,y,y'),$$

are quite common in the context of dynamical systems. The definition of a solution is the obvious extension of the first order case and suitable initial conditions are $y(a) = A_1$, $y'(a) = A_2$. This is a second order equation for one unknown quantity, $y(x)$. An equivalent problem in the standard form (6.4) can be found by introducing two unknown quantities and finding two first order equations satisfied by them. One of the new unknowns has to provide us with the original unknown, so we take $Y_1(x) = y(x)$.

We take the other unknown to be the derivative of the original unknown, $Y_2(x) = y'(x)$. Differentiating the new unknown quantities, we find that

$$\begin{aligned} Y_1' &= y'(x) = Y_2(x), \\ Y_2' &= y''(x) = g(x, y(x), y'(x)) = g(x, Y_1(x), Y_2(x)). \end{aligned}$$

In this way we come to the system of two first order equations in two unknowns:

$$\begin{aligned} Y_1' &= Y_2 \\ Y_2' &= g(x, Y_1, Y_2). \end{aligned}$$

This is in standard form and the theory may be applied to it to conclude the existence of unique functions $Y_1(x)$ and $Y_2(x)$ that satisfy initial conditions

$$\begin{aligned} Y_1(a) &= A_1 \\ Y_2(a) &= A_2. \end{aligned}$$

The solution of the original problem is obtained from $y(x) = Y_1(x)$. To verify this, first notice that one of the equations states that $y'(x) = Y_1'(x) = Y_2(x)$, and the other that

$$y''(x) = Y_2'(x) = g(x, Y_1(x), Y_2(x)) = g(x, y(x), y'(x)).$$

Similarly it is found that the initial conditions are satisfied.

The general mth order equation in one unknown,

$$\begin{aligned} & y^{(m)} = g(x, y, y', \ldots, y^{(m-1)}) \\ & y(a) = A_1, y'(a) = A_2, \ldots, y^{(m-1)}(a) = A_m, \end{aligned}$$

can be put into standard form via the m unknowns $Y_1(x) = y(x)$, $Y_2(x) = y'(x)$, ..., $Y_m(x) = y^{(m-1)}(x)$ and

$$\begin{aligned} F_1(x, Y_1, Y_2, \ldots, Y_m) &= Y_2 \\ F_2(x, Y_1, Y_2, \ldots, Y_m) &= Y_3 \\ &\vdots \\ F_{m-1}(x, Y_1, Y_2, \ldots, Y_m) &= Y_m \\ F_m(x, Y_1, Y_2, \ldots, Y_m) &= g(x, Y_1, Y_2, \ldots, Y_m). \end{aligned}$$

Example 6.4. To convert the initial value problem

$$y'' + (y^2 - 1)y' + y = 0, \; y(0) = 1, \; y'(0) = 4$$

into a system of first order equations, let

$$Y_1(x) = y(x), \; Y_2(x) = y'(x).$$

Then

$$\begin{aligned} Y_1' &= y' = Y_2 \\ Y_2' &= y'' = -(Y_1^2 - 1)Y_2 - Y_1 \end{aligned}$$

and

$$Y_1(0) = 1,\ Y_2(0) = 4.$$

This can be put into the form (6.7) by defining

$$\mathbf{Y} = \begin{pmatrix} Y_1 \\ Y_2 \end{pmatrix},\ \mathbf{A} = \begin{pmatrix} 1 \\ 4 \end{pmatrix},\ \mathbf{F}(x,\mathbf{Y}) = \begin{pmatrix} Y_2 \\ -(Y_1^2-1)Y_2 - Y_1 \end{pmatrix}.$$

■

Example 6.5. Consider the system of second order equations

$$\begin{aligned} u'' + 5v' + 7u &= \sin x, \\ v'' + 6v' + 4u' + 3u + v &= \cos x, \\ u(0) = 1,\ u'(0) &= 2, \\ v(0) = 3,\ v'(0) &= 4. \end{aligned}$$

Let $Y_1(x) = u(x)$, $Y_2(x) = u'(x)$, $Y_3(x) = v(x)$, and $Y_4(x) = v'(x)$. Then the equations are

$$Y_2' + 5Y_4 + 7Y_1 = \sin x, \quad Y_4' + 6Y_4 + 4Y_2 + 3Y_1 + Y_3 = \cos x,$$

which is rearranged as

$$\begin{aligned} Y_1' &= Y_2, \\ Y_2' &= -7Y_1 - 5Y_4 + \sin x, \\ Y_3' &= Y_4, \\ Y_4' &= -3Y_1 - 4Y_2 - Y_3 - 6Y_4 + \cos x, \end{aligned}$$

with initial conditions

$$Y_1(0) = 1,\ Y_2(0) = 2,\ Y_3(0) = 3,\ Y_4(0) = 4.$$

To put this into the form (6.7) define

$$\mathbf{Y} = \begin{pmatrix} Y_1 \\ Y_2 \\ Y_3 \\ Y_4 \end{pmatrix}, \quad \mathbf{A} = \begin{pmatrix} 1 \\ 2 \\ 3 \\ 4 \end{pmatrix},$$

and

$$\mathbf{F}(x,\mathbf{Y}) = \begin{pmatrix} Y_2 \\ -7Y_1 - 5Y_4 + \sin x \\ Y_4 \\ -3Y_1 - 4Y_2 - Y_3 - 6Y_4 + \cos x \end{pmatrix}.$$

Notice that for each unknown in the original set of equations, new unknowns are introduced for the original unknown and each of its derivatives up to an order one less than the highest appearing in the original set of equations. ■

The procedure we have illustrated is the usual way to convert a system of higher order equations to a system of first order equations. There are, however, other ways to do it. For some examples, see Exercise 6.6.

EXERCISES

6.1 As an example of nonuniqueness of solutions, verify that for any constant c, $0 \le c \le b$, the function $y(x)$ defined by

$$y(x) = \begin{cases} 0, & \text{if } 0 \le x \le c \\ \frac{1}{4}(x-c)^2, & \text{if } c < x \le b \end{cases}$$

is a solution of the initial value problem

$$y' = \sqrt{|y|}$$
$$y(0) = 0.$$

6.2 Consider the problem

$$y' = \sqrt{|1-y^2|}$$
$$y(0) = 1.$$

Verify that

(a) $y(x) \equiv 1$ is a solution on any interval containing $x = 0$,

(b) $y(x) = \cosh x$ is a solution on $[0, b]$ for any $b > 0$, and

(c) $y(x) = \cos x$ is a solution on a suitable interval. What is the largest interval containing $x = 0$ on which $\cos x$ is a solution?

6.3 Verify the statement in the text that Dawson's integral is a solution of the initial value problem

$$y' = 1 - 2xy$$
$$y(0) = 0.$$

6.4 For each initial value problem, verify that the given $y(x)$ is a solution.

(a) $y' = -\frac{1}{2}y^3, y(0) = 1; y(x) = 1/\sqrt{1+x}$

(b) $y' = -2xy^2, y(0) = 1; y(x) = 1/(1+x^2)$

(c) $y' = \frac{1}{4}y(1-y/20), y(0) = 1; y(x) = 20/(1+19e^{-x/4})$

(d) $y' = 100(\sin x - y), y(0) = 0; y(x) = [10^2(e^{-100x} - \cos x) + 10^4 \sin x]/(10^4+1)$

(e) $y' = (15\cos 10x)/y, y(0) = 2; y(x) = \sqrt{3\sin 10x + 4}$

6.5 Do the following functions satisfy a Lipschitz condition? If so, give suitable constants.

(a) $f(x,y) = 1 + y^2$ for $0 \le x \le \pi/2$

(b) $f(x,y) = 1 - 2xy$ for $0 \le x \le b$

(c) $f(x,y) = y/x$ for $1 \le x \le 2$

(d) $f(x,y) = y/x$ for $-1 \le x \le 1$

(e) $f(x,y) = \cos x \sin y$ for $-10^6 \le x \le 10^6$

6.6 Verify for each of the following systems of equations

(a) $Y_1' = Y_2, Y_2' = -x^2 Y_1 - xY_2$

(b) $Y_1' = e^{-x^2/2}Y_2, Y_2' = -x^2 e^{x^2/2} Y_1$

(c) $Y_1' = -(x/2)Y_1 + Y_2,\ Y_2' = (1/2 - 3x^2/4)Y_1 - (x/2)Y_2$

that $Y_1(x) = y(x)$, where $y(x)$ satisfies the second order equation

$$y''(x) + xy'(x) + x^2 y(x) = 0.$$

6.7 Put the following problems in standard form. Differentiation is with respect to t.

(a) $u^{(4)} + e^t u' - tu = \cos \alpha t$

(b) $u'' + v' \cos t + u = t,\ v' + u' + v = e^t$

(c) $u'' + 3v' + 4u + v = 8t,\ u'' - v' + u + v = \cos t$

(d) $mx'' = X(t,x,y,z,x',y',z'),$
$my'' = Y(t,x,y,z,x',y',z'),$
$mz'' = Z(t,x,y,z,x',y',z')$

(e) $u^{(6)} + uu' = e^t$

6.2 A SIMPLE NUMERICAL SCHEME

Let us again consider the initial value problem (6.1) and (6.2),

$$y' = f(x,y)$$
$$y(a) = A,$$

on the interval $[a, b]$. The numerical methods we consider generate a table of approximate values for $y(x)$. For the moment we suppose that the entries are for equally spaced arguments in x. That is, we choose an integer N and for $h = (b-a)/N$, we construct approximations at the points $x_n = a + nh$ for $n = 0, 1, \ldots, N$. The notation $y(x_n)$ is used for the solution of (6.1) and (6.2) evaluated at $x = x_n$, and y_n is used for an approximation to $y(x_n)$.

A differential equation has no "memory." If we know the value $y(x_n)$, Theorem 6.1 applies to the problem

$$\begin{aligned} u' &= f(x, u) \\ u(x_n) &= y(x_n) \end{aligned}$$

and says that the solution of this initial value problem on the interval $[x_n, b]$ is just $y(x)$. [After all, $y(x)$ is *a* solution and the theorem says there is only one.] That is, the values of $y(x)$ for x prior to $x = x_n$ do not directly affect the solution of the differential equation for x after x_n. Some numerical methods have memory and some do not. The class of methods known as *one-step* methods have no memory—given y_n, there is a recipe for the value y_{n+1} that depends only on x_n, y_n, f, and h. Starting with the obvious initial value $y_0 = A$, a one-step method generates a table for $y(x)$ by repeatedly taking one step in x of length h to generate successively $y_1, y_2, \ldots$.

The simplest example of a one-step method is Euler's method. We study it because the details do not obscure the ideas and the general case is much the same. A Taylor series expansion of $y(x)$ about $x = x_n$ gives

$$y(x_{n+1}) = y(x_n) + hy'(x_n) + \frac{h^2}{2} y''(\xi_n)$$

with $x_n < \xi_n < x_{n+1}$, provided that $y(x) \in C^2[a, b]$. Using the fact that $y(x)$ satisfies (6.1), this is

$$y(x_{n+1}) = y(x_n) + hf(x_n, y(x_n)) + \frac{h^2}{2} y''(\xi_n). \tag{6.8}$$

For small h,

$$y(x_{n+1}) \approx y(x_n) + hf(x_n, y(x_n)).$$

This relation suggests

Euler's method:

$$\begin{aligned} y_0 &= A \\ y_{n+1} &= y_n + hf(x_n, y_n), \; n = 0, 1, \ldots, N-1. \end{aligned} \tag{6.9}$$

Example 6.6. Tabulate Dawson's integral on [0, 0.5] using Euler's scheme with $h = 0.1$. Recall from Example 6.3 that Dawson's integral is the solution of the initial value problem

$$\begin{aligned} y' &= 1 - 2xy \\ y(0) &= 0. \end{aligned}$$

Taking $y_0 = 0$, we see that

$$y_1 = 0 + 0.1 \times (1 - 2 \times 0 \times 0) = 0.1;$$

similarly,

$$y_2 = 0.1 + 0.1 \times (1 - 2 \times 0.1 \times 0.1) = 0.198.$$

Continuing in this manner, the following table results. The true values of the integral $y(x_n)$ are taken from [7].

x_n	y_n	$y(x_n)$
0.0	0.00000	0.00000
0.1	0.10000	0.09934
0.2	0.19800	0.19475
0.3	0.29008	0.28263
0.4	0.37268	0.35994
0.5	0.44287	0.42444

■

To study the convergence of Euler's method, we relate the error at x_{n+1} to the error at x_n. Subtracting (6.9) from (6.8) gives

$$y(x_{n+1}) - y_{n+1} = y(x_n) - y_n + h[f(x_n, y(x_n)) - f(x_n, y_n)] + \frac{h^2}{2} y''(\xi_n).$$

Denoting the error at x_n by $E_n = y(x_n) - y_n$, the Lipschitz condition on f and this equation imply that

$$|E_{n+1}| \leq |E_n| + hL|y(x_n) - y_n| + \frac{h^2}{2} |y''(\xi_n)|.$$

With the definition

$$M_2 = \max_{a \leq x \leq b} |y''(x)|,$$

we obtain

$$|E_{n+1}| \leq |E_n|(1 + hL) + \frac{h^2}{2} M_2, \; n = 0, 1, \ldots, N-1. \tag{6.10}$$

Here the term $h^2 M_2 / 2$ bounds the error made in the current step and the other term bounds the error propagated from preceding steps.

To prove convergence, we bound the worst error that can arise as we step from $x_0 = a$ to $x_N = b$ and then show that it tends to zero as h does. The first order of business is to see how rapidly the inequality (6.10) allows the error to grow. To do this we establish a general result for later use. Suppose there are numbers $\delta > 0$ and $M > 0$ such that the sequence $d_0, d_1, \ldots$ satisfies

$$d_{n+1} \leq (1 + \delta) d_n + M, \; n = 0, 1, \ldots.$$

The case $n = 0$,

$$d_1 \le (1+\delta)d_0 + M,$$

can be combined with the case $n = 1$ to obtain

$$d_2 \le (1+\delta)d_1 + M \le (1+\delta)^2 d_0 + M[1 + (1+\delta)].$$

Similarly,

$$d_3 \le (1+\delta)d_2 + M \le (1+\delta)^3 d_0 + M[1 + (1+\delta) + (1+\delta)^2].$$

At this point we might guess that

$$d_n \le (1+\delta)^n d_0 + M[1 + (1+\delta) + (1+\delta)^2 + \cdots + (1+\delta)^{n-1}]. \tag{6.11}$$

To prove this, we use induction. The inequality (6.11) certainly holds for $n = 1, 2, 3$. Suppose the inequality is true for the case $n = k$. Then

$$\begin{aligned} d_{k+1} &\le (1+\delta)d_k + M \\ &\le (1+\delta)^{k+1} d_0 + M[1 + (1+\delta) + \cdots + (1+\delta)^k], \end{aligned}$$

which establishes the result for $n = k+1$ and completes the induction argument.

Lemma 6.1. *Suppose there are numbers $\delta > 0$ and $M > 0$ such that the sequence $d_0, d_1, \ldots$ satisfies*

$$d_{k+1} \le (1+\delta)d_k + M, \ k = 0, 1, \ldots.$$

Then for any $n \ge 0$,

$$d_n \le e^{n\delta} d_0 + M\frac{e^{n\delta} - 1}{\delta}. \tag{6.12}$$

Proof. Using the identity

$$(x-1)\sum_{p=0}^{n-1} x^p = \sum_{\ell=1}^{n} x^\ell - \sum_{p=0}^{n-1} x^p = x^n - x^0 = x^n - 1,$$

with $x = 1+\delta$, we see that the right-hand side of (6.11) can be rewritten in the form

$$(1+\delta)^n d_0 + M\frac{(1+\delta)^n - 1}{\delta}. \tag{6.13}$$

Expansion of the exponential function about zero gives for $\delta > 0$,

$$e^\delta = 1 + \delta + \frac{\delta^2}{2}e^\zeta, \ 0 < \zeta < \delta.$$

It then follows that

$$1 + \delta \le e^\delta$$

and

$$(1+\delta)^n \le e^{n\delta}.$$

This implies that (6.13) is bounded by

$$e^{n\delta}d_0 + M\frac{e^{n\delta}-1}{\delta},$$

which establishes (6.12). ■

Returning now to Euler's method, we apply Lemma 6.1 to (6.10) and arrive at

$$|E_n| \le e^{nhL}|E_0| + \frac{hM_2}{2L}(e^{nhL}-1).$$

However, $nh = x_n - a$ and $E_0 = y_0 - A = 0$, so

$$|y(x_n) - y_n| \le \frac{hM_2}{2L}(e^{L(x_n-a)}-1). \tag{6.14}$$

Using $x_n - a \le b - a$, this implies that

$$\max_{0\le n\le N}|y(x_n) - y_n| \le \frac{hM_2}{2L}(e^{L(b-a)}-1). \tag{6.15}$$

It is seen that the error of Euler's method is bounded by a constant times h. When the value of the constant is immaterial, such expressions are written as $O(h)$.

Generally speaking, we try to ignore the effects of finite precision arithmetic. When the tolerance corresponds to a relative accuracy comparable to the word length of the computer, this is not possible. Also, if the solution is very hard to approximate accurately at x_n, the step size necessary may be so small that the precision must be considered. To gain some insight, note that we do not obtain $f(x_n,y_n)$ from a subroutine, but rather $f(x_n,y_n)+\varepsilon_n$. Similarly, in computing $y_{n+1} = y_n + h[f(x_n,y_n)+\varepsilon_n]$ an additional error ρ_n is made. The sequence generated computationally is then

$$y_{n+1} = y_n + hf(x_n,y_n) + h\varepsilon_n + \rho_n.$$

Let us suppose that $|\rho_n| \le \rho$ and $|\varepsilon_n| \le \varepsilon$ for all $h \le h_0$. Then the analysis can be modified to yield

$$\max_{0\le n\le N}|y(x_n) - y_n| \le \frac{e^{L(b-a)}-1}{L}\left(\frac{hM_2}{2}+\varepsilon+\frac{\rho}{h}\right).$$

According to this bound, roundoff effects get worse as the step size is reduced in an attempt to get a more accurate solution. Clearly there is a maximum accuracy possible that depends on the problem, the numerical method, and the arithmetic of the computer used. The effects are more complex than this bound shows, but the bound is qualitatively correct. It is easy to show experimentally that as h is decreased, the numerical solution is at first more accurate, reaches a best value, and subsequently is increasingly less accurate.

The convergence analysis just presented is the traditional one. The trouble is that this is not the way modern codes work. Rather than take a step of specified length h, they select a step size automatically that will produce a solution with a specified accuracy. A reasonable model of the step sizes selected by such codes is that at x_n the code selects a step $h_n = \Theta(x_n)H$, where $\Theta(x)$ is a piecewise-continuous function satisfying $0 < \theta \leq \Theta(x) \leq 1$ for $a \leq x \leq b$. With this model it is easy enough to modify the convergence proof just given to account for variable step size. The result is that as the maximum step size H tends to zero, $\max_{0 \leq n \leq N} |y(x_n) - y_n|$ is $O(H)$. It is not hard to see how the model comes about. The user specifies a tolerance τ. In a step of length h from x_n, Euler's method makes an error of approximately $h^2|y''(x_n)|/2$. The largest step size h_n that can be used and still keep the error less than τ is then about

$$h_n \approx \sqrt{\frac{2\tau}{|y''(x_n)|}}.$$

Special rules come into play in the codes when $y''(x_n)$ is nearly zero, so suppose that $y''(x)$ does not vanish in $[a,b]$. If

$$\zeta = \min_{[a,b]} |y''(x)| > 0$$

and

$$H = \sqrt{\frac{2\tau}{\zeta}},$$

then

$$h_n \approx \sqrt{\frac{\zeta}{|y''(x_n)|}} H = \Theta(x_n) H$$

defines $\Theta(x)$. Notice that $H = O(\tau^{1/2})$ so that $\max |y(x_n) - y_n|$ is $O(\tau^{1/2})$ for Euler's method with automatic selection of the step size.

EXERCISES

6.8 Use Euler's method on the following problems using a fixed step size $h = 1.0$, and then $h = 0.5$. In each case calculate the errors at $x = 1.0$.

(a) $y' = -y/(x+1)$ with $y(0) = 1$, so $y(x) = 1/(x+1)$.

(b) $y' = -y^3/2$ with $y(0) = 1$, so $y(x) = 1/\sqrt{1+x}$.

6.9 Implement Euler's method to estimate solutions of the initial value problem in Exercise 6.8b. Use $h = 1/40$ and $h = 1/80$. Compute the errors at $x = 0.5$ and $x = 1.0$ to see if they are roughly halved as h is. How small an h would you estimate is needed in order for the absolute error to be less than 10^{-6} in magnitude?

6.3 ONE-STEP METHODS

Let us now consider general one-step methods and base our assumptions on the successful treatment of Euler's method. The recipe is to be of the form

$$\begin{aligned} y_0 &= A \\ y_{n+1} &= y_n + h\Phi(x_n, y_n, f, h),\ n = 0, 1, \ldots. \end{aligned} \qquad (6.16)$$

The method has no memory, so Φ depends only on the arguments listed. Usually f and h are omitted in the notation. It is assumed that Φ is continuous in x and y. The treatment of Euler's method had $\Phi(x,y) = f(x,y)$ and a Lipschitz condition was used in an important way. So, for the general procedure we assume that

$$|\Phi(x,u) - \Phi(x,v)| \leq L_\Phi |u - v| \tag{6.17}$$

for $a \leq x \leq b$, all $0 < h \leq h_0$ for some h_0, any continuous function f satisfying a Lipschitz condition, and all u, v.

In discussing Euler's method we used as a starting point the fact that the solution $y(x)$ almost satisfies the recipe (6.9) used to define the numerical approximation. The analog here is

$$y(x_{n+1}) = y(x_n) + h\Phi(x_n, y(x_n)) + h\mu_n, \tag{6.18}$$

with μ_n "small." More precisely, if for all x_n in $[a,b]$ and all $h \leq h_0$, there are constants C and p such that

$$|\mu_n| \leq Ch^p, \tag{6.19}$$

then we say that the method is of order p for the equation (6.1). The quantity μ_n is called the *local truncation error.*

Theorem 6.2. *Suppose the initial value problem*

$$y' = f(x,y)$$
$$y(a) = A$$

on the finite interval $[a,b]$ is solved by the one-step method (6.16) and suppose that the hypotheses of Theorem 6.1 are satisfied. If $\Phi(x,y)$ satisfies (6.17) and if the method is of order $p \geq 1$ for $y(x)$, then for any $x_n = a + nh \in [a,b]$,

$$|y(x_n) - y_n| \leq \frac{Ch^p}{L_\Phi}\left(e^{L_\Phi(x_n - a)} - 1\right).$$

Proof. As before, let $E_n = y(x_n) - y_n$ and subtract (6.16) from (6.18) to obtain

$$E_{n+1} = E_n + h[\Phi(x_n, y(x_n)) - \Phi(x_n, y_n)] + h\mu_n.$$

Using the Lipschitz condition (6.17) and the fact that the method is of order p, we see that

$$|E_{n+1}| \leq (1 + hL_\Phi)|E_n| + Ch^{p+1}.$$

The theorem now follows from Lemma 6.1 and the fact that $E_0 = 0$. ■

As with our discussion of Euler's method, the result of this theorem gives convergence of $O(h^p)$. This explains our calling the method of order p for $y(x)$. The term "a method of order p" is used to describe a method that is of order p if f is sufficiently smooth. The order of convergence is lower when f is not so smooth.

As explained in connection with Euler's method, codes select the step size automatically so as to keep the error smaller than a tolerance τ at each step. At the same time they try to use an efficiently large step. A reasonable model of such a step size algorithm leads to a step size h_n at x_n given by

$$h_n = \Theta(x_n)H$$

for a piecewise-continuous function $\Theta(x)$ with $0 < \theta \leq \Theta(x) \leq 1$ on $[a,b]$. With step sizes specified in this way, the convergence proof can be altered easily to conclude that the error is $O(H^p) = O(\tau^{1/p})$.

The most important task now left is to find functions Φ that are inexpensive to evaluate and of order p for smooth f. We need, then,

$$y(x_{n+1}) = y(x_n) + h\Phi(x_n, y(x_n)) + h\mu_n,$$

with $\mu_n = O(h^p)$. A Taylor series expansion of $y(x)$ shows that

$$y(x_{n+1}) = y(x_n) + h\left[y'(x_n) + \cdots + \frac{h^{p-1}}{p!}y^{(p)}(x_n)\right] + \frac{h^{p+1}}{(p+1)!}y^{(p+1)}(\xi_n)$$

if $y(x) \in C^{p+1}[a,b]$. So we find that if the method is of order p, then it must be the case that

$$\Phi(x, y(x)) = y'(x) + \frac{h}{2!}y''(x) + \cdots + \frac{h^{p-1}}{p!}y^{(p)}(x) + \zeta(x),$$

with $\zeta(x) = O(h^p)$. Because $y(x)$ is a solution of the differential equation $y'(x) = f(x, y(x))$, the derivatives of y can be expressed in terms of total derivatives of f. Using the notation $f^{(m)}(x, y(x))$ for the mth total derivative of f and subscripts for partial derivatives, the relation is

$$y^{(m)}(x) = f^{(m-1)}(x, y(x)),$$

where

$$f^{(1)}(x,y) = f_x(x,y) + f_y(x,y)f(x,y)$$
$$f^{(m)}(x,y) = f_x^{(m-1)}(x,y) + f_y^{(m-1)}(x,y)f(x,y), \ m = 1, 2, \ldots.$$

The expression for $\Phi(x, y)$ becomes

$$\Phi(x,y) = f(x,y) + \frac{h}{2!}f^{(1)}(x,y) + \cdots + \frac{h^{p-1}}{p!}f^{(p-1)}(x,y) + O(h^p). \tag{6.20}$$

An obvious choice for Φ is the function $T(x, y)$,

$$T(x,y) = f(x,y) + \frac{h}{2!}f^{(1)}(x,y) + \cdots + \frac{h^{p-1}}{p!}f^{(p-1)}(x,y),$$

which yields a family of one-step methods called the Taylor series methods. Euler's method is the case $p = 1$. When it is possible to evaluate derivatives efficiently, these methods are very effective.

For a simple equation like that satisfied by Dawson's integral, and especially when high accuracy is desired, a Taylor series method may be the best way to proceed. This equation has

$$f(x,y) = 1 - 2xy$$
$$f^{(1)}(x,y) = -2x + (4x^2 - 2)y,$$

and so forth. Exercise 6.13 develops a simple recursion that makes it easy to use a Taylor series method of very high order for this equation.

Runge–Kutta methods use several evaluations of $f(x,y)$ in a linear combination to approximate $y(x)$. The simplest case is Euler's method that uses one evaluation. Let us now derive a procedure using the two evaluations $f(x_n,y_n)$ and $f(x_n + p_1 h, y_n + p_2 h f(x_n,y_n))$, where p_1 and p_2 are parameters. Then for Φ we use the linear combination $R(x,y)$:

$$R(x_n,y_n) = a_1 f(x_n,y_n) + a_2 f(x_n + p_1 h, y_n + p_2 h f(x_n,y_n)).$$

In this expression we are free to choose any useful values for p_1, p_2, a_1, and a_2. The aim is to choose the parameters so that the representation (6.19) holds for as large a value of p as possible. To carry this out we expand all the quantities in Taylor series in h and equate coefficients of like powers. To simplify the notation, arguments are shown only if they are different from (x_n,y_n). The reader familiar with Taylor series expansions of a function of two variables may skip to the result. Otherwise, we can proceed by a succession of familiar one-variable expansions as follows:

$$\begin{aligned}
R &= a_1 f + a_2 f(x_n + p_1 h, y_n + p_2 h f) \\
&= a_1 f + a_2 \Big[f(x_n, y_n + p_2 h f) + p_1 h f_x(x_n, y_n + p_2 h f) \\
&\qquad + \frac{p_1^2 h^2}{2} f_{xx}(x_n, y_n + p_2 h f) + O(h^3) \Big] \\
&= a_1 f + a_2 \Big[f + p_2 h f f_y + \frac{p_2^2 h^2}{2} f^2 f_{yy} + O(h^3) \\
&\qquad + p_1 h f_x + p_1 p_2 h^2 f f_{xy} + O(h^3) + \frac{p_1^2 h^2}{2} f_{xx} + O(h^3) \Big] \\
&= (a_1 + a_2) f + a_2 h [p_2 f f_y + p_1 f_x] \\
&\qquad + \frac{a_2 h^2}{2} [p_2^2 f^2 f_{yy} + 2 p_1 p_2 f f_{xy} + p_1^2 f_{xx}] + O(h^3).
\end{aligned}$$

Now we want to choose the parameters so that

$$R = f + \frac{h}{2} f^{(1)} + \frac{h^2}{6} f^{(2)} + O(h^3)$$

or, writing this out,

$$R = f + \frac{h}{2}(f f_y + f_x) + \frac{h^2}{6}(f^2 f_{yy} + 2 f f_{xy} + f_{xx} + f_x f_y + f f_y^2) + O(h^3).$$

Equating terms involving the same powers of h, it is found that it is possible to obtain agreement only for h^0 and h^1:

$$\begin{aligned} h^0 &: a_1 + a_2 = 1 \\ h^1 &: a_2 p_2 = \frac{1}{2} \\ &\quad a_2 p_1 = \frac{1}{2}. \end{aligned}$$

Let $a_2 = \alpha$. Then for any value of the parameter α,

$$\begin{aligned} a_2 &= \alpha \\ a_1 &= 1 - \alpha \end{aligned}$$

gives a formula with agreement in all terms involving h^0. If we further require that $\alpha \neq 0$, the choice

$$p_1 = p_2 = \frac{1}{2\alpha}$$

gives a formula with agreement in all terms involving h^1. Then

$$R = (1-\alpha) f(x,y) + \alpha f\left(x + \frac{h}{2\alpha}, y + \frac{h}{2\alpha} f(x,y)\right)$$

yields a family of one-step methods of order 2 when $\alpha \neq 0$ and f is sufficiently smooth.

Some of the members of this family of formulas have names. Euler's method has $\alpha = 0$ and the order $p = 1$. Heun's method (the improved Euler method) is the case $\alpha = 1/2$, and the midpoint or modified Euler method is the case $\alpha = 1$. The broad applicability of these formulas is seen when we ask what is needed for the convergence theorem to be valid. The continuity of R obviously follows from that of f. It is a pleasant fact that the Lipschitz condition on R also follows from that on f:

$$\begin{aligned} |R(x,u) - R(x,v)| &= \Big| (1-\alpha)[f(x,u) - f(x,v)] + \alpha \cdot \\ &\quad \left[f\left(x + \frac{h}{2\alpha}, u + \frac{h}{2\alpha} f(x,u)\right) - f\left(x + \frac{h}{2\alpha}, v + \frac{h}{2\alpha} f(x,v)\right)\right]\Big| \\ &\leq |1-\alpha| L |u-v| + |\alpha| L \left| \left[u + \frac{h}{2\alpha} f(x,u)\right] - \left[v + \frac{h}{2\alpha} f(x,v)\right]\right| \\ &\leq |1-\alpha| L |u-v| + |\alpha| L |u-v| + \frac{h}{2} L^2 |u-v| \\ &= \left\{ |1-\alpha| + |\alpha| + \frac{hL}{2} \right\} L |u-v| \end{aligned}$$

for all $0 < h \leq h_0$, and we may take the Lipschitz constant for R to be

$$\left\{ |1-\alpha| + |\alpha| + \frac{h_0 L}{2} \right\} L.$$

Therefore, if the differential equation satisfies the conditions of Theorem 6.1, and if the function f has two continuous derivatives [which implies the solution $y(x) \in C^3[a,b]$], any member of the family with $\alpha \neq 0$ is convergent of order 2.

Higher order procedures involving more substitutions can be derived in the same way, although naturally the expansions become (very) tedious. As it happens, p evaluations of f per step lead to procedures of order p for $p = 1,2,3,4$ but *not* for 5. For this reason, fourth order formulas were often preferred for constant step size integrations. As in the second order case, there is a family of fourth order procedures depending on several parameters. The classical choice of parameters leads to the algorithm

$$y_0 = A,$$

and for $n = 0, 1, \ldots,$

$$\begin{aligned}
k_0 &= f(x_n, y_n) \\
k_1 &= f\left(x_n + \frac{h}{2}, y_n + \frac{h}{2}k_0\right) \\
k_2 &= f\left(x_n + \frac{h}{2}, y_n + \frac{h}{2}k_1\right) \\
k_3 &= f(x_n + h, y_n + hk_2) \\
y_{n+1} &= y_n + \frac{h}{6}(k_0 + 2k_1 + 2k_2 + k_3).
\end{aligned}$$

This is formulated for the first order system of equations

$$\begin{aligned}
\mathbf{Y}(a) &= \mathbf{A} \\
\mathbf{Y}' &= \mathbf{F}(x, \mathbf{Y})
\end{aligned}$$

in a natural way:

$$\mathbf{Y_0} = \mathbf{A},$$

and for $n = 0, 1, \ldots,$

---

the classical Runge–Kutta algorithm is

$$\begin{aligned}
\mathbf{K_0} &= \mathbf{F}(x_n, \mathbf{Y_n}) \\
\mathbf{K_1} &= \mathbf{F}\left(x_n + \frac{h}{2}, \mathbf{Y_n} + \frac{h}{2}\mathbf{K_0}\right) \\
\mathbf{K_2} &= \mathbf{F}\left(x_n + \frac{h}{2}, \mathbf{Y_n} + \frac{h}{2}\mathbf{K_1}\right) \\
\mathbf{K_3} &= \mathbf{F}(x_n + h, \mathbf{Y_n} + h\mathbf{K_2}) \\
\mathbf{Y_{n+1}} &= \mathbf{Y_n} + \frac{h}{6}(\mathbf{K_0} + 2\mathbf{K_1} + 2\mathbf{K_2} + \mathbf{K_3}).
\end{aligned}$$

Another quite similar fourth order procedure is

$$\begin{aligned}
\mathbf{K_0} &= \mathbf{F}(x_n, \mathbf{Y_n}) \\
\mathbf{K_1} &= \mathbf{F}\left(x_n + \frac{h}{2}, \mathbf{Y_n} + \frac{h}{2}\mathbf{K_0}\right) \\
\mathbf{K_2} &= \mathbf{F}\left(x_n + \frac{h}{2}, \mathbf{Y_n} + \frac{h}{4}\mathbf{K_0} + \frac{h}{4}\mathbf{K_1}\right) \\
\mathbf{K_3} &= \mathbf{F}(x_n + h, \mathbf{Y_n} - h\mathbf{K_1} + 2h\mathbf{K_2}) \\
\mathbf{Y_{n+1}} &= \mathbf{Y_n} + \frac{h}{6}(\mathbf{K_0} + 4\mathbf{K_2} + \mathbf{K_3}).
\end{aligned} \tag{6.21}$$

There is little reason to prefer one of these procedures over the other as far as a single step is concerned. In the next section we shall learn how R. England exploited (6.21) to achieve an estimate of the error made in the step.

EXERCISES

6.10 A definite integral $\int_a^b f(x)\,dx$ can be evaluated by solving an initial value problem for an ordinary differential equation. Let $y(x)$ be the solution of

$$\begin{aligned} y' &= f(x),\ a \le x \le b \\ y(a) &= 0. \end{aligned}$$

Then

$$y(b) = \int_a^b f(x)\,dx.$$

Runge–Kutta methods integrate the more general equation $y' = f(x,y)$. In this special case, they assume forms familiar in quadrature. Referring to Chapter 5, identify the familiar procedure to which both the classical fourth order formula and England's formula degenerate.

6.11 Implement Heun's method to estimate the solution of the initial value problem in Exercise 6.8b. Use $h = 1/40$ and $h = 1/80$. Compute the errors at $x = 0.5$ and $x = 1.0$ to see if they are roughly quartered as h is halved. How small an h would you estimate is needed in order for the absolute error to be less than 10^{-6} in magnitude?

6.12 Implement the classical Runge–Kutta algorithm to estimate the solution of the initial value problem in Exercise 6.8b. Use $h = 1/40$ and $h = 1/80$. Compute the errors at $x = 0.5$ and $x = 1.0$. By what factor are the errors reduced as h is halved? Roughly speaking, what is the largest value of h for which the absolute error will be less than 10^{-6} in magnitude?

6.13 Consider the linear equation

$$y' = P_1(x)y + Q_1(x).$$

Show that the derivatives needed in Taylor series methods can be obtained from

$$y^{(r)} = P_r(x)y + Q_r(x),$$

where

$$\begin{aligned} P_r(x) &= P'_{r-1}(x) + P_1(x)P_{r-1}(x) \\ Q_r(x) &= Q'_{r-1}(x) + Q_1(x)P_{r-1}(x),\ r = 2,3,\ldots. \end{aligned}$$

Use this to develop a fifth order formula for computing Dawson's integral.

6.14 An interesting fact about Runge–Kutta methods is that the error depends on the form of the equation as well as on the solution itself. To see an example of this, show that $y(x) = (x+1)^2$ is the solution of each of the two problems

$$\begin{aligned} y' &= 2(x+1), & y(0) &= 1 \\ y' &= 2y/(x+1), & y(0) &= 1. \end{aligned}$$

Then show that Heun's method is exact for the first equation. Prove that the method is *not* exact when applied to the second equation, although it has the same solution.

6.4 ERRORS—LOCAL AND GLOBAL

Modern codes for the initial value problem do not use a fixed step size h. The error made at each step is estimated and h is adjusted both to obtain an approximation that is sufficiently accurate and to carry out the integration efficiently. There is an unfortunate confusion on the part of many users of codes with error estimates as to what is being measured and what its relation to the true error is.

The function $y(x)$ denotes the unique solution of the problem

$$y' = f(x,y)$$
$$y(a) = A.$$

The *true* or *global error* at x_{n+1} is

$$y(x_{n+1}) - y_{n+1}.$$

Unfortunately, it is relatively difficult and expensive to estimate this quantity. This is not surprising, since in the step to x_{n+1} the numerical procedure is only supplied with x_n, y_n and the ability to evaluate f. The *local solution* at x_n is the solution $u(x)$ of

$$u' = f(x,u)$$
$$u(x_n) = y_n.$$

The *local* error is

$$u(x_{n+1}) - y_{n+1}.$$

This is the error made approximating the solution of the differential equation originating at (x_n, y_n) in a single step. These errors are illustrated in Figure 6.1. It is reasonable to ask that a numerical procedure keep this error small. What effect this has on the global error depends on the differential equation itself. After all,

$$y(x_{n+1}) - y_{n+1} = \{y(x_{n+1}) - u(x_{n+1})\} + \{u(x_{n+1}) - y_{n+1}\}. \qquad (6.22)$$

The quantity

$$y(x_{n+1}) - u(x_{n+1})$$

is a measure of the stability of the differential equation since it is a consequence (at x_{n+1}) of the initial difference $y(x_n) - y_n$ at x_n. If this quantity increases greatly, the problem is poorly posed or ill-conditioned or unstable.

Example 6.7. Consider

$$y' = \alpha y$$

for a constant α. A little calculation shows that

$$y(x) = y(x_n)e^{\alpha(x-x_n)}$$
$$u(x) = y_n e^{\alpha(x-x_n)};$$

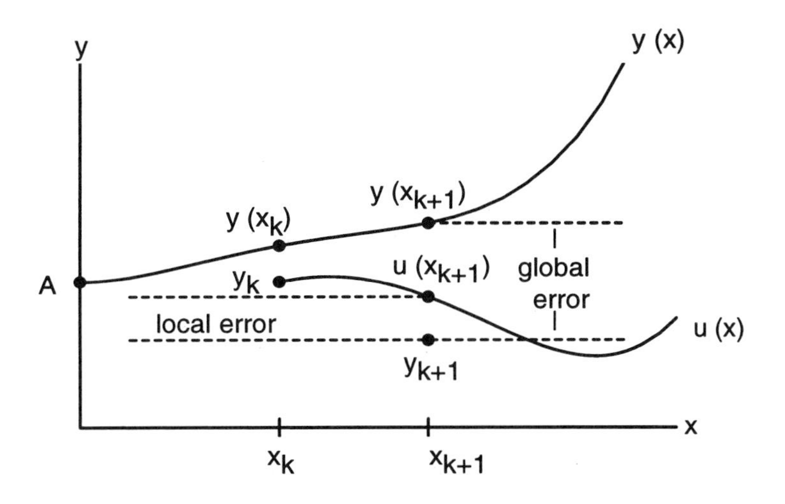

Figure 6.1 Local and global errors.

furthermore,

$$y(x_{n+1}) - u(x_{n+1}) = \{y(x_n) - y_n\}e^{\alpha h}.$$

If $\alpha > 0$, the solution curves spread out (Figure 6.2a), the more so as α is large. From the expression (6.23) it is clear that a small local error at every step does not imply a small global error. On the other hand, if $\alpha < 0$, the curves come together (Figure 6.2b) and (6.23) shows that controlling the local error will control the global error. For general functions $f(x,y)$ the Lipschitz condition alone cannot predict this behavior, since for this example the Lipschitz constant is $|\alpha|$ in either case. ■

The local error is related to the local truncation error. Indeed, it is just h times the local truncation error, μ, for the local solution $u(x)$:

$$\begin{aligned}\text{local error} &= u(x_{n+1}) - y_{n+1}\\ &= (y_n + h\Phi(x_n, y_n) + h\mu_n) - y_{n+1}\\ &= h\mu_n.\end{aligned}$$

For example, when $y(x)$ is a solution of $y' = f(x,y)$, we have seen that Euler's method has

$$\begin{aligned}y(x_{n+1}) = y(x_n) + hf(x_n, y(x_n)) + \frac{h^2}{2}(f(x_n, y(x_n))f_y(x_n, y(x_n))\\ + f_x(x_n, y(x_n))) + O(h^3).\end{aligned}$$

Applying this to $u(x)$, we have

$$\text{local error} = h\mu_n = \frac{h^2}{2}(ff_y + f_x) + O(h^3).$$

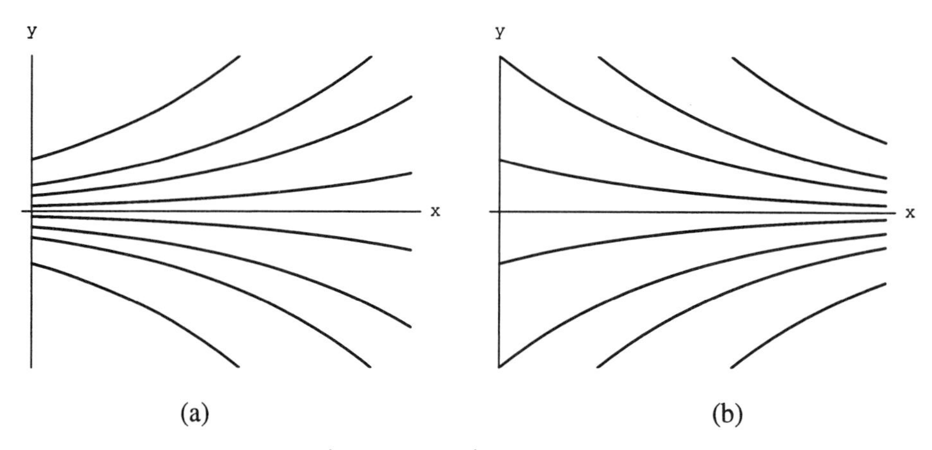

Figure 6.2 Solution curves for (a) $y' = 2y$ and (b) $y' = -2y$.

Similarly, for the Runge–Kutta formulas of order 2 ($\alpha \neq 0$), we have

$$u(x_{n+1}) = y_n + h\left[f + \frac{h}{2}f^{(1)} + \frac{h^2}{6}f^{(2)}\right] + O(h^4)$$

then the numerical approximations satisfy

$$\widehat{y}_{n+1} = y_n + h\left[f + \frac{h}{2}(ff_y + f_x) + \frac{h^2}{8\alpha}(f^2 f_{yy} + 2ff_{xy} + f_{xx})\right] + O(h^4).$$

This leads to

$$\text{local error} = h\widehat{\mu}_n = h^3\left(\frac{1}{6} - \frac{1}{8\alpha}\right)(f^2 f_{yy} + 2ff_{xy} + f_{xx}) + \frac{h^3}{6}(f_x f_y + ff_y^2) + O(h^4).$$

A little reflection about these expressions suggests a way to estimate the local error. Suppose we compute y_{n+1} with Euler's method and we also compute an approximate solution $\widehat{y}_{n+1}$ with one of the second order Runge–Kutta formulas. The expressions above show that

$$\begin{aligned}\widehat{y}_{n+1} - y_{n+1} &= \frac{h^2}{2}(ff_y + f_x) + O(h^3)\\ &= h\mu_n + O(h^3).\end{aligned}$$

That is, the discrepancy between the two values estimates the error in the lower order formula. This is the same principle used in Chapter 5 to estimate quadrature errors. In general, suppose that in addition to the value

$$y_{n+1} = y_n + h\Phi(x_n, y_n)$$

with truncation error $\mu_n = O(h^p)$, we compute another approximation

$$\widehat{y}_{n+1} = y_n + h\widehat{\Phi}(x_n, y_n)$$

with truncation error $\widehat{\mu}_n = O(h^q)$ of higher order, $q > p$. Then by definition

$$\begin{aligned} u(x_{n+1}) &= y_n + h\Phi(x_n, y_n) + h\mu_n \\ &= y_{n+1} + h\mu_n \end{aligned}$$

and, similarly,

$$u(x_{n+1}) = \widehat{y}_{n+1} + h\widehat{\mu}_n,$$

which, on subtracting, shows that

$$\begin{aligned} \widehat{y}_{n+1} - y_{n+1} &= h\mu_n - h\widehat{\mu}_n \\ &= h\mu_n + O(h^{q+1}). \end{aligned}$$

Because $h\widehat{\mu}_n$ goes to zero faster than $h\mu_n$, we can estimate the local error by

$$\text{local error } = h\mu_n \approx \widehat{y}_{n+1} - y_{n+1}.$$

We would like to approximate the local solution $u(x_{n+1})$. In view of the fact that we have a good estimate of the error in y_{n+1}, why not try to improve it by taking out the error? This process, called local extrapolation, is here formally equivalent to advancing the integration with the higher order approximation $\widehat{y}_{n+1}$ because

$$u(x_{n+1}) = y_{n+1} + h\mu_n \approx y_{n+1} + (\widehat{y}_{n+1} - y_{n+1}) = \widehat{y}_{n+1}.$$

This tells us that local extrapolation will raise the effective order of the pair from p to q. Thus we can think of what is happening in two ways. One is that a formula of order p is being used with its result improved by local extrapolation. The other is that a formula of order q is being used with the step size selected conservatively by pretending that the step is being taken with a formula of a lower order p. Because local extrapolation increases the accuracy at no increase in cost, all the current production-grade codes based on explicit Runge–Kutta methods use it.

A Runge–Kutta formula of order 4 requires (at least) four evaluations of **F** per step and a companion formula of order 5 requires at least six. Just as with Gauss–Kronrod quadrature, the trick to being efficient is to derive the formulas as a pair in which function evaluations are used in both formulas. R. England published such a pair of formulas in [5]. To advance from x_n to $x_n + h$, he takes a step of length $h/2$ with (6.21) to get the fourth order result $\mathbf{Y}_{\mathbf{n}+\frac{1}{2}} \approx \mathbf{Y}(x_n + h/2)$ and then another step of length $h/2$ to get the fourth order result $\mathbf{Y}_{\mathbf{n+1}} \approx \mathbf{Y}(x_n + h)$. By working with two half-steps, he has enough function evaluations available that with only one extra evaluation, he is able to form a fifth order approximation $\widehat{\mathbf{Y}}_{n+1}$ to $\mathbf{Y}(x_n + h)$. In this way, one extra function evaluation is made every two half-steps to get an error estimate. An error estimate is used to control the local error and so give some credibility to the computed solution. It also allows the code to select the largest step size that will result in the error test being passed. Except in unusual cases, adaptation of the step size to the solution in this way greatly increases the efficiency of integration. It corresponds to the adaptive quadrature schemes of Chapter 5.

England's formulas are as follows:

$$\mathbf{K_0} = \mathbf{F}(x_n, \mathbf{Y_n})$$

$$\mathbf{K_1} = \mathbf{F}\left(x_n + \frac{h}{4}, \mathbf{Y_n} + \frac{h}{4}\mathbf{K_0}\right)$$

$$\mathbf{K_2} = \mathbf{F}\left(x_n + \frac{h}{4}, \mathbf{Y_n} + \frac{h}{8}(\mathbf{K_0} + \mathbf{K_1})\right)$$

$$\mathbf{K_3} = \mathbf{F}\left(x_n + \frac{h}{2}, \mathbf{Y_n} - \frac{h}{2}\mathbf{K_1} + h\mathbf{K_2}\right)$$

$$\mathbf{Y_{n+\frac{1}{2}}} = \mathbf{Y_n} + \frac{h}{12}(\mathbf{K_0} + 4\mathbf{K_2} + \mathbf{K_3});$$

$$\mathbf{K_4} = \mathbf{F}\left(x_n + \frac{h}{2}, \mathbf{Y_{n+\frac{1}{2}}}\right)$$

$$\mathbf{K_5} = \mathbf{F}\left(x_n + \frac{3}{4}h, \mathbf{Y_{n+\frac{1}{2}}} + \frac{h}{4}\mathbf{K_4}\right)$$

$$\mathbf{K_6} = \mathbf{F}\left(x_n + \frac{3}{4}h, \mathbf{Y_{n+\frac{1}{2}}} + \frac{h}{8}(\mathbf{K_4} + \mathbf{K_5})\right)$$

$$\mathbf{K_7} = \mathbf{F}\left(x_n + h, \mathbf{Y_{n+\frac{1}{2}}} - \frac{h}{2}\mathbf{K_5} + h\mathbf{K_6}\right)$$

$$\mathbf{Y_{n+1}} = \mathbf{Y_{n+\frac{1}{2}}} + \frac{h}{12}(\mathbf{K_4} + 4\mathbf{K_6} + \mathbf{K_7});$$

$$\mathbf{K_8} = \mathbf{F}\Big(x_n + h, \mathbf{Y_n} + \frac{h}{12}(-\mathbf{K_0} - 96\mathbf{K_1} + 92\mathbf{K_2} - 121\mathbf{K_3} + 144\mathbf{K_4}$$
$$+6\mathbf{K_5} - 12\mathbf{K_6})\Big)$$

$$\widehat{\mathbf{Y}}_{\mathbf{n+1}} = \mathbf{Y_n} + \frac{h}{180}(14\mathbf{K_0} + 64\mathbf{K_2} + 32\mathbf{K_3} - 8\mathbf{K_4} + 64\mathbf{K_6} + 15\mathbf{K_7} - \mathbf{K_8}).$$

One drawback of conventional algorithms for solving initial value problems is that they produce a table of approximate values while the mathematical solution $y(x)$ is a continuous function. It is possible to approximate the solution for all x by interpolation. At the beginning of a step we have $\mathbf{Y_n}$ and form

$$\mathbf{K_0} = \mathbf{F}(x_n, \mathbf{Y_n}) \approx \mathbf{Y}'(x_n) = \mathbf{F}(x_n, \mathbf{Y}(x_n)).$$

Similarly, at the start of the second half-step we have $\mathbf{Y_{n+\frac{1}{2}}}$ and

$$\mathbf{K_4} = \mathbf{F}\left(x_n + \frac{h}{2}, \mathbf{Y_{n+\frac{1}{2}}}\right) \approx \mathbf{Y}'\left(x_n + \frac{h}{2}\right).$$

The code Rke does local extrapolation, hence reports the fifth order solution $\widehat{\mathbf{Y}}_{n+1}$ as its approximate solution at x_{n+1}. It will be called $\mathbf{Y}_{n+1}$ on the next step. By programming the procedure so that $\mathbf{F}(x_{n+1}, \widehat{\mathbf{Y}}_{n+1})$ is evaluated in the current step and later used as the $\mathbf{K_0}$ of the next step, we have an approximation to $\mathbf{Y}'(x_{n+1})$. In this way we have approximations to $\mathbf{Y}(x)$ and $\mathbf{Y}'(x)$ at $x_n, x_n + h/2$, and x_{n+1} in the course of taking a step. It is natural to interpolate these data by quintic (degree 5) Hermite interpolation. It turns out that in a certain sense this interpolant is as accurate an approximation to $\mathbf{Y}(x)$ on (x_n, x_{n+1}) as $\mathbf{Y}_{n+1}$ is to $\mathbf{Y}(x_{n+1})$. Notice that only information generated in the current step is needed. Furthermore, the quintic polynomial on $[x_n, x_{n+1}]$ agrees in value and slope at x_{n+1} with the quintic polynomial on $[x_{n+1}, x_{n+2}]$. Thus the piecewise polynomial function resulting from this scheme is continuous and has a continuous derivative on all of $[a, b]$. An interpolation capability can greatly increase the efficiency of an ordinary differential equation solver because the step size can be selected solely for reliability and efficiency.

6.5 THE ALGORITHMS

It is easy enough to write a code based on a pair of Runge–Kutta formulas like England's, but it is not easy to write a code of production quality. A subdiscipline of numerical analysis called *mathematical software* has developed that concerns itself with such tasks. References [1], [2], [4], [6], [8], [9], and [10] discuss the issues for ordinary differential equations at length. The code Rke that we provide is significantly more complex than the codes in other chapters, so its description is far more detailed.

In this section we consider briefly portions of the code with the aim of explaining some of the care needed in converting a pair of formulas into a production code. It is hard to come to a full understanding of such complex codes because decisions about one algorithmic question usually depend on those made in connection with several others. Perhaps a good place to start is with the way error is to be measured.

For simplicity, only the scalar case is described; the vector case is handled similarly. At each step the code attempts to keep the local error less than a tolerance specified by the user:

$$|\text{local error}| \leq \tau.$$

How this error is measured is important. A reasonable error tolerance will depend on the size of the solution. Because this size is usually not known in advance, a good way to proceed is to measure the error relative to the size computed by the code:

$$\frac{|\text{local error}|}{\text{size of } y} \leq \tau.$$

It is not so clear what we should take here as the "size" of y. Besides needing a reasonable definition of size when a solution component vanishes, we need to avoid the technical difficulty of dividing by zero. We have a value y_n at the beginning of the step, an approximate solution of order 4, $y_{n+\frac{1}{2}}$, at half the step, and two approximate solutions of orders 4 and 5, $y_{n+1}, \widehat{y}_{n+1}$, at the end of the step. A reasonable way to define the size of y over the step is to average these magnitudes, taking account of the fact that two of the values approximate $y(x)$ at the same point:

$$wt = \frac{1}{3}\left[|y_n| + |y_{n+\frac{1}{2}}| + \frac{1}{2}(|y_{n+1}| + |\widehat{y}_{n+1}|)\right].$$

With this definition, it is unlikely that a zero value for wt would arise unless the solution underflows identically to zero. The local error is approximated by $\widehat{y}_{n+1} - y_{n+1}$. If $\widehat{y}_{n+1} = y_{n+1}$, the error is estimated to be zero, and there is no need to compute the weighted error. If $\widehat{y}_{n+1} \neq y_{n+1}$, the definition of wt implies that $wt > 0$. Because

$$|\widehat{y}_{n+1} - y_{n+1}| \leq |\widehat{y}_{n+1}| + |y_{n+1}| \leq 6wt,$$

there is then no difficulty in performing the test

$$\frac{|\widehat{y}_{n+1} - y_{n+1}|}{wt} \leq \tau.$$

Proceeding in this way, we have a good measure of "size" and we avoid numerical difficulties. Nonetheless, a pure relative error control may not be appropriate when the solution vanishes or becomes "small." What constitutes "small" necessarily depends on the scale of the problem and must be supplied from the insight of the problem solver or a preliminary computation. The code Rke asks the user to specify a threshold value and measures the error relative to $\max(wt, threshold)$. This tells the code that when the magnitude of the solution drops below the threshold, the user is interested only in an absolute error.

Some attention must be paid to the arithmetic of the computer being used. The error control provided in Rke stresses pure relative error. It makes no sense to ask for a numerical solution more accurate than the correctly rounded true solution. To avoid difficulties the code insists that τ not be smaller than 10 units of roundoff and h not be too small for the precision available. These elementary precautions are very helpful. They are the kinds of things that distinguish mathematical software from research codes.

Suppose we have just tried a step of length h from x_n and formed the local error estimate

$$|\text{local error}| \approx |\widehat{y}_{n+1} - y_{n+1}|.$$

A Taylor series expansion of the local error leads to an expression of the form

$$\text{local error} = h^5 \Phi(x_n, y_n) + O(h^6).$$

Earlier we wrote out Φ explicitly for some low order formulas. If h is small enough,

$$|\widehat{y}_{n+1} - y_{n+1}| \approx h^5 |\Phi(x_n, y_n)|.$$

If the step is a success, meaning that

$$\frac{|\widehat{y}_{n+1} - y_{n+1}|}{wt} \leq \tau,$$

we would like to estimate a suitable step size H for the next step. The largest step size possible has

$$H^5 \frac{|\Phi(x_{n+1}, y_{n+1})|}{wt} \approx \tau.$$

The function Φ is (usually) smooth so that

$$|\Phi(x_{n+1}, y_{n+1})| \approx |\Phi(x_n, y_n)|.$$

Writing $H = \alpha h$ we then find

$$\tau \approx \alpha^5 h^5 \frac{|\Phi(x_n, y_n)|}{wt} \approx \alpha^5 \frac{|\widehat{y}_{n+1} - y_{n+1}|}{wt},$$

so that the "optimal" H is

$$H = h \left(\frac{\tau}{(|\widehat{y}_{n+1} - y_{n+1}|/wt)} \right)^{1/5}.$$

It is much too bold to try this H because if it is even slightly too large, the step will fail and this is expensive. Besides, after making all those approximations, we should not take H too seriously. In practice a fraction of H is tried, or equivalently a fraction of τ is used in computing H, and an efficient fraction determined by experiment. In Rke the tolerance aimed at is 0.6τ. This is equivalent to using about nine-tenths of the "optimal" H.

The same argument is used to obtain the step size for trying again after a failed step. In either case we must program the process with some care. For example, we must deal with the possibility that the estimated local error is zero. This technical difficulty highlights the fact that large increases or decreases cannot be justified by the arguments made. For this reason changes are limited to an order of magnitude. If a large change is truly called for, this action allows a large change to be accumulated over a few steps without the disastrous possibilities opened up by a large change in a single step. In the case of a failed step we must be especially cautious about taking the estimated error at face value. In Rke we try the predicted value once, but if it fails, we simply halve the step size until we get success.

For the numerical solution of ordinary differential equations there are two difficult ranges of tolerances. It is to be expected that tolerances near limiting precision are difficult, but it turns out that nominal tolerances are also difficult. Often users think that because engineering accuracy of 10% will suffice in their use of the results, they can keep their costs down by specifying such a large tolerance. This may result in cheap results that are not reliable because the local error test may not keep the step size small enough to justify the approximations used throughout the code. Even if the results are reliable, they can be far from what is desired because at crude tolerances the true, or global, errors can be much larger than the local errors controlled by the code. It is prudent to ask for accuracy of at least a couple of digits, so that the error control of Rke emphasizes relative error and it is required that the relative tolerance $\tau \leq 0.01$.

EXERCISES

6.15 Implement Euler's method and a local error estimator based on Heun's method. Apply it to the problem

$$y' = 10(y - x),\ y(0) = 1/10$$

and compare the estimated local error to the true local error. Also compare the global error at several points to the general size of the local errors made in the computations up to this point.

6.6 THE CODE RKE

The routine Rke solves the initial value problem for a system of first order ordinary differential equations of the form

$$\frac{d\mathbf{Y}}{dt} = \mathbf{F}(x, \mathbf{Y}), \ \mathbf{Y}(a) = \mathbf{A}.$$

A typical call for Rke in FORTRAN is

```
CALL RKE (F, NEQ, X, Y, H, FIRST, TOL, THRES, FLAG, STEP,
YCOEFF, SCR, NDIM)
```

In C it is

```
Rke (f, neq, &x, y, &h, &first, tol, threshold, &flag, &step, ycoeff);
```

and it is

```
Rke (f, neq, x, y, h, first, tol, threshold, flag, step, ycoeff);
```

in C++.

Input parameters to Rke are F, the name of the routine that defines the differential equations [i.e., $\mathbf{F}(x, \mathbf{Y})$]; NEQ, the number of first order differential equations to be integrated; X, the initial value of the independent variable; Y, an array of dimension NEQ containing the values of the solution components at X; H, step size for the current step (its sign determines the direction of integration); FIRST, a variable indicating whether this is the first or a subsequent step; a scalar TOL and a vector THRES (or threshold in C and C++) are tolerances for the local error control; and NDIM $\geq 6\times$ NEQ, the dimension of the output vector YCOEFF and in FORTRAN of the auxiliary storage vector SCR. Output parameters are X, Y, the integration was advanced to X and Y is the solution there; H, the step size suitable for the next step; FLAG, a flag reporting what the code did; STEP, the actual length of the step taken (output X minus input X); and YCOEFF, an array of coefficient values for quintic Hermite interpolation to be used by the routine Yvalue.

Some of the variables in the call list require more explanation. The initial step size H informs the code of the scale of the problem. It should be small enough to capture changes in the solution near the initial point. Also, the sign of H indicates the direction of integration because the code will try to step to $X + H$. After the first call, the code provides a suitable H for the next call.

The variable FIRST enables the code to initialize itself. The start of a new problem is indicated by input of FIRST = .TRUE. in FORTRAN and first = 1 in C or C++. After the first call, the code sets FIRST = .FALSE. in FORTRAN and first = 0 in C and C++ for subsequent calls. The error parameters TOL and THRES (or threshold in C and C++) tell the code how accurately the solution is to be computed. The vector THRES must have dimension at least NEQ in the calling program. All components of THRES must be nonnegative. The relative error tolerance TOL must satisfy

$$10u \leq \text{TOL} \leq 0.01$$

where u is the unit roundoff of the machine. The tolerances are used by the code in a local error test at each step that requires roughly that

$$|\text{local error}| \leq \text{TOL}\max(|Y(I)|, \text{THRES}(I))$$

for component I of the vector $\mathbf{Y}$. Setting THRES(I) = 0 results in a pure relative error test on the component. On the first call to the code, if some $Y(I) = 0$, the corresponding THRES(I) must be strictly positive. The size of the solution component is carefully defined so that vanishing of the true solution at the current step is very unlikely to cause trouble. Any such trouble can be avoided by a positive value of THRES(I).

The code will not attempt to compute a solution at an accuracy unreasonable for the computer being used. It will report if this situation arises. To continue the integration after such a report, TOL and/or THRES must be increased. Note that Rke is an efficient code for moderate relative accuracies. For more than, say, six-digit accuracy, other methods are likely to be more efficient.

The true (global) error is the difference between the true solution of the initial value problem and the computed one. Nearly all initial value codes, including this one, control only the local error at each step, not the global error. Moreover, controlling the local error in a relative sense does not necessarily result in the global error being controlled in a relative sense. Roughly speaking, the codes produce a solution $\mathbf{Y}(x)$ that satisfies the differential equation with a residual $\mathbf{R}(x)$,

$$\frac{d\mathbf{Y}(x)}{dx} = \mathbf{F}(x, \mathbf{Y}(x)) + \mathbf{R}(x),$$

that is usually bounded in norm by the error tolerances. Usually the true accuracy of the computed $\mathbf{Y}$ is *comparable* to the error tolerances. This code will usually, but not always, deliver a more accurate solution when the problem is solved again with smaller tolerances. By comparing two such solutions a fairly reliable idea of the true error in the solution at the larger tolerances can be obtained.

The principal task of the code is to integrate one step from X toward $X + H$. Routine Rke is organized so that subsequent calls to continue the integration involve little (if any) additional effort. The status of the integration is reported by the value of the FLAG parameter. After a successful step the routine Yvalue is used to approximate the solution at points within the span of the step. A typical call is

```
CALL YVALUE(NEQ, X, STEP, YCOEFF, POINT, U)
```

in the FORTRAN version and

```
Yvalue(neq, x, step, ycoeff, point, u);
```

in the C and C++ versions. Input parameters are NEQ, X, STEP, YCOEFF (as returned from Rke) and POINT, the point at which a solution is desired. The output is U(*), the vector of solution components at P. Routine Yvalue can be used *only* after a successful step by Rke and should be used only to interpolate the solution values on the interval $[X - \text{STEP}, X]$.

Example 6.8. To illustrate Rke, we solve the initial value problem

$$\begin{aligned} Y_1' &= Y_1, & Y_1(0) &= 1 \\ Y_2' &= -Y_2, & Y_2(0) &= 1 \end{aligned}$$

on the interval [0,1] and print the solution at $x = 1$. The problem has the solution $Y_1(x) = e^x$, $Y_2(x) = e^{-x}$. As the solution component Y_1 is increasing, a pure relative error test is appropriate and we set THRES(1) = 0. Y_2, on the other hand, is decreasing and we choose THRES(2) = 10^{-5}, which results in an absolute error test for small $|Y_2|$.

```
At XOUT =       1.00000000000000
The numerical solution is 2.7182755628071  3.6787784616084E-01
The true solution is      2.7182818284590  3.6787944117144E-01
```

■

Example 6.9. This example illustrates the use of Yvalue in conjunction with Rke. The initial value problem

$$\begin{aligned} Y_1' &= Y_2 \\ Y_2' &= -Y_1 - (Y_1^2 - 1) \cdot Y_2 \\ Y_1(0) &= Y_2(0) = 1 \end{aligned}$$

is integrated over the interval [0,10] and the solution tabulated at $x = 0, 1, 2, \ldots, 10$. Note that Rke must be called before Yvalue. The output is as follows.

```
XOUT = 0.00    Y(1) = 1.000000    Y(2) = 1.000000
XOUT = 1.00    Y(1) = 1.298484    Y(2) =-0.367034
XOUT = 2.00    Y(1) = 0.421178    Y(2) =-1.488951
XOUT = 3.00    Y(1) =-1.634813    Y(2) =-1.485472
XOUT = 4.00    Y(1) =-1.743960    Y(2) = 0.568922
XOUT = 5.00    Y(1) =-0.878664    Y(2) = 1.258102
XOUT = 6.00    Y(1) = 1.187072    Y(2) = 2.521700
XOUT = 7.00    Y(1) = 1.933030    Y(2) =-0.406833
XOUT = 8.00    Y(1) = 1.245572    Y(2) =-0.963183
XOUT = 9.00    Y(1) =-0.329599    Y(2) =-2.467240
XOUT =10.00    Y(1) =-2.008260    Y(2) =-0.034172
```

■

EXERCISES

Unless otherwise indicated, use 10^{-5} for tolerance values and 10^{-7} for the threshold values for the computing exercises.

6.16 Use Rke with a tolerance of 10^{-6} and threshold values of 1 to calculate $y(b)$ in the following cases. Check the computed output with the exact answers that are given in Exercise 6.4.

(a) $y' = -\frac{1}{2}y^3$, $y(0) = 1$; $b = 3$

(b) $y' = -2xy^2$, $y(0) = 1$; $b = 1$

(c) $y' = \frac{1}{4}y(1 - y/20)$, $y(0) = 1$; $b = 5$

(d) $y' = 100(\sin x - y)$, $y(0) = 0$; $b = 1$

(e) $y' = (15\cos 10x)/y,\ y(0) = 2;\ \ b = \pi/4.$

Are the true (global) errors within the tolerance on the local errors? Which problems needed the most steps? Why do you think they are more difficult than the others?

6.17 An important equation of nonlinear mechanics is van der Pol's equation:

$$x''(t) + \varepsilon(x^2(t) - 1)x'(t) + x(t) = 0$$

for $\varepsilon > 0$. Regardless of the initial conditions, all solutions of this equation converge to a unique periodic solution (a stable limit cycle). For $\varepsilon = 1$, choose some initial conditions t_0, $x(t_0)$, and $x'(t_0)$ and integrate the equation numerically until you have apparently converged to the limit cycle. A convenient way to view this is to plot $x'(t)$ against $x(t)$, a phase plane plot. In the phase plane, a periodic solution corresponds to a closed curve.

6.18 Deriving the equations for the quintic interpolating polynomial used in Rke/Yvalue is not difficult. Write

$$U(p) = a + bz + cz^2 + dz^3 + ez^4 + fz^5$$

for p in $[x - \Delta, x]$, where

$$z = \frac{p - x}{\Delta} + \frac{1}{2}.$$

(a) Apply the interpolation conditions

$$\begin{aligned} U(x-\Delta) &= U_L \\ U(x-\Delta/2) &= U_M \\ U(x) &= U_R \\ U'(x-\Delta) &= U'_L \\ U'(x-\Delta/2) &= U'_M \\ U'(x) &= U'_R \end{aligned}$$

to generate six linear equations in the six unknowns a, b, c, d, e, and f.

(b) Solve the linear system to get

$$\begin{aligned} a &= U_M,\ b = \Delta \cdot U'_M \\ c &= 4[U_L - 2U_M + U_R] + \frac{1}{2}\Delta[U'_L - U'_R] \\ d &= 10[U_R - U_L] - \Delta[U'_L + 8U'_M + U'_R] \\ e &= -8[U_L - 2U_M + U_R] - 2\Delta[U'_L - U'_R] \\ f &= -24[U_R - U_L] + 4\Delta[U'_L + 4U'_M + U'_R]. \end{aligned}$$

(In the code, $\alpha = \Delta[U'_L - U'_R]$, $\beta = U_L - 2U_M + U_R$, and $\gamma = U_R - U_L$.)

6.19 Use Rke to approximate the solution to the initial value problem

$$P'(t) = 0.001P(t)[1000(1 - 0.3\cos\pi t/6) - P(t)],$$

with $P(0) = 250$; sketch its graph for $0 \le t \le 36$. (If you have worked Exercise 5.10, compare results.)

6.20 The response of a motor controlled by a governor can be modeled by

$$\begin{aligned} s'' + 0.042s' + 0.961s &= \theta' + 0.063\theta \\ u'' + 0.087u' &= s' + 0.025s \\ v' &= 0.873(u - v) \\ w' &= 0.433(v - w) \\ x' &= 0.508(w - x) \\ \theta' &= -0.396(x - 47.6). \end{aligned}$$

The motor should approach a constant (steady-state) speed as $t \to \infty$. Assume $s(0) = s'(0) = u'(0) = \theta(0) = 0$, $u(0) = 50$, $v(0) = w(0) = x(0) = 75$.

(a) Evaluate $v(t)$ for $t = 0, 25, 50, 75, 100, 150, 200, 250, 300, 400, 500$.

(b) What does $\lim_{t\to\infty} v(t)$ appear to be? You can check this by working out the steady-state solution (the constant solution of the differential equation).

6.21 Consider the initial value problem

$$\begin{aligned} y''' - y''\sin x - 2y'\cos x + y\sin x &= \ln x, \\ y(1) &= A_1, \\ y'(1) &= A_2, \\ y''(1) &= A_3. \end{aligned}$$

Show that the solution $y(x)$ satisfies the first integral relation

$$y''(x) - y'(x)\sin x - y(x)\cos x = c_2 + x\ln x - x$$

and the second integral relation

$$y'(x) - y(x)\sin x = c_1 + c_2 x + \frac{1}{2}x^2\ln x - \frac{3}{4}x^2.$$

What are c_1, c_2 in terms of A_1, A_2, A_3? Choose values for A_1, A_2, and A_3 and integrate this problem numerically. Monitor the accuracy of your solution by seeing how well it satisfies the integral relations. Argue that if the integral relations are nearly satisfied, then the numerical solution may or may not be accurate, but that if they are not satisfied, the numerical solution must be inaccurate.

6.22 The Jacobian elliptic functions $sn(x)$, $cn(x)$, and $dn(x)$ satisfy the initial value problem

$$\begin{aligned} y_1' &= y_2y_3, \quad y_1(0)=0 \\ y_2' &= -y_1y_3, \quad y_2(0)=1 \\ y_3' &= -k^2y_1y_2, \quad y_3(0)=1 \end{aligned}$$

where k^2 is a parameter between 0 and 1 and $y_1(x) = sn(x)$, $y_2(x) = cn(x)$, and $y_3(x) = dn(x)$.

Evaluate these functions numerically. Check your accuracy by monitoring the relations

$$\begin{aligned} sn^2(x)+cn^2(x) &\equiv 1 \\ dn^2(x)+k^2sn^2(x) &\equiv 1 \\ dn^2(x)-k^2cn^2(x) &\equiv 1-k^2. \end{aligned}$$

Argue that if these relations are well satisfied numerically, you cannot conclude that the computed functions are accurate, rather that their errors are correlated. If the relations are not satisfied, the functions must be inaccurate. Thus, this test is a necessary test for accuracy but it is not sufficient.

The Jacobian elliptic functions are periodic. You can get the true solutions for $k^2 = 0.51$ from the fact that the period is $4K$, where $K =$ 1.86264 08023 32738 55203 02812 20579 $\cdots$. If $t_j = jK$, $j = 1,2,3,\ldots$, the solutions are given by the relation

$$y_i(t_{j+4}) = y_i(t_j)$$

and the following table:

j	$y_1(t_j)$	$y_2(t_j)$	$y_3(t_j)$
0	0	1	1.0
1	1	0	0.7
2	0	−1	1.0
3	−1	0	0.7

6.23 A simple model of the human heartbeat gives

$$\begin{aligned} \varepsilon x' &= -(x^3 - Ax + c) \\ c' &= x, \end{aligned}$$

where $x(t)$ is the displacement from equilibrium of the muscle fiber, $c(t)$ is the concentration of a chemical control, and ε and A are positive constants. Solutions are expected to be periodic. This can be seen by plotting the solution in the phase plane (x along the horizontal axis, c on the vertical), which should produce a closed curve. Assume that $\varepsilon = 1.0$ and $A = 3$.

(a) Calculate $x(t)$ and $c(t)$ for $0 \le t \le 12$ starting with $x(0) = 0.1$, $c(0) = 0.1$. Sketch the output in the phase plane. What does the period appear to be?

(b) Repeat (a) with $x(0) = 0.87$, $c(0) = 2.1$.

6.24 Devise a step size strategy for Euler's method with a local error estimator based on Heun's method. Implement it in a code for a single equation. Test it on some of the problems of this section and compare it to Rke.

6.7 OTHER NUMERICAL METHODS

The explicit Runge–Kutta methods discussed in this chapter have no memory of what has happened prior to the current point of the integration. Other methods take advantage of previously computed solution values. The Adams methods furnish a very important example that is easily understood. On reaching x_n with the approximate solution $y_n \approx y(x_n)$, there are (usually) available values $y_{n+1-i} \approx y(x_{n+1-i})$ for $i = 2,3,\ldots,k$. From the differential equation $y' = f(x,y)$, approximations to the derivatives $y'(x_{n+1-i})$ can be obtained as

$$f_{n+1-i} \approx y'(x_{n+1-i}) = f(x_{n+1-i}, y(x_{n+1-i})).$$

Knowledge of solution values prior to the current point x_n are exploited by means of the integrated form of the differential equation:

$$y(x_{n+1}) = y(x_n) + \int_{x_n}^{x_{n+1}} y'(t)\,dt = y(x_n) + \int_{x_n}^{x_{n+1}} f(t,y(t))\,dt.$$

This is done with ideas used throughout this book: interpolate $y'(t)$ by a polynomial and approximate the integral by integrating the polynomial. Let $P(t)$ be the polynomial that interpolates f_{n+1-i} at x_{n+1-i} for $i = 1,2,\ldots,k$. A numerical approximation y_{n+1} to the exact solution $y(x_{n+1})$ is then defined by

$$y_{n+1} = y_n + \int_{x_n}^{x_{n+1}} P(t)\,dt.$$

This is called the *Adams–Bashforth formula* of order k. When $P(t)$ is written in Lagrange form, this formula becomes

$$y_{n+1} = y_n + \sum_{i=1}^{k} \left(\int_{x_n}^{x_{n+1}} L_i(t)\,dt \right) f_{n+1-i}.$$

In terms of the current step size $h_n = x_{n+1} - x_n$ and the coefficients

$$\alpha_{k,i} = \frac{1}{h_n} \int_{x_n}^{x_{n+1}} L_i(t)\,dt,$$

this is

$$y_{n+1} = y_n + h_n \sum_{i=1}^{k} \alpha_{k,i} f_{n+1-i}.$$

The integration coefficients $\alpha_{k,i}$ depend on the spacing of the mesh points $x_{n+1}, x_n, \ldots$ and in general must be computed at each step. It is easy to verify that they depend only on the *relative* spacing, so when the step size is a constant h, they can be computed once and for all. Using the theory of interpolation developed in Chapter 3, it is not difficult to show that if the "memorized" values y_{n+1-i} are sufficiently accurate and f satisfies a Lipschitz condition, then this formula is of order k (hence the name). An Adams–Bashforth formula involves only one evaluation of f per step. Given y_n and previously computed values $f_{n-1}, f_{n-2}, \ldots$, the value $f_n = f(x_n, y_n)$ is formed; if necessary, the coefficients $\alpha_{k,i}$ are computed, and then y_{n+1} is evaluated by the formula to advance to x_{n+1}. An attractive feature of this approach is that it naturally provides a polynomial approximation to $y(x)$ that can be used to obtain values between mesh points:

$$y(x) \approx y_n + \int_{x_n}^{x} P(t)\,dt.$$

An Adams–Bashforth formula is so much cheaper than a Runge–Kutta formula of the same order that it is natural to ask how Runge–Kutta codes can possibly be competitive. It seems that by recycling previously computed values we get something (high order) for almost nothing (only one new f evaluation per step). Unfortunately, we do not. All methods with memory have certain difficulties. One is getting started: Where do the "previously computed" values come from on the first few steps? A related difficulty is the recurring one of changing the step size. When previously computed values are recycled, it is natural to wonder if a "feedback" of errors might cause the computed results to "explode." This instability *can* occur, and some accurate formulas that resemble the Adams–Bashforth formulas cannot be used at all because the integration is unstable even for arbitrarily small step sizes. Fortunately, if the step size is small enough, integration with Adams–Bashforth formulas is stable. Returning to the striking difference in cost of the formulas, it is important to realize that it is not merely the cost per step that is the issue but also how big a step can be taken and still achieve a desired accuracy. The popular Runge–Kutta formulas cost much more per step, but offset this by taking larger steps. Which method proves more efficient in practice depends on the problem, the accuracy desired, and the particular formulas being compared. There are many issues to be considered when selecting a method and unfortunately, there is no choice that is best for all problems.

The Adams–Moulton formulas arise when the polynomial $P(t)$ interpolates f_{n+1-i} for $i = 0, 1, \ldots, k-1$. Proceeding analogously to the Adams–Bashforth case, we are led to the formula

$$y_{n+1} = y_n + h_n \alpha^*_{k,0} f(x_{n+1}, y_{n+1}) + h_n \sum_{i=1}^{k-1} \alpha^*_{k,i} f_{n+1-i}.$$

The term involving interpolation to f_{n+1} at x_{n+1} has been extracted from the sum to emphasize that y_{n+1} is only defined *implicitly* by this formula. It is not obvious that y_{n+1} is even well defined. To establish that it is, we will show how to solve the nonlinear equations defining y_{n+1} for all sufficiently small step sizes. This is accomplished by first "predicting" a value using an explicit formula such as an Adams–Bashforth formula. Let $y^{(0)}_{n+1}$ denote this predicted value. "Simple" or "functional" iteration improves or "corrects" the mth approximation according to the explicit recipe

$$y^{(m+1)}_{n+1} = y_n + h_n \alpha^*_{k,0} f(x_{n+1}, y^{(m)}_{n+1}) + h_n \sum_{i=1}^{k-1} \alpha^*_{k,i} f_{n+1-i}$$

for $m = 0, 1, \ldots$. If L is a Lipschitz constant for f and the step size is small enough that for some constant ρ,

$$|h_n \alpha^*_{k,0}| L \le \rho < 1,$$

it is not difficult to show that there is a unique solution y_{n+1} to the algebraic equations and that the error of each iterate is decreased by a factor of ρ at each iteration. For "small" step sizes, the predicted value is close to y_{n+1} and the iteration converges quickly.

An implicit formula like an Adams–Moulton formula is more trouble and more expensive to evaluate than an explicit formula like an Adams–Bashforth formula. Why bother? For one thing, the Adams–Moulton formula of order k is considerably more accurate than the Adams–Bashforth formula of the same order so it can use a bigger step size. For another, the Adams–Moulton formula is much more stable. When all factors are considered, the Adams–Moulton formulas are advantageous. A modern code based on such formulas is more complicated than a Runge–Kutta code because it must cope with the difficulties mentioned above concerning starting values and changing the step size. It is *much* more complicated than even this brief discussion suggests. With sufficiently many memorized values, we can use whatever order formula we wish in the step from x_n. Modern codes attempt to select the most efficient formula at each step. Unfortunately, the art of computing has run ahead of the theory in this regard—there is an adequate theoretical understanding of variation of step size with a fixed formula, but little has been proven about variation of order. Nevertheless, years of heavy usage of codes that vary the order have demonstrated that they do "work" and that the variation of the order is very important to the efficiency of such codes.

Another natural approach to approximating the solutions of differential equations is based on numerical differentiation. Again using a basic idea of this book, we start by interpolating previously computed solution values $y_n, y_{n+1}, \ldots, y_{n+1-k}$ as well as the new one y_{n+1} by a polynomial $P(t)$. The derivative of the solution at x_{n+1} is then approximated by $P'(x_{n+1})$. This approximation is tied to the differential equation at x_{n+1} by requiring that

$$P'(x_{n+1}) = f(x_{n+1}, P(x_{n+1})) = f(x_{n+1}, y_{n+1}).$$

A formula for y_{n+1} is obtained by writing $P(t)$ in Lagrange form and using it in the $P'(x_{n+1})$ term of the equation. For certain practical reasons it is usual with this family of formulas to work with a constant step size h. Making this assumption, carrying out the substitution, and scaling by h lead to a formula of the form

$$\alpha_0 y_{n+1} + \alpha_1 y_n + \cdots + \alpha_k y_{n+1-k} = hf(x_{n+1}, y_{n+1}).$$

This is a member of a family known as backward differentiation formulas, or just BDFs. They were popularized by Gear [6] and are sometimes known as Gear's formulas. Obviously, these formulas are implicit like the Adams–Moulton formulas. They are not nearly as accurate as the Adams–Moulton formulas of the same order, and formulas of orders 7 and up cannot be used because they are not stable (hence not convergent) as the step size tends to zero. The reason they are interesting is that at the orders for which they are stable, they are *much* more stable than explicit Runge–Kutta and Adams formulas. Before discussing their usage, some general remarks about the step size are necessary.

The selection of the step size is affected by a number of issues. The one that receives the most attention is choosing the step size sufficiently small to obtain the desired accuracy. We have also seen that for some methods the step size might have to be reduced to produce an answer at a desired point. There are other less obvious constraints on the step size. Earlier we touched on the matter that the step size might have to be restricted so as to evaluate an implicit formula efficiently and also alluded to the issue of restricting the step size in order to make the integration stable. There are problems of great practical interest called *stiff* for which these other restrictions will cause an explicit Runge–Kutta method or an Adams–Moulton formula evaluated by simple iteration to need a step size very much smaller than that permitted by the accuracy of the formula. The excellent stability properties of the BDFs have made them the most popular formulas for solving such problems. They cannot be evaluated by simple iteration because it restricts the step size too much. In practice, a modification of the Newton iteration described in Chapter 4 is used to solve the nonlinear algebraic equations for y_{n+1}. This has many disagreeable consequences due to the necessity of approximating partial derivatives and solving systems of linear equations. Each step is very expensive compared to a Runge–Kutta or Adams method, but when the problem is stiff, the steps can be so much larger that this is a bargain. Indeed, the solution of a problem that is quite stiff is impractical with codes not specifically intended for such problems. As with the Adams formulas, modern codes based on the BDFs vary the order as well as the step size. Stiff problems are difficult technically as well as practically and how to solve them is an active area of research.

There is a large literature on the solution of the initial value problem for a system of ordinary differential equations. References [4], [8], and [9] provide useful orientation, especially in regard to the software available. State-of-the-art codes are available from netlib: RKSUITE [1] makes available explicit Runge–Kutta formulas of three different orders. ODE/STEP/INTRP [10] is a variable step, variable order Adams–Bashforth–Moulton code. The methods of RKSUITE and ODE/STEP/INTRP are not appropriate for stiff problems. Both suites of codes diagnose stiffness when it is responsible for unsatisfactory performance. VODE [2] is a variable step, variable order code that makes available two kinds of methods, Adams–Moulton formulas and a variant of the BDFs. The computing environments MATLAB and MATHCAD provide codes based on a variety of methods, including some not mentioned here, but the code that is to be tried first (assuming that the problem is not stiff) is an explicit Runge–Kutta code. *Mathematica* provides a single code that, like VODE, makes available both Adams–Moulton methods and the BDFs. It is unusual in that the code attempts to recognize stiffness and select an appropriate method automatically.

6.8 CASE STUDY 6

The restricted three-body problem is obtained from Newton's equations of motion for the gravitational attraction of three bodies when one has a mass infinitesimal in comparison to the other two. For example, the position (x,y) of a spaceship or satellite moving under the influence of the earth and the moon in a coordinate system rotating so as to keep the positions of the earth and moon fixed changes according to

$$x'' = 2y' + x - \frac{\mu^*(x+\mu)}{r_1^3} - \frac{\mu(x-\mu^*)}{r_2^3}$$

$$y'' = -2x' + y - \frac{\mu^* y}{r_1^3} - \frac{\mu y}{r_2^3}.$$

Here

$$r_1 = \sqrt{(x+\mu)^2 + y^2}$$

$$r_2 = \sqrt{(x-\mu^*)^2 + y^2}$$

and $\mu = 1/82.45$, $\mu^* = 1 - \mu$. More insight is possible when the general equations of motion are reduced to those of the restricted three-body problem, but it is still not possible to determine orbits analytically. A search using high precision computation identified several periodic orbits. One has initial conditions $x(0) = 1.2$, $x'(0) = 0$, $y(0) = 0$, and $y'(0) = -1.04935750983031990726$. The period of this orbit is about $T = 6.19216933131963970674$. Integration of this problem with Rke is straightforward after the equations are written as a first order system by introducing the vector of unknowns $\mathbf{y}(t) = (x(t), y(t), x'(t), y'(t))^T$. The orbit displayed in Figure 6.3 was computed using 10^{-6} for TOL and all components of THRESHOLD. Although the

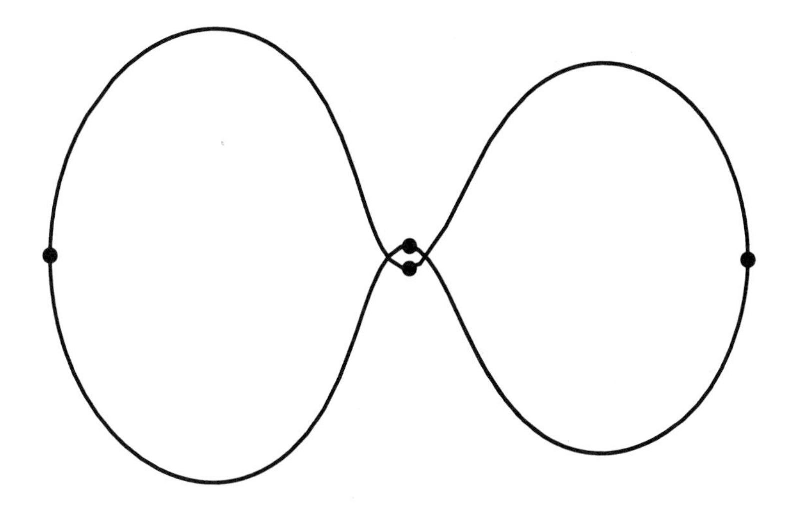

Figure 6.3 Rke solution of the satellite problem.

analytical solution is not known, we do know $\mathbf{y}(T) = \mathbf{y}(0) = \mathbf{Y}_0$ because the orbit has period T. Using this known value, we measured the global error of the approximation $\mathbf{Y}_N$ to $\mathbf{y}(T)$ and tested whether the computed orbit is periodic by computing $\|\mathbf{Y}_N - \mathbf{Y}_0\|$. The discrepancy turned out to be about 6.1×10^{-5}, which is about what we would expect for the local error tolerances given the code. The figure shows the "natural" output, the values at each step, connected by straight lines in the manner typical of plotting packages. At this tolerance the natural output is sufficiently dense that the curve appears to be smooth. However, at less stringent tolerances it was found that in portions of the integration, the step size is so large that the curve is visibly composed of straight lines. This unsatisfactory situation can be remedied easily by computing inexpensively with Yvalue the additional output values needed for a smooth graph. A less efficient alternative that is acceptable in this particular instance is to limit the step size so as to force Rke to take more steps.

Conservation of energy has a special form for the restricted three-body problem. The Jacobi integral is

$$J(x(t),y(t),x'(t),y'(t)) = 0.5\left(x'(t)^2 + y'(t)^2 - x(t)^2 - y(t)^2\right) - \frac{\mu^*}{r_1(t)} - \frac{\mu}{r_2(t)}.$$

A little calculation shows that the derivative dJ/dt is zero when it is evaluated at arguments $x(t)$, $y(t)$ that satisfy the differential equations. This leads to the conservation law

$$G(t) = J(x(t),y(t),x'(t),y'(t)) - J(x(0),y(0),x'(0),y'(0)) = 0,$$

expressing the fact that the Jacobi integral is constant along a solution of the restricted three-body problem. We monitored the integration by computing $G(t)$ at each step. For the tolerances specified, it was never larger in magnitude than about 1.8×10^{-5}. Many differential equations have conservation laws that arise naturally from physical principles, but others satisfy laws that are not so readily interpreted. Recall, for example, the conservation law we used in Case Study 4 to solve the Lotka–Volterra equations in the phase plane. Conservation laws are consequences of the form of the differential equations. It does not follow that numerical approximations will satisfy the laws, and generally speaking they do not. However, they must satisfy the laws approximately. To see this, suppose that $G(t) = G(t,\mathbf{y}(t)) = 0$ for any solution $\mathbf{y}(t)$ of a differential equation $\mathbf{y}' = \mathbf{F}(t,\mathbf{y})$. If $\mathbf{y}_n \approx \mathbf{y}(t_n)$, then linearization tells us that

$$\begin{aligned} G(t_n,\mathbf{y}_n) &\approx G(t_n,\mathbf{y}(t_n)) + \frac{\partial G}{\partial \mathbf{y}}(t_n,\mathbf{y}(t_n))\,(\mathbf{y}_n - \mathbf{y}(t_n)) \\ &= \frac{\partial G}{\partial \mathbf{y}}(t_n,\mathbf{y}(t_n))\,(\mathbf{y}_n - \mathbf{y}(t_n)). \end{aligned}$$

Evidently the residual of the numerical solution in the conservation law is of the same order as the global error of the numerical solution, $\mathbf{y}(t_n) - \mathbf{y}_n$. This observation helps us understand the size of the residual we found in integrating the periodic earth–moon orbit. It is worth remarking that solutions of a system of differential equations might satisfy several conservation laws.

The conservation laws mentioned so far are nonlinear, but others arising from conservation of mass, charge balance, and the like are linear. A linear conservation law for the equation $\mathbf{y}' = \mathbf{F}(t,\mathbf{y})$ arises mathematically when there is a constant vector $\mathbf{v}$ such that $\mathbf{v}^T\mathbf{F}(t,\mathbf{u}) = 0$ for all arguments $(t,\mathbf{u})$. If $\mathbf{y}(t)$ is a solution of the equation, then

$$\frac{d}{dt}\mathbf{v}^T\mathbf{y}(t) = \mathbf{v}^T\mathbf{F}(t,\mathbf{y}(t)) = 0,$$

implying that $G(t) = \mathbf{v}^T\mathbf{y}(t) - \mathbf{v}^T\mathbf{y}(0) = 0$. A simple example is provided by a system of equations that describes a certain radioactive decay chain:

$$\begin{aligned} y_1' &= -y_1 \\ y_k' &= (k-1)y_{k-1} - ky_k \;\text{ for } k = 2,\dots,9 \\ y_{10}' &= 9y_9. \end{aligned}$$

The right-hand sides here sum to zero, hence the system satisfies a linear conservation law with $\mathbf{v}^T = (1,1,\dots,1)^T$. Figure 6.4 shows the solution of this system with initial condition $\mathbf{y}(0) = (1,0,\dots,0)^T$ computed using Rke with TOL equal to 10^{-3} and all components of THRESHOLD equal to 10^{-10}. Despite the modest relative accuracy requested, the error in the conservation law was at the roundoff level, specifically a maximum of 4.4×10^{-16} in the MATLAB environment on the workstation we used. This illustrates an interesting fact: all standard numerical methods produce approximations that satisfy linear conservation laws *exactly*. This is easy to show for explicit Runge–Kutta formulas. In advancing from x_n to $x_n + h$, such a formula has the form

$$\mathbf{Y}_{n+1} = \mathbf{Y}_n + h(b_0\mathbf{K}_0 + b_1\mathbf{K}_1 + \cdots + b_s\mathbf{K}_s).$$

Each stage $\mathbf{K}_i$ has the form $\mathbf{F}(x^*,\mathbf{Y}^*)$ for arguments $(x^*,\mathbf{Y}^*)$ that are defined in terms of x_n, h, and the previous stages. The details do not matter, for all we need here is that $\mathbf{v}^T\mathbf{K}_i = 0$ because $\mathbf{v}^T\mathbf{F}(t,\mathbf{u}) = 0$ for *all* arguments $(t,\mathbf{u})$. It then follows immediately

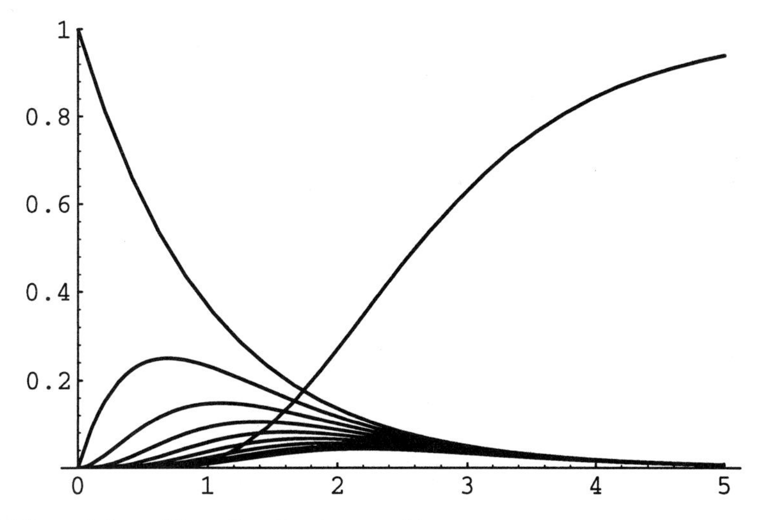

Figure 6.4 Rke solution of the radioactive decay problem.

that $\mathbf{v}^T\mathbf{Y}_{n+1} = \mathbf{v}^T\mathbf{Y}_n$ for all n. We start the integration with the given initial values so that $\mathbf{Y}_0 = \mathbf{y}(0)$. This implies that for all n,

$$\mathbf{v}^T\mathbf{Y}_n = \mathbf{v}^T\mathbf{Y}_0 = \mathbf{v}^T\mathbf{y}(0).$$

Put differently, $G(x_n, \mathbf{Y}_n) = 0$ for all n. Accordingly, it was no accident that the numerical solution computed by Rke satisfied the conservation law to roundoff error, that is what we should expect of a *linear* conservation law.

Returning now to the satellite problem, suppose we would like to know when the satellite is nearest and farthest from earth. The distance to the satellite is

$$d(t) = \sqrt{x(t)^2 + y(t)^2},$$

so we look for the extrema of this function by finding the times t for which $d'(t) = 0$. Let us avoid square roots by working with the square of the distance, $D(t)$. The derivative of this function is

$$\begin{aligned} D'(t) &= 2\left(x(t)x'(t) + y(t)y'(t)\right) \\ &= 2\left(y_1(t)y_3(t) + y_2(t)y_4(t)\right) \end{aligned}$$

in the original variables and those of the first order system, respectively. Notice that for the orbit we study, the initial distance $d(0) = 1.2$ is an extremum because $D'(0) = 0$. This will afford a check on our computations because the orbit is periodic and the same must be true of $d(T)$. We want to compute the roots of $F(t) = D'(t) = 0$. At each step we have an approximation to the solution that allows us to evaluate $F(t)$, so when we reach x after a step of size *step*, we ask if $F(x - step)F(x) \leq 0$. If so, we have found just the kind of bracket for a root that we need for applying Zero to locate a root precisely. Evaluation of $F(t)$ is easy enough; we just invoke Yvalue to get an approximation to $\mathbf{y}$ at t, and then use these approximations in the expression for $D'(t)$. There is one snag, which is a very common one when combining items of mathematical software. We invoke Zero with the name of a function F of just one argument t. However, to evaluate $F(t)$ we must invoke Yvalue, and it requires three other arguments, namely x, $step$, and $ycoeff$, the array of coefficients defining the interpolating polynomial over $[x - step, x]$, returned by Rke. Somehow we must communicate this output from Rke in the main program to the function for evaluating F. There can be more than one way to do this, depending on the language, but it is always possible to do it by means of global variables. As a specific example, we coded the function in MATLAB as

```
function Ft = F(t)
global x step ycoeff
yt = Yvalue(t,x,step,ycoeff);
Ft = 2*(yt(1:2)'*yt(3:4));
```

The quantities listed in the second line of this code are computed by Rke in the main program. By duplicating the line there, the quantities are made accessible from the function F. Proceeding in the manner described with the same tolerances in both Rke and Zero, we found

time	distance
1.45857	0.03338
3.09606	1.26244
4.73354	0.03338
6.19210	1.19998

The last extremum occurs at a time that agrees to about five figures with the period and agrees with the initial distance to about the same accuracy, which is quite reasonable given the tolerances. In Figure 6.3 the points nearest the earth and farthest away that we found in this way are indicated by circles.

REFERENCES

1. R. Brankin, I. Gladwell, and L. Shampine, "RKSUITE: a suite of Runge–Kutta codes for the initial value problem for ODEs," Softreport 92-S1, Math. Dept., Southern Methodist University, Dallas, Tex., 1992.
2. P. Brown, G. Byrne, and A. Hindmarsh, "VODE: a variable-coefficient ODE solver," *SIAM J. Sci. Stat. Comput.*, 10 (1989), pp. 1038–1051.
3. E. Coddington, *An Introduction to Ordinary Differential Equations*, Prentice Hall, Englewood Cliffs, N.J., 1961.
4. W. Cowell, ed., *Sources and Development of Mathematical Software*, Prentice Hall, Englewood Cliffs, N.J., 1984.
5. R. England, "Error estimates for Runge–Kutta type solutions to systems of ordinary differential equations," *Computer Journal*, 12 (1969), pp. 166–170.
6. C. W. Gear, *Numerical Initial Value Problems in Ordinary Differential Equations*, Prentice Hall, Englewood Cliffs, N.J., 1971.

7. *Handbook of Mathematical Functions*, M. Abramowitz and I. Stegun, eds., Dover, Mineola, N.Y., 1964.
8. L. Shampine, *Numerical Solution of Ordinary Differential Equations*, Chapman & Hall, New York, 1994.
9. L. Shampine and C. W. Gear, "A user's view of solving stiff ordinary differential equations," *SIAM Review*, 21 (1979), pp. 1–17.
10. L. Shampine and M. Gordon, *Computer Solution of Ordinary Differential Equations: The Initial Value Problem*, Freeman, San Francisco, 1975.

MISCELLANEOUS EXERCISES FOR CHAPTER 6

Unless otherwise indicated, use 10^{-5} for tolerance values and 10^{-7} for the threshold values for the computing exercises.

6.25 Consider the problem

$$y' = 2|x|y,\ y(-1) = 1/e$$

on the interval $[-1, 1]$. Verify that the existence and uniqueness theorem (Theorem 6.1) applies to this problem. Verify that the solution to this problem is

$$y(x) = \begin{cases} e^{x^2}, & x \geq 0 \\ e^{-x^2}, & x < 0 \end{cases}$$

and further that $y(x)$ has one continuous derivative on $[-1, 1]$ but does not have two. Is Euler's method convergent and $O(h)$ for this problem? What about "higher order" Runge–Kutta methods? What are the answers to these questions for the two problems

$$y' = 2|x|y,\ y(-1) = 1/e$$

on $[-1, 0]$ and

$$y' = 2|x|y,\ y(0) = 1$$

on $[0, 1]$? Show that solving the original problem on $[-1, 1]$ with a mesh point at $x = 0$ is the same as solving these two problems.

Explain the following numerical results: A fourth order Runge–Kutta code was used to integrate

$$y' = 2|x|y,\ y(-1) = 1/e$$

from $x = -1$ to $x = 1$ using a fixed step size h and the true error at $x = 1$ computed from the analytical solution. Two computations were done. One used $h = 2/2^k$ and the other $h = 2/3^k$. The results in the following table were obtained:

k	error $h = 2/2^k$	error $h = 2/3^k$
1	1.0E−1	2.4E−1
2	1.0E−2	2.3E−2
3	8.2E−4	2.5E−3
4	5.6E−5	2.8E−4
5	3.6E−6	3.1E−5
6	2.3E−7	3.4E−6
7	1.4E−8	3.8E−7
8	9.0E−10	4.2E−8
9	5.6E−11	4.7E−9
10	3.5E−12	5.2E−10

6.26 In modeling circuits containing devices with electrical properties that depend on the current, differential equations of the form

$$A(\mathbf{x})\frac{d\mathbf{x}}{dt} = f(t, \mathbf{x})$$

occur. For the case where

$$A(\mathbf{x}) = \begin{pmatrix} 3 - x_1^2/40 & 0 & 1 - x_3^2/8 \\ 0 & 4 & 1 - x_3^2/8 \\ 1 - x_1^2/40 & 2 & 4 - x_3^2/8 \end{pmatrix},$$

$$f(t, \mathbf{x}) = \begin{pmatrix} 30\cos t - 4x_1 + 5x_3 \\ 2x_1 - 3x_2 \\ 3x_2 - 3x_3 \end{pmatrix},$$

and

$$\mathbf{x}(0) = \begin{pmatrix} 0 \\ 0 \\ 0 \end{pmatrix},$$

compute $\mathbf{x}(t), t = 0.4, 0.8, 1.2, \ldots, 16$. Plot $x_1(t)$, $x_2(t)$, and $x_3(t)$, $0 \leq t \leq 16$, on separate graphs. (Hint: Use Factor/Solve in conjunction with Rke.)

6.27 Approximate the solution to the nonlinear two-point

boundary value problem

$$y'' = e^y - 1,\ y(0) = 0,\ y(1) = 3$$

at $x = 0.1, 0.2, \ldots, 0.9$. First use Rke and Zero to find the missing initial condition $y'(0)$. Then use Rke to solve the resulting initial value problem at the desired x values. (Hint: Denoting the solution of

$$y'' = e^y - 1,\ y(0) = 0,\ y'(0) = s$$

by $y(t;s)$, the problem is to find s so that $y(1;s) = 3$. Use Zero to find the root of $G(s) = y(1;s) - 3 = 0$.) If you have worked Exercise 5.12, compare results.

6.28 The following set of differential equations arises in semiconductor theory:

$$\begin{aligned} \varepsilon^2 E' &= p - n + 1 \\ \varepsilon p' &= pE - \varepsilon \\ \varepsilon n' &= -nE + \varepsilon. \end{aligned}$$

Typical side conditions are $n(0) = p(0)$, $p(1) = 0$, and $n(1) = 1$. For $\varepsilon = 1$ find $E(1)$ such that $n(0) = p(0)$. Print out your final value for $E(1)$. What happens if $\varepsilon = 10^{-2}$?

6.29 The motion of a wind-blown balloon can be approximated by the solution of the initial value problem

$$\begin{aligned} \frac{d}{dt}a(t) &= -10^{-3}t + 10^{-6}a(t)\frac{d}{dt}x(t) \\ \frac{d}{dt}x(t) &= v(a) \\ a(0) &= A \\ x(0) &= 0, \end{aligned}$$

where for each time t, $a(t)$ is the altitude of the balloon above ground, $x(t)$ is the horizontal distance of the balloon from a fixed starting point, and $v(a)$ is the velocity of the wind as a function of altitude. Assumptions are that the wind direction is constant, the ground is level, and the balloonist is only allowed to coast. Lengths are measured in feet and time in seconds. The following data have been collected for $v(a)$:

a	0	50	200	500	1000
v	0	3	5	10	15
a	1500	2000	3000	4000	5000
v	20	20	18	15	12

The problem is to find the initial altitude, $a(0) = A$, so that the balloon lands at $x = 35{,}000$ ft from its point of origin, and the time of flight t_f. At t_f we know that $a(t_f) = 0$ and $x(t_f) = 35{,}000$ so the system can be integrated backwards in time until $t = 0$. The value t_f is to be chosen so that $x(0) = 0$; the initial altitude A will be the value of $a(0)$ for this t_f.

To solve the problem, perform the following steps in the order indicated.

(a) Assuming that $v(a) = 20$ (constant wind velocity) and $t_f = 2000$, use Rke to calculate $x(0)$ and $a(0)$. Check your results with an analytical solution of the differential equations.

(b) Fit the $v(a)$ data using the complete cubic interpolating spline. Plot v as a function of a.

(c) Using $v(a)$ from part (b), again calculate $x(0)$ and $a(0)$, this time for $t_f = 1000, 1500, 2000, 2500$. Tabulate t_f, $x(0)$, and $a(0)$ for each case.

(d) Modify your program in (c) to use Zero to find the time of flight t_f so that $x(0) = 0$. What is A?

APPENDIX A

NOTATION AND SOME THEOREMS FROM THE CALCULUS

We assume that the reader is familiar with the topics normally covered in the undergraduate analytical geometry and calculus sequence. For purpose of reference, we present some standard notation used in this book and a few theorems from the calculus.

A.1 NOTATION

$[a,b]$, the interval consisting of the real numbers x such that $a \leq x \leq b$.

(a,b), the interval consisting of the real numbers x such that $a < x < b$.

$x \in [a,b]$, x lies in the interval $[a,b]$.

$f'(x) = \frac{d}{dx} f(x)$

$f^{(n)}(x) = \frac{d^n}{dx^n} f(x)$

$f \in C^1[a,b]$, f belongs to the class of functions having a first derivative continuous on the interval $[a,b]$.

$f \in C^n[a,b]$, f belongs to the class of functions having an nth derivative continuous on $[a,b]$.

The norm $||f||$ of a function $f(x)$ continuous on the interval $[a,b]$ is the maximum value of $|f(x)|$ over the interval.

$g_x(x,y) = \frac{\partial}{\partial x} g(x,y)$, $g_y(x,y) = \frac{\partial}{\partial y} g(x,y)$, $g_{xy}(x,y) = \frac{\partial^2}{\partial x \partial y} g(x,y)$, and so on, denote partial differentiation.

$\sum_{i=0}^{n} a_i = a_0 + a_1 + \cdots + a_n$

$\prod_{i=0}^{n} a_i = a_0 \times a_1 \times \cdots \times a_n$

$0.d_1 d_2 \cdots d_n(e)$ stands for $0.d_1 d_2 \cdots d_n \times 10^e$; for example, 0.123(4) means 0.123×10^4. More often, we use 0.123E4 for this.

$\approx$, is approximately equal to.

$f(x^+) =$ limit of $f(x+\eta)$ as $\eta \to 0$ with $\eta > 0$, the limit from the right.

$f(x^-) =$ limit of $f(x-\eta)$ as $\eta \to 0$ with $\eta > 0$, the limit from the left.

$\ll$, much less.

$\mathbf{v} = (v_1 v_2 \cdots v_n)$, notation for a vector.

The norm $||\mathbf{v}||$ of a vector $\mathbf{v}$ is the maximum magnitude component of the vector.

$q(h) = 0(h^k)$ as $h \to 0$ means that there are (unknown) constants h_0, K such that $|q(h)| \leq Kh^k$ for all $0 < h \leq h_0$.

A.2 THEOREMS

Theorem. Intermediate Value Theorem. *Let $f(x)$ be a continuous function on the interval $[a,b]$. If for some number α and for some $x_1, x_2 \in [a,b]$ we have $f(x_1) \leq \alpha \leq f(x_2)$, then there is a point $c \in [a,b]$ such that $\alpha = f(c)$.*

Theorem. Rolle's Theorem. *Let $f(x)$ be continuous on the finite interval $[a,b]$ and differentiable on (a,b). If $f(a) = f(b) = 0$, there is a point $c \in (a,b)$ such that $f'(c) = 0$.*

Theorem. Mean Value Theorem for Integrals. *Let $g(x)$ be a nonnegative function integrable on the interval $[a,b]$. If $f(x)$ is continuous on $[a,b]$, then there is a point $c \in [a,b]$ such that*

$$\int_a^b f(x)g(x)\,dx = f(c)\int_a^b g(x)\,dx.$$

Theorem. Mean Value Theorem for Derivatives. *Let $f(x)$ be continuous on the finite interval $[a,b]$ and differentiable on (a,b). Then there is a point $c \in (a,b)$ such that*

$$\frac{f(b)-f(a)}{b-a} = f'(c).$$

Theorem. Taylor's Theorem (for $f(x)$). *Let $f(x)$ have $n+1$ continuous derivatives on (a,b) for some $n \geq 0$ and let $x, x_0 \in (a,b)$. Then*

$$f(x) = P_n(x) + R_{n+1}(x),$$

where

$$P_n(x) = f(x_0) + \frac{(x-x_0)}{1!}f'(x_0) + \cdots + \frac{(x-x_0)^n}{n!}f^{(n)}(x_0)$$

and

$$R_{n+1}(x) = \frac{(x-x_0)^{n+1}}{(n+1)!}f^{(n+1)}(z)$$

for some z between x_0 and x.

Theorem. Taylor's Theorem (for $f(x,y)$). *Let (x_0,y_0) and $(x_0+\xi,y_0+\eta)$ be given points and assume that $f(x,y)$ is $n+1$ times continuously differentiable in some neighborhood of the line $L(x_0,y_0;x_0+\xi,y_0+\eta)$ connecting (x_0,y_0) and $(x_0+\xi,y_0+\eta)$. Then*

$$f(x_0+\xi,y_0+\eta) = f(x_0,y_0) + \left\{ \sum_{j=1}^{n} \frac{1}{j!} \left(\xi \frac{\partial}{\partial x} + \eta \frac{\partial}{\partial y} \right)^j f(x,y) \bigg|_{\substack{x=x_0 \\ y=y_0}} \right\}$$
$$+ \frac{1}{(n+1)!} \left(\xi \frac{\partial}{\partial x} + \eta \bigg| \frac{\partial}{\partial y} \right)^{n+1} f(x,y) \bigg|_{\substack{x=x_0+\theta\xi \\ y=y_0+\theta\eta}}$$

for some $0 \le \theta \le 1$. [The point $(x_0+\theta\xi,y_0+\theta\eta)$ is an unknown point on the line L.] For $n=2$, the formula without remainder becomes

$$f(x_0+\xi,y_0+\eta) = f(x_0,y_0) + \xi \frac{\partial f}{\partial x}(x_0,y_0) + \eta \frac{\partial f}{\partial y}(x_0,y_0)$$
$$+ \frac{1}{2!} \left[\xi^2 \frac{\partial^2 f}{\partial x^2}(x_0,y_0) + 2\xi\eta \frac{\partial^2 f}{\partial x \partial y}(x_0,y_0) + \eta^2 \frac{\partial^2 f}{\partial y^2}(x_0,y_0) \right].$$

Theorem. *Let $f(x)$ be a continuous function on the finite interval $[a,b]$. Then $f(x)$ assumes its maximum and minimum values on $[a,b]$; that is, there are points $x_1,x_2 \in [a,b]$ such that*

$$f(x_1) \le f(x) \le f(x_2)$$

for all $x \in [a,b]$.

Theorem. Integration by Parts. *Let $f(x)$ and $g(x)$ be real-valued functions with derivatives continuous on $[a,b]$. Then*

$$\int_a^b f'(t)g(t)\,dt = f(t)g(t)|_{t=a}^{t=b} - \int_a^b f(t)g'(t)\,dt.$$

Theorem. Fundamental Theorem of the Integral Calculus. *Let $f(x)$ be continuous on the interval $[a,b]$, and let*

$$F(x) = \int_a^x f(t)\,dt \quad \text{for all} \quad x \in [a,b].$$

Then $F(x)$ is differentiable on (a,b) and

$$F'(x) = f(x).$$

Theorem. Common Maclaurin Series. *For any real number x,*

$$e^x = 1 + x + \frac{x^2}{2!} + \cdots + \frac{x^n}{n!} + \cdots$$

$$\sin x = x - \frac{x^3}{3!} + \frac{x^5}{5!} - \cdots + (-1)^n \frac{x^{2n+1}}{(2n+1)!} + \cdots$$

$$\cos x = 1 - \frac{x^2}{2!} + \frac{x^4}{4!} - \cdots + (-1)^n \frac{x^2 n}{(2n)!} + \cdots.$$

For $|x| < 1$,

$$\frac{1}{1-x} = 1 + x + x^2 + \cdots + x^n + \cdots.$$

For discussions and proofs of these theorems, see A. Taylor, Advanced Calculus, Wiley, New York, 1983.

ANSWERS TO SELECTED EXERCISES

Exercise Set 1.1 (Page 11)

1.2 (a) In six-digit decimal rounded arithmetic $U = 0.00001 = 10^{-5}$.
(b) Same as those in part (a).

1.3 (a) With $n = 3$ in six-digit decimal chopped arithmetic $x - y = 0.000456$.

1.5 $[F(x+\varepsilon x) - F(x)]/F(x) = [(1+2\varepsilon+\varepsilon^2)x^2 - 2(1+\varepsilon)x + 1 - x^2 + 2x - 1]/(x-1)^2 = \varepsilon^2 x^2/(x-1)^2$. The last quantity becomes infinite as x approaches 1.

Exercise Set 1.2 (Page 23)

1.7 With $s = \sqrt{1-c^2} = F(c)$, $\delta s \approx F'(c)\delta c = -(c/s)\,\delta c$. For $\pi/4 \le \theta \le \pi/2$, the factor $|c/s| = |\cot(\theta)| \le 1$. In this range, the value of s is as accurate as the value of c, and it will be of comparable accuracy for θ not greatly smaller than $\pi/4$. However, for $\theta \ll 1$, the factor $\cot(\theta) \approx 1/\theta$ is very large and s computed in this manner is much less accurate than c. The relative error

$$\frac{\delta s}{s} \approx -\frac{1}{s}\left(\frac{c}{s}\right)\delta c = -\left(\frac{c}{s}\right)^2 \frac{\delta c}{c}.$$

Because the relative error in s is related to the relative error in c by a factor that is the square of that arising for absolute errors, conclusions about the usefulness of the scheme for the various θ are essentially the same as for absolute errors.

1.9 In four-digit decimal chopped $(0.8717 + 0.8719)/2 = 1.743/2 = 0.8715$, in four-digit decimal rounded the midpoint is $1.744/2 = 0.8720$; neither of these is even inside the desired interval. A good alternative is to compute $h = b - a$ and then the midpoint is $a + h/2$. For this example, we get $h = 0.0002000$ and the midpoint is 0.8718, the exact value.

1.11 (a) In IEEE single precision the final $k = 46$ with $S = 0.00008138$; its relative error is -0.793 or 79%.
(c) The series is a better algorithm for positive x because there is no cancellation error.

1.13 Some information is lost in the subtraction of $L_N^2/4$ from 1 because $L_N \to 0$ as $N \to \infty$, $\sqrt{1 - L_N^2/4} \approx 1 - L_N^2/8$, showing that there is severe cancellation when this quantity is subtracted from 1. The information lost is small in magnitude, but it becomes important due to cancellation, especially after it is multiplied by the large value of N to form the approximation $NL_{2N}/2$ to π. The rearranged form avoids the cancellation in the original form of the recurrence.

Miscellaneous Exercises for Chapter 1 (Page 29)

1.15 (a) Let $x = 8.01, y = 1.25, z = 80.8$, then $(x \otimes y) \otimes z = 808.$ while $x \otimes (y \otimes z) = 809.$
(c) Let $x = 200., y = -60.0, z = 6.03$, then $x \otimes (y \oplus z) = -10700.$ while $(x \otimes y) \oplus (x \otimes z) = -10800.$

1.17 Use

$$\begin{aligned} \sin nx &= \sin(1+n-1)x \\ &= \sin x \cos(n-1)x + \cos x \sin(n-1)x \end{aligned}$$

and

$$\begin{aligned} \cos nx &= \cos(1+n-1)x \\ &= \cos x \cos(n-1)x - \sin x \sin(n-1)x. \end{aligned}$$

Then

$$\begin{aligned} \varepsilon_n &= s_n - \hat{s}_n \\ &= s_1(c_{n-1} - \hat{c}_{n-1}) \\ &\quad + c_1(s_{n-1} - \hat{s}_{n-1}) \end{aligned}$$

$$= s_1\tau_{n-1} + c_1\varepsilon_{n-1}$$

and

$$\begin{aligned}\tau_n &= c_n - \hat{c}_n \\ &= c_1(c_{n-1} - \hat{c}_{n-1}) \\ &\quad - s_1(s_{n-1} - \hat{s}_{n-1}) \\ &= c_1\tau_{n-1} - s_1\varepsilon_{n-1}.\end{aligned}$$

Hence,

$$\begin{aligned}\varepsilon_n^2 + \tau_n^2 &= s_1^2\tau_{n-1}^2 \\ &\quad + 2c_1s_1\varepsilon_{n-1}\tau_{n-1} + c_1^2\varepsilon_{n-1}^2 \\ &\quad + c_1^2\tau_{n-1}^2 - 2c_1s_1\varepsilon_{n-1}\tau_{n-1} \\ &\quad + s_1^2\varepsilon_{n-1}^2 \\ &= \varepsilon_{n-1}^2 + \tau_{n-1}^2 = \cdots \\ &= \varepsilon_2^2 + \tau_2^2.\end{aligned}$$

Exercise Set 2.1 (Page 42)

2.1 (a) nonsingular; $x_1 = -1, x_2 = 1/2, x_3 = 3$
(c) singular (but consistent)
(e) singular (but consistent)

2.3 $R_E = 223, R_A = 177, C_h = 56.7607,$
$D_h = -56.7607, C_v = 29.3397, D_v = -252.340,$
$B_v = -147.660, B_h = 56.7607$

Exercise Set 2.2 (Page 48)

2.4 (a) $x_1 = 9, x_2 = -36, x_3 = 30$
(c)

$$U = \begin{bmatrix} 1.0 & 0.50 & 0.33 \\ 0 & 0.080 & 0.090 \\ 0 & 0 & 0.001 \end{bmatrix},$$

$$L^{-1}b = \begin{bmatrix} 1.0 \\ -0.50 \\ 0.22 \end{bmatrix},$$

so $x_3 = 220, x_2 = -230$, and $x_1 = -37$.
(e) $x_1 = 500/9, x_2 = -2500/9, x_3 = 2300/9$

2.5 (a)

$$L = \begin{bmatrix} 1 & 0 & 0 \\ -1 & 1 & 0 \\ 2 & -2/3 & 1 \end{bmatrix},$$

$$U = \begin{bmatrix} -1 & 2 & 1 \\ 0 & 6 & -2 \\ 0 & 0 & -7/3 \end{bmatrix}$$

(c)

$$L = \begin{bmatrix} 1.0 & 0 & 0 & 0 \\ 0.5 & 1 & 0 & 0 \\ 1.5 & 13 & 1 & 0 \\ 0.5 & 5 & 4 & 1 \end{bmatrix}$$

$$U = \begin{bmatrix} 2 & -3 & 2 & 5 \\ 0 & 0.5 & 0- & 0.5 \\ 0 & 0 & -1 & 0 \\ 0 & 0 & 0 & -1 \end{bmatrix}$$

Exercise Set 2.3 (Page 61)

2.7 $r = (0.000772, 0.000350)$ and
$s = (0.000001, -0.000003)$, so the more accurate answer has the larger residual.

2.9 (a) The uncertainty in each x_i is $\pm||\mathbf{x}||$ times the right side of the Condition Inequality, that is, ± 0.0075386.

Exercise Set 2.4 (Page 63)

2.11 Factor/Solve produces the exact answers $R_1 = 51.67$, $R_2 = 26.66$, and $R_3 = 31.67$ with minor perturbations at the roundoff level. The value of COND is 1.50.

2.13 COND = 1.438E+4

$$A^{-1} = \begin{bmatrix} 98.331 & -199.329 & 99.998 \\ -198.662 & 399.658 & -199.995 \\ 99.998 & -199.995 & 99.998 \end{bmatrix}$$

2.15 COND = 112.9; for $V = 50$ we have
$v = (35, 26, 20, 15.5, 11, 5)$ with minor perturbations at the roundoff level.

2.17 (a) COND = 6.44

$$X = \begin{bmatrix} 3.14 & 22.38 & 0.21 & 7.05 \\ -0.014 & 0.19 & 0.62 & 0.83 \\ 6.25 & 3.46 & 8.85 & 18.93 \end{bmatrix}$$

For exact data the answers are very reliable; however, if the entries in A are known only to ± 0.0005 and those in B to ± 0.005, then from the condition number inequality

$$||\Delta X_k||/||X_k|| \leq 6.44(0.0011 + 0.005/||B_k||),$$

where X_k is the kth column of X. For example, when $k = 1$ we have $||\Delta X_k||/||X_k|| \leq 0.018$. In particular, x_{21} is probably incorrect; it certainly makes no sense physically.
(c) The analog of (2.27) here is $\Delta x_{ij} \approx \alpha_{ip} \Delta b_{pj}$. For this problem, α_{23} is much smaller than the other entries, but the rest are about the same. The x_3 values are most sensitive to changes in b since the third row of A^{-1} contains the largest entries. When b_{11} is changed to 1.43, that is, $\Delta b_{11} = -0.01$, then $x_{21} = 0.016$; hence, $\Delta x_{21} = 0.03 \approx \alpha_{21} \Delta b_{11}$ as predicted by the theory.

Exercise Set 2.5 (Page 72)

2.19 There are $n-1$ divisions for the elimination and n for each b; there are $n-1$ multiplications for the elimination and $2n-2$ for each b. Hence, there is a total of $(m+1)n-1$ divisions and $(2m+1)(n-1)$ multiplications. The number of additions/subtractions equals the number of multiplications.

Miscellaneous Exercises for Chapter 2 (Page 79)

2.21 x_1^T = (0.5500E−5, −0.1653E−3, −0.3717E−5, −0.4737E−4, 0.3714E−4, −0.1212E−3, 0.6434E−4, 0.6362E−4) and COND = 18.01. Then f_1^T = (−0.6190, −0.9217E−1, 0.1202E−1, 0.3714, 0.2720, −0.7209E−2, −0.1325, −0.4654, 0.1656, −0.1421).

2.23 (a) det A = 1,500,000

Exercise Set 3.1 (Page 89)

3.1 No, coefficients of higher degree terms may be zero. For example, for the data $x_i = i, y_i = 2i, 1 \leq i \leq 4$, the interpolating polynomial is clearly $P_4(x) = 2x$ which has degree only 1.

3.3

$$P_3(x) = 2\frac{(x-2)(x-3)}{(-1)(-2)} + 4\frac{(x-1)(x-3)}{(1)(-1)} + c\frac{(x-1)(x-2)}{(2)(1)}$$

interpolates (1, 2), (2, 4), and (3, c) for any c. The formula for $P_3(x)$ simplifies to

$$\frac{1}{2}(c-6)x^2 + (11 - 3c/2)x + (c-6)$$

so that we get exact degree 2 as long as $c \neq 6$. This does not contradict Theorem 3.1 since the degree of P_2 is too large for that theorem.

3.5 The standard algorithm requires $1 + 2 + \cdots + N - 1$ multiplications for a total of $N(N-1)/2$. The nested version requires only $N - 1$.

3.7

$$w_9(x) = \prod_{n=1}^{9}(x-n)$$

Near the ends of the interpolating points $w_9(x)$ is large in magnitude, for example, $-w_9(0) = w_9(10) = 362,880$. In the middle, for example, $w(5.5) = 193.80$, it is smaller.

3.9 $H(x_n) = a = f_n, H'(x_n) = b = f_n'$, $H(x_{n+1} = a + bh + ch^2 + dh^3 =$ $f_n + hf_n' + [3(f_{n+1} - f_n) - 2hf_n' - hf_{n+1}'] + [hf_n' + hf_{n+1}' - 2(f_{n+1} - f_n)] = f_{n+1}$. Similarly, $H'(x_{n+1}) = f_{n+1}'$.

Exercise Set 3.2 (Page 93)

3.11 The errors on $[-5, 5]$ are diverging, but they appear to be converging in $[-1, 1]$. These data came from a sample of 1001 points.

m	N	Error on $[-5,5]$	Error on $[-1,1]$
7	15	7.19	0.019716
10	21	59.77	0.003280
13	27	538.17	0.000748

3.13

N	Error
15	4.06
21	95.11
27	2933.04

Error increases as N does.

Exercise Set 3.3 (Page 98)

3.15 (a) $P_4(x) =$

$$2\frac{x(x-1)(x-2)}{6}+2\frac{(x+1)(x-1)(x-2)}{2}$$
$$+2\frac{(x+1)x(x-2)}{-2}+5\frac{(x+1)x(x-1)}{6}$$

(c)

x	f			
−1	2			
0	2	0		
1	2	1	0	
2	5	8	1.5	0.5

Hence $P_4(x) = 2+\frac{1}{2}(x+1)x(x-1)$.

Exercises for Section 3.5 (Page 115)

3.17

x	f		
0	0.0		
$\pi/2$	1.0	$2/\pi$	
π	0.0	$-2/\pi$	$-4/\pi^2$

Hence $P_3(x) = 2x/\pi-(4/\pi^2)x(x-\pi/2) = -4x^2/\pi^2+4x/\pi$ as before.

3.21 For $S''(x_1) = f''(x_1)$ use $c_1 = f''(x_1)$; for $S''(x_N) = f''(x_N)$ use $c_N = f''(x_N)$.

3.23 The results for $S(x)$ are the same (to the displayed digits) for the three different sets of $\{x_i\}$.

x	$f(x)$	$S(x)$
3375	0.400	0.447
3625	0.449	0.424
3875	0.769	0.741
4125	0.750	0.741
4375	0.315	0.316
4625	0.144	0.137
4875	0.252	0.262

3.25

f	$A(f)$	(a) S	(b) S
63	0.070	0.070	0.070
200	0.359	0.340	0.343
800	0.935	0.984	0.959
2000	2.870	2.796	2.835
10000	53.478	54.761	54.343

The results from (b) are better than those from (a) but neither is especially accurate (particularly for large f).

3.27 Using all the data but those at $\{21, 22.6, 22.8, 23.0, 23.2, 23.4\}$ produces an $S(x)$ for which

x	Exact	$S(x)$
21.0	503	500
22.6	550	548
22.8	565	570
23.0	590	570
23.2	860	767
23.4	944	966

This is good for small x, but deteriorates eventually. For this choice of interpolating points there are 10 sign changes in the $\{c_i\}$ sequence indicating 10 inflection points, not 1. Hence, there must be a lot of undesired oscillation; however, a graph of $S(x)$ would show that, for the most part, the amplitudes of the oscillations are small enough to not be visible.

3.31 $S' = 0$ in $[x_n, x_{n+1}]$ if and only if $b_n+2c_n(z-x_n)+3d_n(z-x_n)^2 = 0$ for $x_n \leq z \leq x_{n+1}$. Using the quadratic formula this reduces to the statement in the text. Checking $b_nb_{n+1} < 0$ will not detect all zeros of S', since S'

(a piecewise quadratic) may have two zeros in a particular $[x_n, x_{n+1}]$ and consequently $S'(x_n)S'(x_{n+1}) > 0$.

3.33 For the data used in Exercise 3.15, the resulting $S(x)$ had a local maximum at $(4001.3, 0.8360)$, and local minima at $(3514.9, 0.3811)$ and $(4602.5, 0.1353)$.

3.35 For the choice of the 12 data points in Exercise 3.20, there was one critical point at $r_e = 5.5544$ for which $S(r_e) = -12.036$; there were two inflection points at $(6.199, -8.979)$ and $(9.685, -0.6798)$. The second inflection point is spurious.

Exercises for Section 3.6 (Page 127)

3.37 The four coefficients are

$$a = f_{11}$$
$$b = (f_{21} - f_{11})/(x_2 - x_1)$$
$$c = (f_{12} - f_{11})/(y_2 - y_1)$$
$$d = (f_{11} + f_{22} - f_{12} - f_{21})/[(x_2 - x_1)(y_2 - y_1).$$

Miscellaneous Exercises for Chapter 3 (Page 132)

3.39 $\{t_i\} = \{0.00, 1.54, 2.81, 3.65, 4.49, 5.23, 5.78, 6.13, 6.46, 6.76, 7.00\}$. The graph is a spiral in the xy plane.

Exercise Set 4.1 (Page 149)

4.1 If a given $F(x)$ has residual ε at $x = r$, then the scaled function $f(x) = MF(x)$ has residual $M\varepsilon$ at $x = r$. Hence, a small residual (ε) can be scaled up by a large M while a large residual can be scaled down by a tiny M; consequently, a single residual tells us very little about accuracy.

4.3 (a) The next bracket is $[0.5, 1.0]$, the second is $[0.75, 1.0]$, and the third is $[0.75, 0.875]$.
(c) Newton's method:

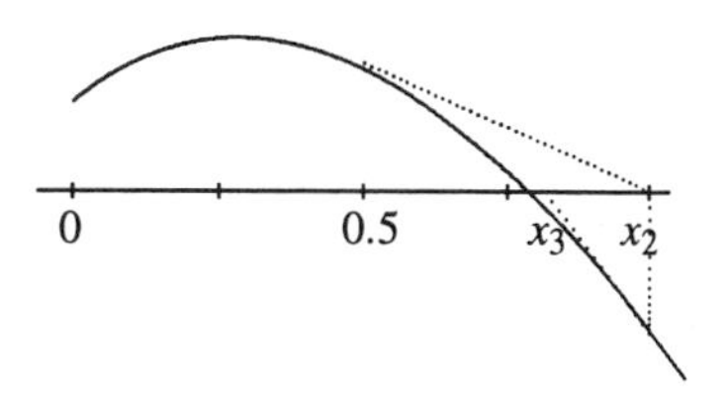

4.5 (a) There are many possible brackets; here $\pi/2$ and π were used.
(c) $x_3 = 1.928478$ and $x_4 = 1.897313$
(e) $B = 1.895494$

4.7 Let $\varepsilon = \max(\varepsilon_0, \varepsilon_1) < 1$. Then $\varepsilon_i \le \varepsilon^{\delta_i}$, where

$$\delta_i = \frac{1}{\sqrt{5}}\left[\left(\frac{1+\sqrt{5}}{2}\right)^{i+1} - \left(\frac{1-\sqrt{5}}{2}\right)^{i+1}\right].$$

For $i = 0$, $\delta_0 = 1$, so $\varepsilon_0 = \varepsilon$. Assume that $\varepsilon_{n-1} \le \varepsilon^{n-1}$; then

$$\varepsilon_n \le \varepsilon_{n-1}\varepsilon_{n-2} \le \varepsilon^{\delta_{n-1}+\delta_{n-2}}.$$

But, after some algebra,

$$\delta_{n-1} + \delta_{n-2} = \delta_n;$$

hence $\varepsilon_n \le \varepsilon^{\delta_n}$, which, by induction, gives the desired result.

Exercise Set 4.2 (Page 152)

4.9 (a) One step of the bisection method reduces the width of a bracket by a factor of 2, so n steps reduce the width by 2^N. To get from a width of 10^{10} to one of 10^{-5} then requires

$$\frac{10^{10}}{2^N} \le 10^{-5}$$

or $N \ge 15 \log 10 / \log 2 = 49.8 \approx 50$. Technically, this is the number of midpoint evaluations required; you may want to add two more to get the function values at the endpoints of the initial bracket.
(b) The worst case for Zero is four wasted secant iterations for every three bisections. Hence

$$N \ge \frac{7}{3} \cdot 50 = 116.27 \approx 117.$$

Exercise Set 4.3 (Page 155)

4.10 (a) The output $B = 0.7853982$, $Flag = 0$, Residual $= -2.1E-11$, and there were 7 calls to the f function. There was just one root in $[0,1]$ which Zero found quickly and accurately.
(d) The output $B = 0.8000003$, $Flag = 0$, Residual $= 5.3E-46$, and there were 44 calls to the f function. The high multiplicity root is ill-conditioned, yet, with this form of f, Zero was able to compute it accurately after many function evaluations. The small residual is in accord with the flatness of the graph at the root.
(h) There is no root in the input interval $[0,1]$; Zero correctly reported the lack of a bracket through the Flag $= -2$ value.

4.11 $C = 633.162$ and $T = 353.878$

4.13 The three smallest positive roots are 1.30654, 3.67319, and 6.58462.

4.15 There are two roots: one at $T = 456.9975$, the other at $T = 12{,}733.77$ is physically dubious.

4.17 With $f(k) = \omega_0 \ln(1+k)/(1-k) - 2k$, it is easily seen that $f(-k) = -f(k)$, $f(0) = 0$, $f(1^-) = +\infty$, and $f''(k) > 0$ on $[0,1)$, so the mathematical conclusions follow. The three k values are 0.99933, 0.95750, and 0.77552.

4.19 $E = 1.1903\text{E}-11$

Exercise Set 4.4 (Page 160)

4.23 $m = 2$ and $\alpha = 1/3$, so the right side of (4.14) is $10^{-6}/\sqrt{87} \approx 10^{-7}$.

Exercise Set 4.5 (Page 162)

4.25

n	x	y	z
0	0.5000	1.0000	0.0000
1	0.2808	0.6278	0.0365
2	0.2138	0.4635	0.0565
3	0.2113	0.4305	0.0622
4	0.2114	0.4293	0.0623
5	0.2114	0.4293	0.0623

There is also an answer $(0.2114, -0.4293, 0.0623)$.

Miscellaneous Exercises for Chapter 4 (Page 167)

4.27

x	t
0.1	0.656578
0.2	1.236000
0.3	1.918448
0.4	2.694180
0.5	3.556555

4.31 (a) The expression for a_0 follows from $a_0 = P(0)$ and the factorization of $P(x)$. The expression

$$\left|\frac{P(\sigma)}{a_0}\right|^{1/n} = \left|\prod_{i=1}^{n}\left(\frac{\sigma - r_i}{r_i}\right)\right|^{1/n}$$

is the geometric mean (average value) of the factors $(\sigma - r_i)/r_i$, and the bound merely states that the least value of a factor is no more than the average value. If some factor is much larger than the smallest one, the average will be substantially larger than the smallest factor and the bound will be rather poor. If σ is approximately a large root, say r_j, then the factor $(\sigma - r_j)/r_j$ is small. But if some root, say r_k, is much smaller than σ, that is, $|r_k| \gg |\sigma|$, then the factor

$$\left|\frac{\sigma - r_k}{r_k}\right| \approx \left|\frac{\sigma}{r_k}\right|$$

is quite large and the bound is poor.

Exercise Set 5.1 (Page 183)

5.1 $A_1 = A_4 = \frac{1}{4}$ and $A_2 = A_3 = \frac{3}{4}$. From $f(x) = x^4$ we get $d = 3$ and $c = -2/405$. (The name $\frac{3}{8}$-rule comes from the coefficient values for the interval $[0,1]$ instead of $[-1,1]$.)

5.3 With $h = \pi/N$ we have

N	N-panel trapezoid
60	0.811148922853
100	0.811155733422
140	0.811155735194

Note: the routine Adapt from Section 5.3 requires 441 function evaluations for comparable accuracy.

Exercise Set 5.3 (Page 189)

5.4 (a) This integrand is very smooth, so Adapt has no trouble getting an accurate answer in only 21 function calls. The output is Answer $= 3.14159122$, Flag $= 0$, Errest $= 1.431\text{E}{-5}$. Since the exact answer is π, the actual error is $1.431\text{E}{-5}$, which was what was estimated by Adapt.
(d) This integrand is highly oscillatory, yet Adapt has little trouble getting an accurate answer in 49 function calls. The output is ANSWER $= 1.5000000$, FLAG $= 0$, ERREST $=$ $-1036\text{E}{-16}$. Since the exact answer is $3/2$, Adapt's error estimate is quite accurate.
(f) This integrand has a vertical asymptote at 0.25 but it still is integrable with exact answer of $1+\sqrt{3}$. Adapt has great difficulty with this problem, even when f is defined to be finite, say 1.E+8, at 0.25 to avoid the division by zero. The output is ANSWER $= 2.7319539$, FLAG $= 2$, ERREST $= 9.264\text{E}{-5}$. The actual error is $9.264\text{E}{-5}$, which is close to the actual error of $9.691\text{E}{-5}$. The Flag of 2 indicates that Adapt was unable to compute a sufficiently accurate answer in the 3577 function calls allowed.

5.5 ANSWER $= 1.500000$ and ERREST $= -2.4\text{E}{-18}$ are comparable to the results from Exercise 5.4d. The number of integrand evaluations was 21, which is faster than for Exercise 5.4d.

5.7

x	Answer	Flag	F calls
0.0	0.0000000	0	7
0.1	0.0993360	0	7
0.2	0.1947510	0	7
0.3	0.2826317	0	7
0.4	0.3599435	0	7
0.5	0.4244364	0	7

5.9 $T(0.50) = 3.736767$, $T(0.75) = 3.904747$, $T(1.00) = 4.187184$.

5.11 For a machine with $u = 1.4 \times 10^{-17}$, the first $n = 4$ and the second is 12. Exercise 1.3 suggests that for the first n, 4 digits of accuracy (out of 17) might be lost due to the noise. For the second n, 12 digits might be lost. This idea was used with integrand (b) of Exercise 5.4. The output ANSWER $= 0.909090912$, with ERREST $= -7.82\text{E}{-9}$ that required 49 function calls. The noise level was not magnified by Adapt, so the code appears to be stable. When $n = 12$ we got ANSWER $= 0.909099105$ in 2807 function calls with ERREST $= -7.82\text{E}{-6}$. The accuracy was contaminated by the noise after the expected 5 digits; the number of function evaluations required also increased due to the lack of smoothness.

Exercise Set 5.4 (Page 198)

5.13

Part	Answer	Flag	F calls
(a)	0.6205370	0	147
(b)	0.6205362	0	35
(c)	0.6205366	0	49

For the first method in (b) we used

$$\int_0^1 \frac{\sin x}{\sqrt{x}}\,dx = \int_0^{\varepsilon} \frac{x - x^3/6}{\sqrt{x}}\,dx + \int_{\varepsilon}^1 f(x)\,dx.$$

with $\varepsilon = 0.23$; the first integral was done analytically and the second by Adapt. For (c), we used the change of variable $t = x^2$.

5.15 (a) For ABSERR $= 10^{-9}$ Adapt returned ANSWER $= 0.2865295$ and an error estimate of $-0.34\text{E}{-9}$ with 1897 function calls.
(c) Method (a) is faster at the larger tolerances and is more accurate at ABSERR $= 10^{-9}$.

5.17 The singularity at $x = 0$ causes no problems.

t	$E_1(t)$ estimate
1.0	0.2193838
2.0	0.0489004
3.0	0.0130475

5.19

T	$C_0(T)$	F calls
100	0.067688	21
200	0.521380	21
300	1.41696	7
400	2.37916	7
500	3.17812	21
1000	5.00724	49
total		126

This function has the form $C_v(T) = g(T)\int_0^{1900/T} f(x)\,dx$ so it is most efficiently evaluated as a reverse iteration. Let

$$F(T_i) = F(T_{i+1}) + \int_{1900/T_{i+1}}^{1900/T_i} f(x)\,dx$$

for $i = 5, 4, \ldots, 1$ with the integral for $F(T_6)$ taken over the interval $[0, 1900/T_6]$ to get the iteration started. Then $C_v(T) = g(T)F(T)$.

Exercise Set 5.5 (Page 201)

5.21

Method	(a)	(b)
Glucose	6930.9	6927.1
Insulin	1905.1	1904.6
Glucagon	92319.4	92688.3
Gr. Hormone	256.5	256.7

Exercise Set 5.6 (Page 203)

5.24 (a) $A_1 = A_2 = A_3 = 1/6$; hence,

$$\iint f(x,y)\,dx\,dy \approx [f(0,1) + f(0,0) + f(1,0)]/6$$

(b)

$$\begin{aligned}\iint f(x,y)\,dx\,dy \approx\ & (1/12)\{[f(0,1)+f(0,1/2)+f(1/2,1/2)] \\ & +[f(0,1/2)+f(0,0)+f(1/2,0)] \\ & +[f(1/2,1/2)+f(1/2,0)+f(1,0)] \\ & -[f(1/2,0)+f(1/2,1/2)+f(0,1/2)]\} \\ & = [f(0,0)+f(1/2,0)+f(0,1/2) \\ & +f(0,1)+f(1,0)+f(1/2,1/2)]/12\end{aligned}$$

Miscellaneous Exercises for Chapter 5 (Page 208)

5.25 For convenience in notation, let

$$h(r,\theta,\theta') = \frac{1-r^2}{1-2r\cos(\theta-\theta')+r^2}.$$

Since

$$1 = \frac{1}{2\pi}\int_0^{2\pi} h(r,\theta,\theta')\,d\theta,$$

we have, after some algebra

$$\phi(r,\theta) = \frac{1}{2\pi}\int_0^{2\pi} h(r,\theta,\theta')[f(\theta')-f(\theta)],$$

which should have better numerical properties for r near 1. This is because as $\theta' \to \theta$, both the numerator $f(\theta') - f(\theta)$ and the denominator $1 - 2r\cos(\theta-\theta') + r^2$ become small. Since the integral acts as a small correction to the term $f(\theta)$ and since the integrand is of one sign, the numerator must balance out the effect of the small denominator. Moreover, even if the integral is grossly in error, it has little effect on the result of the sum as $r \to 1$. To illustrate this, let $\theta = \pi/2$ with $f(\theta) = \sin\theta$ so that the solution becomes $\phi(r,\theta) = r$. The function ϕ was evaluated using both forms, the original and the modified, and function counts were used as a measure of the computational effort. For example, at $r = 0.9921875$ we got for the original ϕ, a value of 0.99218012 requiring 1057 function evaluations. The modified ϕ value was 0.99218750 requiring 329 function evaluations. The modified version is clearly faster and a little more accurate.

5.27 For $n = 3$ we have $q_n = 7.08638$ so Adapt returned $\int_0^a rf(r)J_0(q_n r)\,dr = 0.000049746$ with Errest $= -7.21\text{E}{-9}$; hence, $A_n = 0.011046$.

5.29 The main program is the usual driver to produce the answer T_O. The function f for Zero has independent variable T_O and output

$$f = \frac{m}{\pi D}\text{Answer} - L,$$

where ANSWER is the output from Adapt that is called inside F since the upper limit of integration is T_O. The result is $T_O = 246.73$.

Exercise Set 6.1 (Page 216)

6.1 Clearly $y(0) = 0$ and $y'(x) = 0$ for $0 \le x \le c$ and $y' = (x-c)/2$ for $c < x \le b$. The limits as $x \to c$ are the same in both cases, so y' is continuous on $[0,b]$. Also, for $0 \le x \le c$,

$$y' = 0 = \sqrt{|0|} = \sqrt{|y|}$$

and for $c < x \le b$,

$$y' = \frac{1}{2}(x-c) = \sqrt{|\frac{1}{4}(x-c)^2|} = \sqrt{|y|},$$

so that $y(x)$ satisfies the differential equation.

6.3 Recall that Dawson's integral is the expression

$$y(x) = e^{-x^2} \int_0^x e^{t^2}\, dt.$$

Now, $y(0) = 0$ and differentiation shows that

$$\begin{aligned} y' &= e^{-x^2} e^{x^2} + (-2x) e^{-x^2} \int_0^x e^{t^2}\, dt \\ &= 1 - 2xy. \end{aligned}$$

Obviously, y' is continuous on $[0,b]$ for any finite b.

6.5 (a) $f_y = 2y$; the partial derivative is not bounded for all y and so does not satisfy a Lipschitz condition for $0 \le x \le \pi/2$.

(c) $f_y = 1/x$; since $|f_y| \le 1/1 = 1$ for x in $[1,2]$, the function f does satisfy a Lipschitz condition with constant $L = 1$.

(e) $f_y = \cos x \cos y$; since $|f_y| \le 1 \cdot 1$ for any x and y, the function f does satisfy a Lipschitz condition with constant $L = 1$.

6.7 (a) One solution is $Y_1 = u$, $Y_2 = u'$, $Y_3 = u''$, and $Y_4 = u'''$; then $Y_1' = Y_2$, $Y_2' = Y_3$, $Y_3' = Y_4$, and $Y_4' = \cos \alpha t + tY_1 - e^t Y_2$.

(c) One solution is $Y_1 = u$, $Y_2 = u'$, and $Y_3 = v$; then solving the system for u'' and v', we have

$$\begin{aligned} u'' &= 2t + \frac{3}{4}\cos t - \frac{7}{4}u - v \\ v' &= 2t - \frac{\cos t}{4} - \frac{3}{4}u. \end{aligned}$$

Hence $Y_1' = Y_2$, $Y_2' = 2t + (3\cos t)/4 - 7Y_1/4 - Y_3$, and $Y_3' = 2t - (\cos t)/4 - 3Y_1/4$.

(e) One solution is $Y_1 = u$, $Y_2 = u'$, $Y_3 = u''$, $Y_4 = u^{(3)}$, $Y_5 = u^{(4)}$, and $Y_6 = u^{(5)}$; then $Y_1' = Y_2$, $Y_2' = Y_3$, $Y_3' = Y_4$, $Y_4' = Y_5$, $Y_5' = Y_6$, and $Y_6' = e^t - Y_1 Y_2$.

Exercise Set 6.2 (Page 221)

6.9 For $h = 1/40$ the error at 0.5 is 0.0021; at 1.0 the error is 0.0010. For $h = 1/80$ the error at 0.5 is 0.0023; at 1.0 the error is 0.00116. For both x values the error does drop by a factor of 2 as h is halved. To achieve an absolute error of 10^{-6} in magnitude would require $h \approx 0.000011$.

Exercise Set 6.3 (Page 227)

6.11 For $h = 1/40$ the error at 0.5 is $-0.1434\text{E}{-4}$; at 1.0 the error is $-0.1395\text{E}{-4}$. For $h = 1/80$ the error at 0.5 is $-0.3564\text{E}{-5}$; at 1.0 the error is -0.3471. For both x values the error does drop by a factor of 4 as h is halved. To achieve an absolute error of 10^{-6} in magnitude would require $h \approx 0.0067$.

6.13 To show that $y^{(r)}$ can be computed via the formula $y^{(r)} = P_r y + Q_r$ we use induction. The result is true for $r = 2$ since

$$\begin{aligned} y'' &= P_1' y + P_1 y' + Q_1' \\ &= P_1' y + P_1(P_1 y + Q_1) + Q_1' \\ &= (P_1' + P_1 \cdot P_1) y + (Q_1' + Q_1 P_1) \\ &= P_2 y + Q_2. \end{aligned}$$

Assume it to be true for $r = k$; then

$$\begin{aligned} y^{(k+1)} &= P_k' y + P_k y' + Q_k' \\ &= P_k' y + P_k(P_1 y + Q_1) + Q_k' \\ &= (P_k' + P_1 P_k) y + (Q_k' + Q_1 P_k) \\ &= P_{k+1} y + Q_{k+1}. \end{aligned}$$

A fifth order formula is obtained by dropping the remainder term in the expression

$$y(x_{n+1}) = y(x_n) + h_n \Big(y'(x_n) + \frac{h_n y''(x_n)}{2!} + \frac{h_n^2 y^{(3)}(x_n)}{3!}$$

$$+\frac{h_n^3 y^{(4)}(x_n)}{4!}+\frac{h_n^4 y^{(5)}(x_n)}{5!}\Bigg)$$

$$+\frac{h_n^6 y^{(\xi)}(x_n)}{6!},$$

where $h_n = x_{n+1} - x_n$, $y' = 1 - 2xy$, and $Y^{(r)} = P_r y + Q_r$. The polynomials P_r and Q_r are given recursively by

$$\begin{aligned} P_r &= P'_{r-1} + P_1 P_{r-1}, \\ Q_r &= Q'_{r-1} + Q_1 P_{r-1}, \end{aligned}$$

with $P_1 = -2x$ and $Q_1 = 1$.

Exercise Set 6.6 (Page 238)

6.16 (a) The solution is quite smooth on $[0,3]$ and Rke computes an accurate solution very efficiently. The output $y = 0.49999991$ with 8 calls to Rke has absolute error 0.00000009 well within the requested tolerance.

(d) The large Lipschitz constant of 100 is a clue that this problem may be difficult. Rke produces $y = 0.83598456$ requiring 63 calls. Thge actual global error is -0.0000002, which is within the requested tolerance. This problem is "stiff" in the sense of Section 6.7.

6.17 The equation $x'' + (x^2 - 1)x' + x = 0$ is of the form treated by Liénhard with $f(x) = x^2 - 1$. The indefinite integral $G(x) = x^3/3 - x$, so the Liénard variables are

$$\begin{aligned} Y_1(t) &= x(t) \\ Y_2(t) &= x'(t) + G(x(t)) \end{aligned}$$

and we have

$$\begin{aligned} Y_1'(t) &= Y_2(t) + Y_1(t) - Y_1^3(t)/3 \\ Y_2'(t) &= -Y_1(t). \end{aligned}$$

To plot the solution in the phase plane, it is necessary to plot $x'(t) = Y_2(t) - G(x(t)) = Y_2(t) - G(Y_1(t))$ against $x(t) = Y_1(t)$. With $Y_1(0) = -1$ and $Y_2(0) = 1$ for $0 \le t \le 15$ the resulting phase plane plot is shown below. The closed curve was traced over in the computation so that (to plotting accuracy) the limit cycle was obtained.

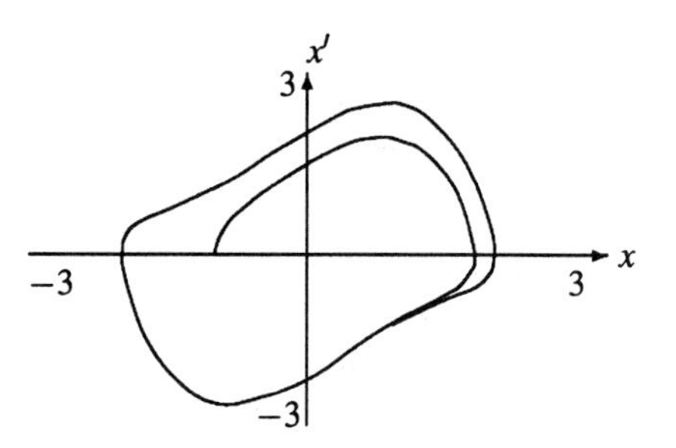

6.19 The results are comparable to those of Exercise 5.15. For example, $P(12) = 778.030$, $P(21) = 1112.06$, and $P(30) = 1246.86$.

6.21 Note that the equation can be written in the form

$$\frac{d}{dx}(y'' - y' \sin x - y \cos x) = \ln x.$$

Integrate this to obtain

$$y'' - y' \sin x - y \cos x = x \ln x - x + c_2,$$

the first integral relation. Write this in the form

$$\frac{d}{dx}(y' - y \sin x) = x \ln x - x + c_2,$$

which can be integrated to get the second integral relation

$$y' - y \sin x = c_1 + c_2 x + \frac{1}{2}x^2 \ln x - \frac{3}{4}x^2.$$

To determine c_2, evaluate the first relation at $x = 1$ and use the initial conditions to get

$$c_2 = A_3 - A_2 \sin 1 - A_1 \cos 1 + 1.$$

Evaluate the second integral relation at $x = 1$ to get

$$c_1 = A_2 - A_1 \sin 1 - c_2 + \frac{3}{4}.$$

The following sample results were generated by Rke with the residuals computed for each integral relation.

x	1.500000	2.000000
$y(x)$	1.647253	2.642695
$y'(x)$	1.629417	2.316041
$y''(x)$	1.468284	1.010749
First res.	1.94E−6	7.32E−6
Second res.	−3.92E−7	−4.80E−6

6.23 (a)

t	$x(t)$	$c(t)$
0.00	0.10000	0.10000
2.00	1.39122	1.75478
4.00	−1.95327	2.57504
6.00	−1.58142	−0.11364
8.00	0.39677	−3.14858
10.00	1.71102	0.40587
12.00	0.66878	3.04524

The period appears to be approximately 9.

Miscellaneous Exercises for Chapter 6 (Page 249)

6.25 Clearly $f(x,y) = 2|x|y$ is continuous on $[-1,1]$; also, $f_y = 2|x| \leq 2$, so f satisfies a Lipschitz condition with constant $L = 2$. The hypotheses of Theorem 1 are satisfied. Let $y(x)$ be defined by

$$y = \begin{cases} e^{x^2}, & x \geq 0 \\ e^{-x^2}, & x < 0 \end{cases}$$

so that $y(-1) = e^{-1}$ and

$$y' = \begin{cases} 2xe^{x^2}, & x \geq 0 \\ -2xe^{-x^2}, & x < 0. \end{cases}$$

Thus, y' is continuous on $[-1,1]$. Also, for x positive or negative $y' = 2|x|y$, so that y does satisfy the differential equation. Since

$$y'' = \begin{cases} 2e^{x^2} + 4x^2e^{x^2}, & x \geq 0 \\ -2e^{-x^2} + 4x^2e^{-x^2}, & x < 0, \end{cases}$$

we have $y''(0^+) = 2$ while $y''(0^-) = -2$, so y'' is not continuous at $x = 0$. Euler's method is convergent for this problem but not $O(h)$. Higher order Runge–Kutta methods will not improve convergence past $O(h^2)$. If the problem is split at $x = 0$, then $y(x)$ is infinitely differentiable on each of $[-1,0]$ and $[0,1]$. If, in the original problem, a mesh point is placed at $x = 0$, this is equivalent to solving the two problems separately. By using a fixed step $h = 2/2^k$ we guarantee that $x = 0$ is a mesh point, so convergence is $O(h^4)$ with the fourth order Runge–Kutta code; in contrast, $x = 0$ cannot be a mesh point for $h = 2/3^k$, so convergence will occur at a slower rate.

6.27 The missing initial condition is $y'(0) = 2.155796$; the following table gives the solution at a few values.

x	$y(x)$	$y'(x)$
0.0	0.000000	2.155580
0.2	0.434337	2.205839
0.4	0.891567	2.396098
0.6	1.408763	2.830332
0.8	2.055972	3.763435
1.0	3.000000	6.067751

6.29 (a) The exact solution is $x(t) = 20t - 5000$, $a(t) = -2.6 \times 10^6 \exp[2 \times 10^{-5}(t - 2000)] + 50t + 0.25 \times 10^7$. Rke produces

$$x(0) = -5000.00$$

and

$$a(0) = 1947.46,$$

which are both correct.

(c) If FORTRAN is used, the program should be written so that Spcoef is called only once by the driver. The vectors X, F, B, C, and D should be passed through a COMMON statement to the routine defining the differential equation. In C and C++ these will have to be global variables. The output is given in the following table.

t_f	$a(0)$	$x(0)$
1000	497.29	27884.60
1500	1109.42	16736.78
2000	1950.42	1656.81
2500	3030.94	−7811.70

INDEX